EMERGING CONCEPTS IN ANALYSIS AND APPLICATIONS OF HYDROGELS

Edited by **Sutapa Biswas Majee**

Emerging Concepts in Analysis and Applications of Hydrogels
http://dx.doi.org/10.5772/61692
Edited by Sutapa Biswas Majee

Contributors

Morteza Bahram, Naimeh Mohseni, Mehdi Moghtader, Juan Manuel Lazaro Martinez, Viviana Campodall'Orto, Xiangdong Bi, Aiye Liang, Gabriela Ionita, Lavinia Vlaia, Georgeta Coneac, Ioana Olariu, Vicenţiu Vlaia, Dumitru Lupuleasa, Juntao Wu, Li Zhang, Gaetano Lamberti, Anna Angela Barba, Sara Cascone, Diego Caccavo, Annalisa Dalmoro, Dorota Kołodyńska, Sutapa Biswas Majee

Published by InTech

Janeza Trdine 9, 51000 Rijeka, Croatia

Publishing Process Manager Andrea Koric

Technical Editor SPi Global

Cover InTech Design team

First published August, 2016

Printed in Croatia

Additional hard copies can be obtained from orders@intechopen.com

Emerging Concepts in Analysis and Applications of Hydrogels, Edited by Sutapa Biswas Majee

p. cm.

Print ISBN 978-953-51-2509-9

Online ISBN 978-953-51-2510-5

For EU product safety concerns:
IN TECH d.o.o., Prolaz Marije Krucifikse Kozulić 3, 51000 Rijeka, Croatia,
info@intechopen.com or visit our website at intechopen.com.

Meet the editor

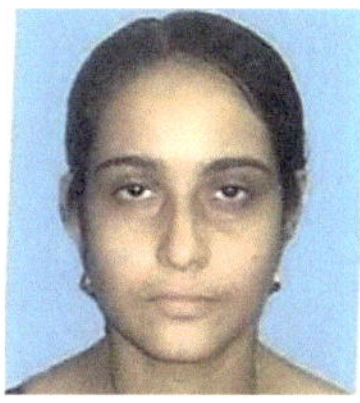

The editor Dr. Sutapa Biswas Majee is actively engaged in exploring the role of natural and synthetic polymers in modulating drug release from various dosage forms. She is also keenly involved in mathematical modeling of different diseases. She acquired MPharm (Pharmaceutics) and PhD degrees from Jadavpur University, India. She carried out her post-doctoral research on Parasitology, Phytochemistry and Microbiology from Bose Institute and Jadavpur University, India and published peer-reviewed research articles and reviews in various national and international journals. She authored/co-authored book chapters and complete books on various topics pertaining to modulation of drug release and study of release kinetics from drug delivery devices and also on mathematical modeling.

Contents

Preface

The field of "Hydrogels" is ever expanding with newer applications being explored always. Hydrogels act as unique platform for desired biomedical applications and applications in the field of drug delivery, guided by different cross-linking chemistry. They are usually found to vary in their intrinsic physicochemical characteristics such as hydrophilicity, swellability, gelation, mechanical strength, porosity, biocompatibility, and biodegradability. Molecular constitution of the hydrogels can be tuned by physical or chemical cross-linking, influencing their in vitro and in vivo characteristics and ultimately their performance. An ideal hydrogel for biomedical applications in the area of regenerative medicine, tissue adhesion, and design of 3D tissue engineering scaffolds for cellular growth should promote cell-cell interactions, cell-polymer bioadhesion, cellular penetration, proliferation, differentiation, and migration. Hydrogels are being increasingly investigated for their utilization in oral, parenteral, ocular, topical, and transdermal drug delivery in order to achieve tailor-made, pre-programmed controlled drug release. Porosity, rheological behavior and in vivo erosion rate are the primary factors governing drug release from hydrogel-based dosage forms. Stimuli-responsive hydrogels offer great potential in fabrication of specialized drug delivery systems. Growing interest in this class of polymeric macromolecules necessitates a holistic study covering every aspect, from synthesis, characterization, mathematical modeling, performance improvement to embarking opportunities. The intent of this book is to equip academia and industry with adequate basic and applied principles related to hydrogels. The book is an up-to-date, authoritative, and multi-authored treatise on hydrogels and has been divided into eight Chapters. Chapter 1 focuses on the key features, classification, and synthesis of hydrogels with a comprehensive review on stimulus-responsive hydrogels and their diverse applications. Chapter 2 provides an insight into various forms of spectroscopy as analytical tool for hydrogel characterization, in addition to microscopic and rheological studies. Chapters 3 4 have attempted to bring together the emerging concepts, theories of sorption, characterization, and applications of the hydrogel in the area of uptake of metal ions and fertilizers. The central theme of Chapters 5 through 7 is the mathematical modeling of diffusion into and release from hydrogel-based devices and their biomedical and pharmaceutical applications. The last chapter deals with a novel class of hydrogels, endowed with self-healing attribute. The book will be of interest to academicians and researchers of biomedical and pharmaceutical science as well as practitioners of cognate disciplines with demand for hydrogel.

Dr. Sutapa Biswas Majee,
NSHM College Of Pharmaceutical Technology,
NSHM Knowledge Campus, Kolkata-Gropu Of Institutions,
Kolkata
India

Chapter 1

Introductory Chapter: An Overview of Hydrogels

Sutapa Biswas Majee

Additional information is available at the end of the chapter

http://dx.doi.org/10.5772/64302

Hydrogels are hydrophilic, water-insoluble polymeric macromolecules represented as semi-open network systems comprising of entangled chains or short strands of varying lengths joined together by cross-links. On being exposed in a thermodynamically compatible solvent (water or any biological fluid), they can entrap a large fraction of solvent within the pores or interstitial spaces and can achieve a fully swollen state. Swelling is accompanied by dimensional change, which leads to a drastic change in the rheological characteristics and ultimately phase transition [1]. These molecules of natural or synthetic origin are endowed with intrinsic physicochemical characteristics such as hydrophilicity, swellability, gelation, mechanical strength, porosity, biocompatibility and biodegradability conferring on them the capability of being utilized in different industries in the area of water purification, ion exchange chromatography, enhanced oil recovery, sensor development, removal of azo-dye pollutants or toxic materials, development of immobilized enzyme systems, agriculture, food processing, pharmaceutical, medical and biomedical fields [2, 3]. Stiffness and water-absorbing capacity of hydrogels are attributed to the presence of hydrophilic pendant moieties attached to the backbone such as alcohols, carboxylic acids and amides, whereas the presence of cross-links makes it resistant to dissolution in the aqueous medium [1]. Some of the common examples of hydrogels include agarose, alginate, chitosan, collagen, fibrin, gelatin, hyaluronic acid, poly(vinyl alcohol), poly(acrylic acid), poly(acrylamide), poly(propylene fumarate-co-ethylene glycol), poly-2-hydroxyethyl methacrylate (PHEMA), polypeptides, etc. [2, 4].

The science of hydrogels is incomplete without a discussion on mathematical models and equations describing swelling thermodynamics and swelling kinetics. Swelling involves the sorption of solvent molecules into the pores or voids in the macromolecular structure of hydrogel. Therefore, porosity constitutes an essential physicochemical parameter, on the basis of which hydrogels may be classified as non-porous, microporous, macroporous and superporous [2, 5]. During hydrogel swelling, two forces come into play: osmotic force and a counter-elastic force, which resists the elongation of macromolecular chains and thus solvent-induced deformation. Florey and Rehner's equilibrium swelling theory proposed in 1943 laid the foundation stone for mathematical description of the swelling process of polymer networks. Equilibrium or steady state is attained when the counteracting forces of osmotic force and elastic force balance each other and no further swelling occurs and maximum dimensional change has taken place [5, 6]. Several attempts have been made to establish a correlation between network structure of polymeric hydrogel, swelling behaviour and desirable mechanical characteristics, gelation properties and drug-release profile [7].

During adsorption process of metal ions by the non-porous or porous hydrogels, three processes are assumed to occur in succession according to intra-particle diffusion model as proposed by Weber and Morris. They can be described as external mass transfer of the adsorbate on the surface of the adsorbent (film diffusion), internal mass transfer (intraparticle diffusion) of the adsorbate into the pores and capillaries of the adsorbent and chemical-binding reactions. Any of the above-mentioned steps can be the rate-limiting step [8, 9]. Kinetics of adsorption by most of the hydrogels has been successfully described by Freundlich's, Temkin's and Dubinin-Radushkevich's isotherms [10].

Molecular constitution of the hydrogels affects their in vitro and in vivo performance during any conceivable application of hydrogel in any sphere of life. Chemical or physical cross-linking produces irreversible or reversible gels, respectively, with improved spatial and temporal control of cross-linking, thereby producing gels of tunable mechanical strength, elasticity, gelation characteristic and in vivo degradation/erosion rate [1, 11]. Introduction of different substituent groups or chemical cross-links between identical/different monomers or similar/dissimilar polymers involves the formation of permanent covalent bonds, induces configurational changes and imparts greater stability than physical gels held together by ionic bonds, hydrogen bonds or hydrophobic forces and chain entanglements. Conformational changes are manifested in physical gels. For cross-linking, chemicals such as carbodiimides, formaldehyde and glutaraldehyde have been frequently used. Different reactions involved in producing chemically cross-linked hydrogels include free radical polymerization, addition and condensation polymerization, classical organic reactions between functional groups (viz. Michael addition, click reaction, Schiff base formation, epoxide coupling, etc.), enzymatic reactions and gamma and electron beam polymerization as well as grafting. Polymerization can be carried out in solution, suspension or emulsion phase [2, 12]. Gelatin matrix has been cross-linked with oxidized cellulose nanowhiskers for enhanced mechanical strength and better thermal stability [13–15]. The parameters that control the micro-architecture and hydrogel attributes include the concentration of the cross-linking agent, structure and concentration of the monomers [1, 11]. It is much easier to initiate disintegration of physical or reversible gels through alteration of adjacent environmental parameters such as pH, temperature, solvent composition, ionic strength and electric field. Physical hydrogels, which are capable of responding to changes in the above-mentioned internal or external stimuli, are termed as 'smart' or 'intelligent' polymers. Swelling and volume phase transition of these 'smart' polymers may be either continuous or discontinuous over a range of level of stimulus or at a threshold level of the stimulus. Since, synthetic hydrogels usually contain ionic or ionizable groups, they are known as polyelectrolyte gels and are reported to possess higher swelling capacity than the nonionic gels[1]. For these gels, interaction with mobile counterions in the medium has an effect on the hydrogel behaviour and performance. High cross-link density in polyelectrolyte gels may not lead to spatial inhomogeneity or network imperfections as in the case of nonionic gels [1, 3, 4]. At this juncture, it is worth mentioning that the change in environmental pH does not act as a stimulus for nonionic gels [4].

Various instrumental techniques such as dynamic contact angle analysis, tensile analysis and thermogravimetry are employed to study surface topography, mechanical properties and swelling behaviour. Mechanical properties of a hydrogel are time-dependent where poroelasticity is attributed to solvent imbibition and its movement whereas viscoelasticity results from the rearrangement of polymer chains bound together by reversible and irreversible cross-links. Mechanical characterization of hydrogels through extensiometry, parallel plate compression test, bulge test and indentation test provides a precise idea about gel parameters such as Young's modulus, yield strength, tensile strength, viscoelastic and poroelastic properties [16, 17]. Rheological behaviour is analysed with the help of oscillatory shear rheometry through dynamic frequency sweep tests [18]. Modulated differential scanning calorimetry (DSC) is utilized in the characterization of temperature-dependent changes in hydrogel properties [4]. Hydrogels with specific ligands or chelating agents attached to the polymeric backbone are known to adsorb or absorb metal ions or organic compounds such as fertilizers and can be used to either remove heavy metal ions from a particular system or release the encapsulated phosphorous-containing fertilizers slowly over a prolonged time period, thereby facilitating plant nutrition. The adsorbed/absorbed components and state of metal ion coordination can be studied by different spectroscopic techniques. Spectroscopic methods, if properly utilized, can provide valuable information on the micro-architecture and interactions occurring between polymer and water, mesh-size distribution and also about the phenomena of solvent diffusion and release of the embedded/entrapped molecules [19–21].

For biomedical applications in the area of regenerative medicine, tissue adhesion and design of three-dimensional (3D) tissue engineering scaffolds for cellular growth, an ideal hydrogel should possess excellent biocompatibility, low immunogenicity, capability of integrating and encoding biologically active motifs or functional groups in the network structure, in addition to optimum network architecture, porosity and desirable mechanical properties and stiffness to promote cell-cell interactions, cell-polymer bioadhesion, cellular penetration, proliferation, differentiation and migration [11, 14]. In situ-forming hydrogels are gaining popularity among scientists engaged in biomedical research. Extreme precaution should be taken during selection of monomers, cross-linking agent, initiator and catalyst as well as the reaction conditions should be highly specific in order to minimize cytotoxicity and immunogenicity. The formation of hydrogel in the presence of specific ions, inclusion complexes with β-cyclodextrin, stereocomplexation and complementary-binding reactions are some of the methods to produce in situ-forming hydrogels [22]. The ultimate objective of hydrogel-based tissue constructs is to mimic native tissue and extracellular matrix as closely as possible, that is, they should be bio-mimetic. Moreover, the hydrogel should retain its integrity after administration by parenteral routes and encapsulated cells should maintain their viability and phenotype. Natural polymer gelatin possesses Arg-Gly-Asp (RGD) sequences, which act as sites for adhesion to cell surfaces through integrins [13]. However, native gelatin fails to perform efficiently as scaffold material owing to its low mechanical strength and shape stability at body temperature. Therefore, numerous studies have been carried out to produce cross-linked gelatin for the development of engineered tissues such as cartilage, bone and smooth muscle [3, 13, 15]. Another study reported the effect of elastic and viscous modulus of modified alginate hydrogels on the rate of proliferation and differentiation of embedded neural stem cells. The modulus value of the alginate gel is controlled by the molecular weight distribution of the polymer, divalent ion concentration and the α-L-guluronic acid content in the gel [18, 23]. Recently, different three-dimensional bioprinting techniques are being exploited to design hydrogel-based bio-scaffolds for hard and soft tissues containing viable cells [24]. Self-healing hydrogels are gaining popularity because of their unique capability of autonomous healing and repair on damage. They can be either linear polymers or supramolecular networks, and recent advances have been made in inducing self-healing property in hydrogels with permanent cross-links [25].

Alterations in the degree and density of cross-linking of the hydrogels and hydrogels with different hydrophilicity/hydrophobicity ratios have enabled the development of drug-delivery devices with tailor-made and preprogrammed release profiles. Different hydrogels thus obtained show different solubilities in aqueous medium and also characteristic swelling behaviour. Among the synthetic polymers commercially available, hydroxypropylmethyl cellulose (HPMC) is the most commonly used polymer in the development of different dosage forms because of its availability in different grades varying in their degree and type of substitution, cross-linking density, sensitivity to pH and temperature, rheological and swelling behaviour as well as drug-release profiles [4]. Drug entrapment and its subsequent release from hydrogels are governed by porosity and degradability of the gels. Solvent influx or absorption and drug release from the hydrogel matrices or reservoirs occurs primarily by the process of diffusion, although swelling and erosion/biodegradation may also contribute to the release phenomenon depending upon the chemical structure of the hydrogel [3, 6, 23]. The devices may exhibit Fickian or non-Fickian diffusion kinetics. Although diffusion is the primary mechanism for solvent influx and efflux of entrapped molecules, convection may play a significant role in the transport of water molecules into or drug molecules or out of the hydrogel matrix [6]. Hydrogels have shown great promise in metal-based therapy as intelligent carriers for anticancer drugs like cisplatin where a complex of well-defined stoichiometry resulted and the encapsulated platinum was released obeying near zero-order kinetics demonstrating remarkable cytotoxic activity [26]. Similar cytotoxic effect has also been observed with herceptin loaded into hydrogel exhibiting improved therapeutic efficacy and better retention inside the tumour [27].

By virtue of their unique property of being stimuli-responsive, thermo-sensitive hydrogels have found wide spectrum of applications in the field of parenteral drug delivery since they can undergo gelation or sol-gel transformation readily at body temperature, especially at sites of Inflammation having elevated temperature [28]. Thermoresponsive hydrogels exhibit a lower critical solution temperature (LCST) where there is a sudden change in aqueous solubility and gelation characteristics of hydrogels and this is often termed as volume-phase transition temperature (VPTT). Swelling occurs below LCST and gel network collapses due to dehydration above the critical temperature. Temperature-induced volume-phase transition behaviour is thus governed by water-polymer interactions [4, 29, 30]. Temperature sensitivity of hydrogels has also been exhibited by hydrophobically modified hydrogels where a phenomenon known as re-entrant swelling transition along with change in size is manifested in aqueous solution of non-polar solvents as observed in the case of poly(N-isopropyl acrylamide) (PNIPAAm) [1]. Linear copolymer of PNIPAAm and N-hydroxymethylacrylamide (HMAAm) has been synthesized as cross-linked thermoresponsive polymer and investigated as microparticulate carrier for drug molecules. Drug diffusion through the matrix of the microspheres depended highly on the presence of salt in the medium and its temperature. Above VPTT, the collapse of the gel network, exposure of the hydrophobic domains and subsequent shrinkage retarded drug release although some of the dissolved drug is mechanically expelled out. Below VPTT, diffusion was faster ultimately leading to pulsatile drug release from cross-linked copolymer microspheres over the temperature range studied [4, 29].

Physical hydrogels exhibiting shear-thinning and self-healing properties during and after injection respectively can be exploited for controlled and minimally invasive delivery of biologically active molecules, proteins and peptides in the form of injectable delivery systems. They show lower viscosity or fluidity on the application of shear stress rendering syringeability, which is recovered on relaxation at the site of application, when shear is absent. Therefore, these systems exhibit shear-dependent sol-gel transformation. The desirable qualities of an injectable drug delivery system include customized mechanical strength, stability, biodegradation rate, erosion in addition to minimum toxicity and maximum biocompatibility [31]. Hydrogels have also gained attention in the field of topical and transdermal drug delivery as also in ocular drug delivery and in the manufacture of different types of contact lenses differing in their flexibility, water vapour transmission and gas permeability. Fabrication of gastroretentive dosage forms for oral administration through the use of superporous hydrogels has offered new possibilities in achieving sustained drug release [3].

Constant efforts are being made to improve mechanical properties of hydrogels and also to obtain fast response to various exogenous stimuli. Topological gels, double-network gels or interpenetrating network consisting of two polymeric entities with different degree of stiffness, nanocomposite gels containing PNIPAAm and inorganic clay, tetrapoly(ethylene glycol) gels submicrometer-sized gel particles, gels having dangling chains and macroporous gels have been designed with this objective [1, 32]. In biomedical field, hydrogels can be employed in controlling the process of vascularization through controlled release of vascular endothelial growth factor (VEGF) and other angiogenic factors at the desired target site. However, hydrogel-based tissue scaffolds are yet to be commercially viable because of the huge expenditure involved [23, 33].

Therefore, modifications in the chemistry and architecture of hydrogels and advances in the field of 'smart' or stimuli-responsive hydrogels and self-healing hydrogels have opened up new avenues in the fabrication of bio-mimetic scaffolds, artificial tissues and organs, drug delivery systems, injectable hydrogels and controlled release systems for fertilizers. Physicochemical characterization of hydrogel parameters, mathematical modelling of polymer-solvent interaction during hydrogel swelling and sorption phenomena will enable in the future evolution of hydrogel materials.

Author details

Sutapa Biswas Majee

Address all correspondence to: sutapabiswas2001@yahoo.co.in

Division of Pharmaceutics, NSHM College of Pharmaceutical Technology, NSHM Knowledge Campus, Kolkata Group of Institutions, Kolkata, West Bengal, India

References

[1] Okay O (2009). General properties of hydrogels. *In* Hydrogel sensors and actuators. G. Gerlach and K.F. Arndt (eds.). Springer Series on Chemical Sensors and Biosensors, 6, Springer–Verlag Berlin Heidelberg. (http://www.springer.com/978-3-540-75644-6).

[2] Ahmed EM (2015). Hydrogel: Preparation, characterization, and applications: A review. *Journal of Advanced Research* 6, 105–121.

[3] Calo E and Khutoryanskiy VV (2015). Biomedical applications of hydrogels: A review of patents and commercial products. *European Polymer Journal* 65, 252–267.

[4] Omidian H and Park K (2012). Hydrogels. *In* Fundamentals and applications of controlled release drug delivery, J.Siepmann, R. Siegel and M. Rathbone (eds.). Controlled Release Society, Springer New York, 75–106 (DOI: 10.1007/978-1-4614-0881-9).

[5] Ganji F, Vasheghani-Farahani S and Vasheghani-Farahani E (2010). Theoretical description of hydrogel swelling: A review. *Iranian Polymer Journal* 19, 375–398.

[6] Porter TL, Stewart R, Reed J and Morton K (2007). Models of hydrogel swelling with applications to hydration sensing. *Sensors* 7,1980–1991.

[7] Brannon-Peppas L and Peppas NA (1991). Equilibrium swelling behavior of pH-sensitive hydrogels. *Chemical Engineering Science* 46, 715–722.

[8] Chen Y, Zhang Q, Chen L , Bai H and Li L (2013). Basic aluminum sulfate@graphene hydrogel composites: Preparation and application for removal of fluoride. *Journal of Materials Chemistry A* 1, 13101–13110.

[9] Medina RP, Ballesteros FC and Rodrigues DF (2015). The influence of physical configuration on the performance of chitosan–poly(acrylic acid) hydrogel beads as adsorbent

for lead(II). *Proceedings of the 14th International Conference on Environmental Science and Technology*, Rhodes, Greece.

[10] Saeed R and Ul Abdeen Z (2014). Sorption and kinetics study of potassium chloride in polyvinyl alcohol-borate hydrogel. *International Journal of Scientific and Technology Research* 3, 321–327.

[11] Zhang YN, Avery RK, Vallmajo-Martin Q, Assmann A, Vegh A, Memic A, Olsen BD, Annabi N and Khademhosseini A (2015). A highly elastic and rapidly crosslinkable elastin-like polypeptide-based hydrogel for biomedical applications. *Advanced Functional Materials* 25, 4814–4826.

[12] Maitra J and Shukla VK (2014). Cross-linking in hydrogels—A review. *American Journal of Polymer Science* 4, 25–31.

[13] Xing Q, Yates K, Vogt C, Qian Z, Frost MC and Zhao F (2014). Increasing mechanical strength of gelatin hydrogels by divalent metal ion removal. *Scientific Reports* 4, 4706.

[14] Yue K, Trujillo-de Santiago G, Moises Alvarez M, Tamayol A, Annabi N and Khademhosseini A (2015). Synthesis, properties, and biomedical applications of gelatin methacryloyl (GelMA) hydrogels. *Biomaterials* 73, 254–271.

[15] Cao L, Cao B, Lu C, Wang G, Yu L and Ding J (2015). An injectable hydrogel formed by *in situ* cross-linking of glycol chitosan and multi-benzaldehyde functionalized PEG analogues for cartilage tissue engineering. *Journal of Materials Chemistry B* 3, 1268–1280.

[16] Ahearne M, Yang Y and Liu KK (2008). Mechanical characterisation of hydrogels for tissue engineering applications. *In* Topics in tissue engineering, Vol. 4. N. Ashammakhi, R Reis and F Chiellini (eds.).Available from : http://www.oulu.fi/spareparts/ebook-topics-in-t-e-vol4/published-chapters.html.

[17] Wang QM, Mohan AC, Oyen ML and Zhao XH (2014). Separating viscoelasticity and poroelasticity of gels with different length and time scales. *Acta Mechanica Sinica* 30, 20–27.

[18] Banerjee A, Arha M, Choudhary S, Ashton RS, Bhatia SR, Schaffer DV and Kane RS (2009). The influence of hydrogel modulus on the proliferation and differentiation of encapsulated neural stem cells. *Biomaterials* 30, 4695–4699.

[19] Wallace M, Adams DJ and Iggo JA (2013). Analysis of the mesh size in a supramolecular hydrogel by PFG-NMR spectroscopy. *Soft Matter* 9, 5483–5491.

[20] Bäckström S, Benavente J, Berg RW, Stibius K, Larsen MS, Bohr H and Hélix-Nielsen C (2012). Tailoring properties of biocompatible PEG-DMA hydrogels with UV light. *Materials Science Applications* 3, 425–431.

[21] Ferro M, Castiglione F , Punta C, Melone L, Panzeri W, Rossi B, Trotta F and Mele A(2014). Anomalous diffusion of ibuprofen in cyclodextrin nanosponge hydrogels: An HRMAS NMR study. *Beilstein Journal of Organic Chemistry* 10, 2715–2723.

[22] Choi B, Loh XJ, Tan A, Loh CK, Ye E, Joo MK and Jeong B (2015). Introduction to in situ forming hydrogels for biomedical applications. *In* In-situ gelling polymers, Series in BioEngineering X.J. Loh (ed.) Springer Science+Business Media.(http://www.springer.com/978-981-287-151-0).

[23] Silva EA and Mooney DJ (2007). Spatiotemporal control of vascular endothelial growth factor delivery from injectable hydrogels enhances angiogenesis. *Journal of Thrombosis and Haemostasis* 5, 590–598.

[24] Tabriz AG, Hermida MA, Leslie NR and Shu W (2015). Three-dimensional bioprinting of complex cell laden alginate hydrogel structures. *Biofabrication* 7, 045012.

[25] Phadke A, Zhang C, Arman B, Hsu CC, Mashelkar RA, Lele AK, Tauber MJ, Arya G and Varghese S (2012). Rapid self-healing hydrogels. *PNAS*,109, 4383–4388.

[26] Casolaro M, Cini R, Del Bello B, Ferrali M and Maellaro E (2009). Cisplatin/hydrogel complex in cancer therapy. *Biomacromolecules* 10, 944–949.

[27] Lee ALZ, Ng VWL, Gao S, Hedrick JL and Yang YY (2014). Injectable hydrogels from triblock copolymers of vitamin e-functionalized polycarbonate and poly(ethylene glycol) for subcutaneous delivery of antibodies for cancer therapy. *Advanced Functional Materials* 24, 1538–1550.

[28] Sinha MK, Gao J, Stowell CET and Wang Y (2015). Synthesis and biocompatibility of a biodegradable and functionalizable thermo-sensitive hydrogel. *Regenerative Biomaterials* 2, 177–85.

[29] Constantin M, Cristea M, Ascenzi P, Fundueanu G (2011). Lower critical solution temperature versus volume phase transition temperature in thermoresponsive drug delivery systems. *eXPRESS Polymer Letters* 5, 839–848.

[30] Mah E and Ghosh R (2013). Thermo-responsive hydrogels for stimuli-responsive membranes. *Processes* 1, 238–262.

[31] Guvendiren M, Lu HD and Burdick JA (2012). Shear-thinning hydrogels for biomedical applications. *Soft Matter* 8, 260–272.

[32] Noda Y, Hayashi Y and Ito K (2014). From topological gels to slide-ring materials. *Journal of Applied Polymer Science.* 131, 40509–517

[33] Corbin EA, Millet LJ, Pikul JH, Johnson CL, Georgiadis JG, King WP and Bashir (2013). Micromechanical properties of hydrogels measured with MEMS resonant sensors. *Biomedical Microdevices.* 15, 311–19.

Chapter 2

An Introduction to Hydrogels and Some Recent Applications

Morteza Bahram, Naimeh Mohseni and
Mehdi Moghtader

Additional information is available at the end of the chapter

http://dx.doi.org/10.5772/64301

Abstract

Hydrogels have existed for more than half a century, and today they have many applications in various processes ranging from industrial to biological. There are numerous original papers, reviews, and monographs focused on the synthesis, properties, and applications of hydrogels. This chapter covers the fundamental aspects and several applications of hydrogels based on the old and the most recent publications in this field.

Keywords: Hydrogels, Classification, Synthesis, Applications, Drug delivery, Spinal cord injury, Supercapacitor

1. Introduction

A hydrogel is a three-dimensional (3D) network of hydrophilic polymers that can swell in water and hold a large amount of water while maintaining the structure due to chemical or physical cross-linking of individual polymer chains. Hydrogels were first reported by Wichterle and Lím (1960) [1]. By definition, water must constitute at least 10% of the total weight (or volume) for a material to be a hydrogel. Hydrogels also possess a degree of flexibility very similar to natural tissue due to their significant water content. The hydrophilicity of the network is due to the presence of hydrophilic groups such as $-NH_2$, -COOH, -OH, $-CONH_2$, - CONH -, and $-SO_3H$.

Hydrogels undergo a significant volume phase transition or gel-sol phase transition in response to certain physical and chemical stimuli. The physical stimuli include temperature, electric and magnetic fields, solvent composition, light intensity, and pressure, while the chemical or biochemical stimuli include pH, ions, and specific chemical compositions. However, in most cases such conformational transitions are reversible; therefore, the hydrogels are capable of returning to their initial state after a reaction as soon as the trigger is removed. The response of hydrogels to external stimuli is mainly determined by the nature of the monomer, charge density, pendant chains, and the degree of cross-linkage. The magnitude of response is also directly proportional to the applied external stimulus.

There are numerous original papers, reviews, and monographs focused on the synthesis, properties, and applications of hydrogels. This chapter covers the fundamental aspects and application areas of hydrogels.

2. Classifications of hydrogels

The literature reports a number of classifications of hydrogels and presents several views. Hydrogels are mainly formed from biopolymers and/or polyelectrolytes. Concerning definitions of hydrogel types, according to the source, hydrogels can be divided into those formed from natural polymers and those formed from synthetic polymers [2]. Depending on the ionic charges on the bound groups, hydrogels may be cationic, anionic, or neutral. The types of cross-linking agents also can be the criteria for classification.

Hydrogels can be physical, chemical, or biochemical. Physical gels can undergo a transition from liquid to a gel in response to a change in environmental conditions such as temperature, ionic concentration, pH, or other conditions such as mixing of two components. Chemical gels use covalent bonding that introduces mechanical integrity and degradation resistance compared to other weak materials. In biochemical hydrogels, biological agents like enzymes or amino acids participate in the gelation process.

It is also possible to divide hydrogels into groups based on their structure: amorphous, semicrystalline, crystalline, and hydrocolloid aggregates [3]. **Figure 1** clearly represents the classification of hydrogels based on their source and properties, along with detailed classifi-

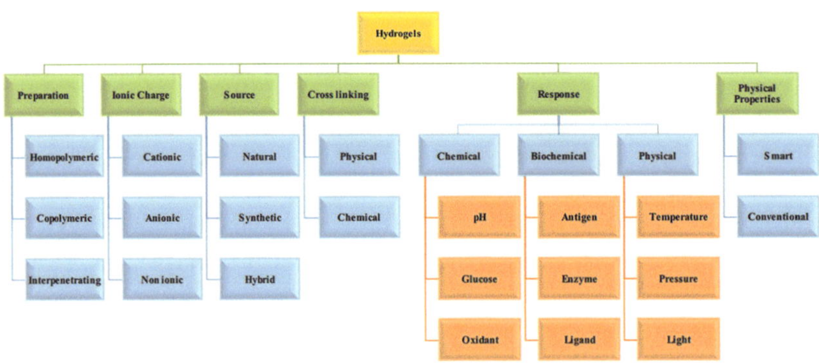

Figure 1. Classification of hydrogels based on the different properties.

http://dx.doi.org/10.5772/64301

cations based on their response, that is, physically, chemically, and biochemically responsive hydrogels (**Figures 2** and **3**).

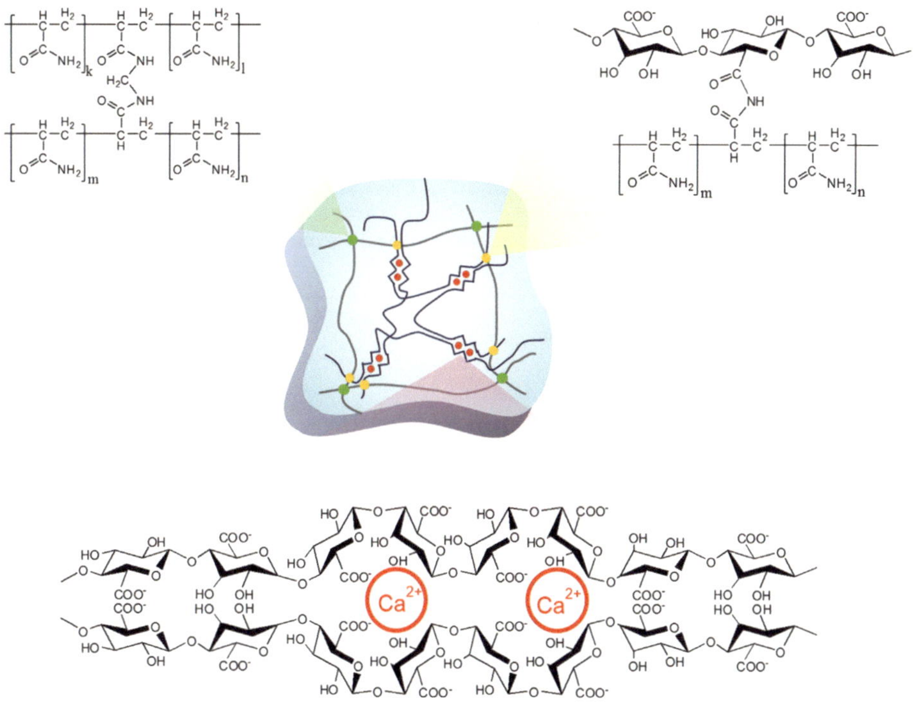

Figure 2. In situ hydrogel formation using chemical cross-linking and ionic interaction between alginate and calcium ions [61, 62].

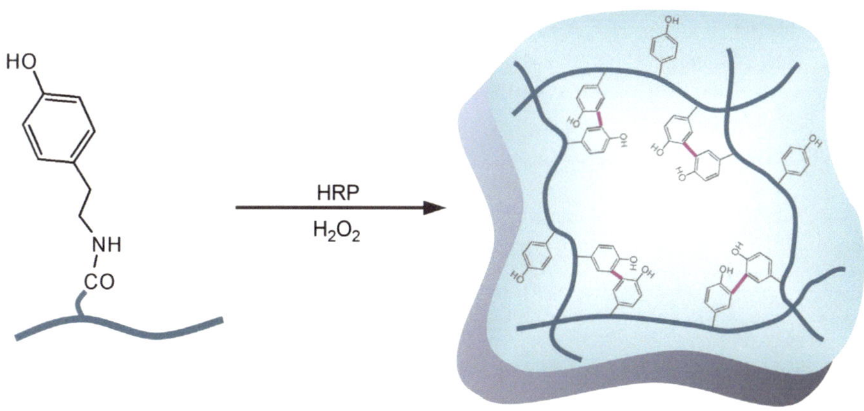

Figure 3. In situ hydrogel formation using an enzymatic cross-linking reaction with horseradish peroxidase (HRP) and H_2O_2 [62].

3. Synthesis of hydrogels

Based on the methods of preparation, hydrogels may be classified as homopolymer, copolymer, semi-interpenetrating network (semi-IPN) and interpenetrating network (IPN). **Table 1** indicates some examples.

Type of hydrogel	Monomer	Cross-linker	Specific reaction conditions	References	Applications
Homopolymer	Poly(2-hydroxyethyl methacrylate) (PHEMA) 2-Hydroxyethyl methacrylate (HEMA) Polyethylene glycol (PEG)	Polyethylene glycol dimethacrylate TEGDMA (triethylene glycol dimethacrylate)	Presence of benzoin isobutyl ether as the UV-sensitive initiator	[4, 5]	Drug delivery systems, contact lenses, scaffolds for protein recombination Wound healing and functional tissues production
Copolymer	Methacrylic acid (MAA) PEG-PEGMA Carboxymethyl cellulose (CMC) Polyvinylpyrrolidone (PVP)	Tetra(ethylene glycol) dimethacrylate	Free-radical photopolymerization	[4–7]	Drug delivery, hydrogel dressing material
Semi-inter penetrating network	Acrylamide/acrylic acid copolymer Linear cationic polyallylammonium chloride	*N,N′*-methylene bisacrylamide	Template copolymerization	[75]	Drug delivery
Inter penetrating network	Poly(*N*-isopropyl acrylamide) (PNIPAM) Chitosan	*N,N′*-methylene bisacrylamide	*N,N,N′*,N′-tetramethylethylenediamine (TEMED), ammonium persulphate (APS) and Presence of UV light	[76]	Drug delivery
Self -assembling peptide systems	Acrylate-modified PEG and acrylate-modified hyaluronic acid Heparin Amine end-functionalized 4-arm star-PEG	No cross -linking agent	Presence of UV light and a photo-initiator EDC/sulfo-NHS solution and low temperature	[63, 49]	Tissue regeneration

Table 1. Some examples of synthesis methods and applications of hydrogels.

http://dx.doi.org/10.5772/64301

Homopolymers contain only one type of monomer in their structure, and based on the nature of the monomer and the technique used for polymerization, they may have a cross-linked structure (**Figure 4**).

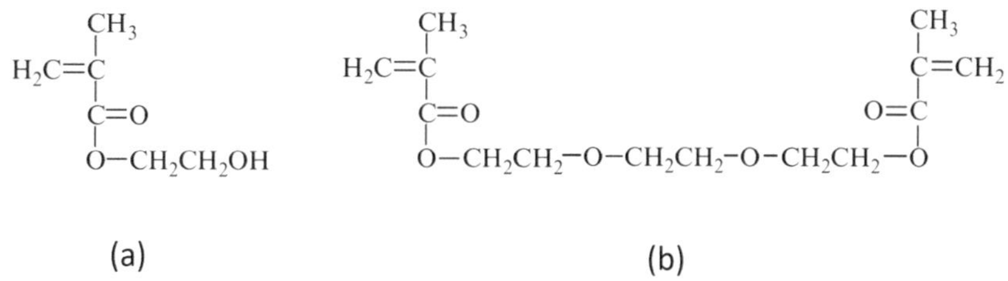

Figure 4. Structures of (a) HEMA and (b) TEGDMA.

Copolymeric hydrogels are composed of two types of monomers, of which at least one is hydrophilic in nature (**Figures 5–7**).

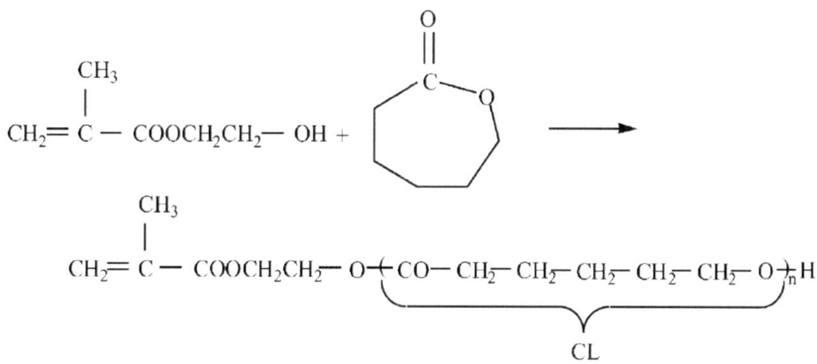

Figure 5. Synthesis of the poly(e-caprolactone)-HEMA macromonomer.

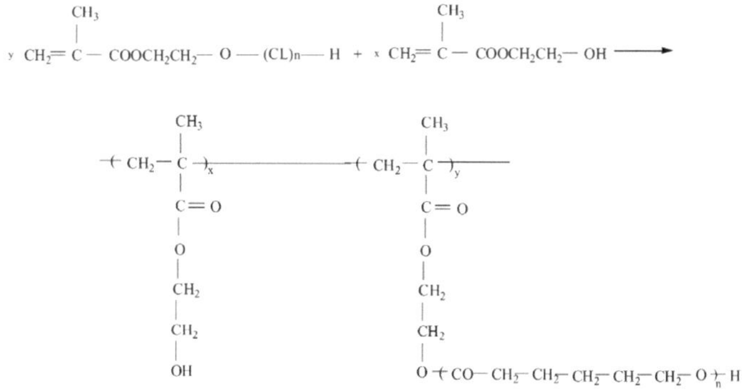

Figure 6. Synthesis of the poly(2-hydroxyethyl methacrylate)-graft-poly(e-caprolactone) copolymer.

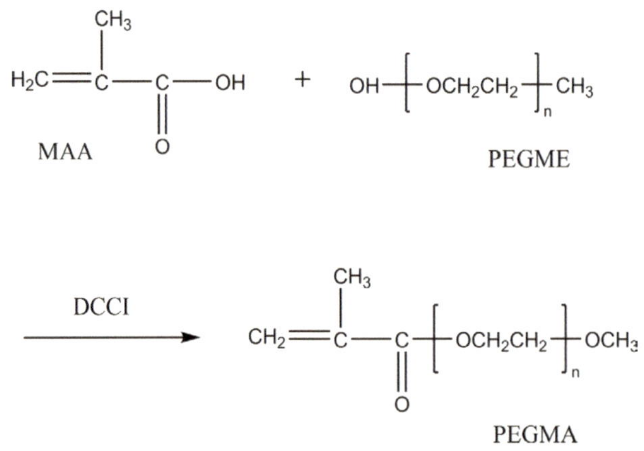

Figure 7. Dicyclohexylcarbodiimide (DCCI) method to synthesize PEG-containing macromonomers.

A semi-IPN forms when a linear polymer penetrates into another cross-linked network without any other chemical bonds between them. Semi-IPNs can more effectively preserve rapid kinetic response rates to pH or temperature due to the absence of a restricting interpenetrating elastic network while still providing the benefits like modified pore size, slow drug release, etc.

Combining of two polymers can lead to the formation of IPNs provided that one of them is already present in the solution and the other is synthesized or cross-linked in situ. This process is done by preparing a solution of monomers and initiators and then immersing a pre-polymerized hydrogel into this solution. The pore size and surface properties of an IPN can be modified to control the kinetics of drug release, environmental interactions of the hydrogel, and its mechanical features.

It is worth to point here the self-assembling peptide systems that are synthetic amino acid-based molecules which undergo a sol-gel transition when brought to neutral pH and ionic concentration. These systems do not use cross-linking agents; hence, they can safely encapsulate cells and/or drugs without exposing them to toxic agents [49] (**Figures 8** and **9**).

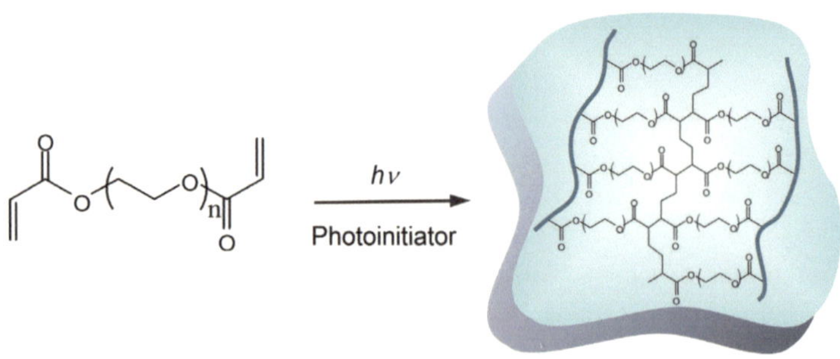

Figure 8. In situ hydrogel formation using photo-cross-linking.

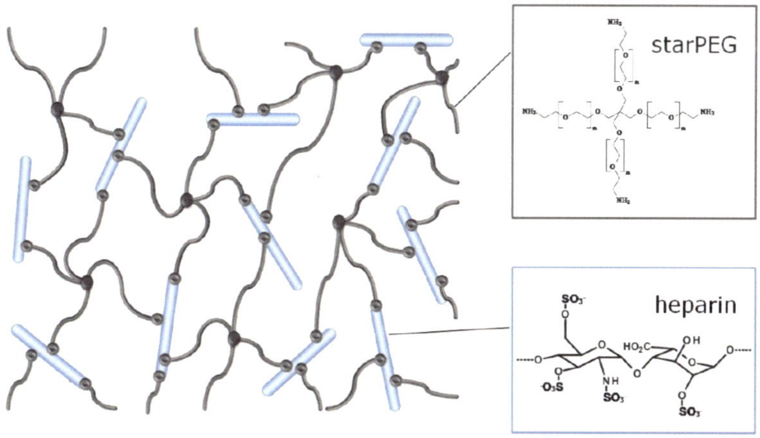

Figure 9. Hydrogel formation by cross-linking of star-PEG and heparin.

4. Some common types of hydrogels

The environment-sensitive hydrogels, also called "intelligent" or "smart" hydrogels, are currently the subject of considerable scientific research in various fields including biomedical, biotechnology, pharmaceutical, and separation science. In this section, we will introduce four classes of most used hydrogels.

4.1. pH-sensitive hydrogels

Any pH-sensitive polymer structurally contains hanging acidic (e.g. carboxylic and sulfonic acids) or basic (e.g. ammonium salts) groups that respond to the pH changes in their environment by gain or loss of protons. Polyelectrolytes are polymers that have a large number of such ionizable groups. Anionic polyelectrolytes such as poly(acrylic acid) (PAA) are deprotonated in basic environmental conditions and then electrostatic repulsions between the chains strongly increase, which allow water molecules to penetrate causing drastic swelling of the hydrogel. However, in an acidic media, the acidic polymer protonates resulting in a decrease of charge density and polymer volume collapse (**Figure 10**). In contrast, cationic polyelectrolytes such as poly(*N,N* 9-diethylaminoethyl methacrylate) become ionized and swell in acidic pH (**Figure 10**). Amphiphilic hydrogels contain both acidic and basic moieties; therefore, they exhibit two-phase transitions in both acidic and basic environments, rather than neutral media. **Figure 11** clearly demonstrates phase transition behavior of polyelectrolyte hydrogels. Worth noting is that the phase transition from collapsed state to expanded state occurs in a small range close to the apparent dissociation constant pK_a of the hydrogel which is mostly identical with the pK_a of the ionizable groups. Approximately at the apparent pK_a of the polymer, the ionization begins and the electrostatic repulsions of the same charges present in the polymer network cause a drastic swelling of the hydrogel. If the ionization of the ionizable component is complete, the swelling process stops and further pH increase only increases the ionic strength [7, 8]. The phase transition pH range can be modulated by selecting the ionizable moiety with a pK_a matching the desired pH range or by incorporating hydrophobic moieties into the polymer backbone [10].

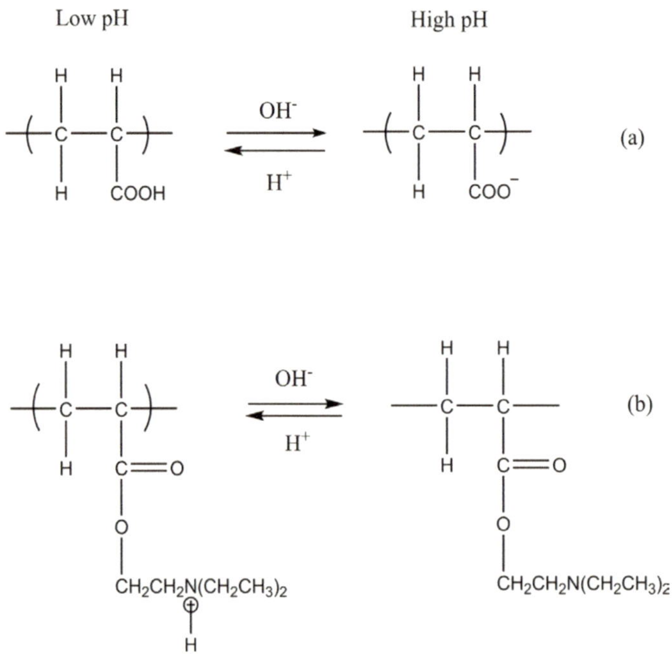

Figure 10. pH-dependent ionization of polyelectrolytes. (a) Poly(acrylic acid) and (b) poly(*N,N*′-diethylaminoethyl methacrylate).

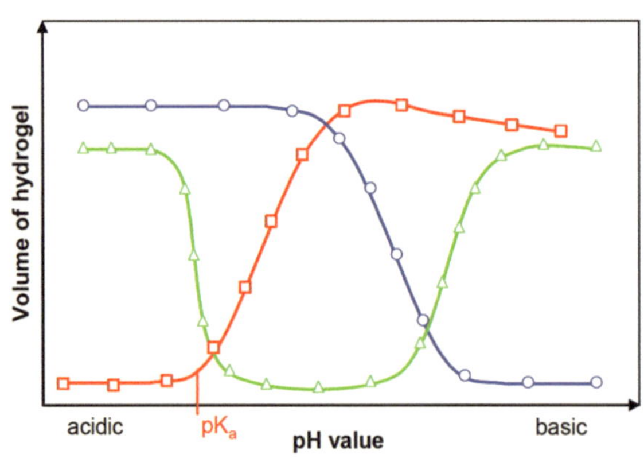

Figure 11. Phase transition behavior of polyelectrolyte hydrogels. Acidic hydrogels (□) are ionized by deprotonation in basic solutions. Basic hydrogels (○) swell in acidic solutions due to the ionization of their basic groups by protonation. Amphiphilic hydrogels (Δ) contain both acidic and basic groups, therefore they show two-phase transitions.

Ionization of a polyelectrolyte takes place similar to acidic or basic groups of monoacids or monobases, but due to the electrostatic effects of neighboring groups, it will have a different dissociation constant (K_a) from corresponding monoacids or monobases.

The extent of swelling is influenced by any factor that alters this electrostatic repulsion including pH, ionic strength, and the type of counterions. **Figure 12** shows this phenomenon. In this figure, hydrogel has two phases: one phase is separated, gel-like, and formed by polymer-polymer interactions. In this condition, the maximum hydrophobicity takes place and the shrinkage of hydrogel occurs. In the second phase, interactions between the solvent and the polymer create a mixed phase in which the polymer and the aqueous solution are mixed well. The maximal value of hydrophilicity and swelling occurs in this second phase [7].

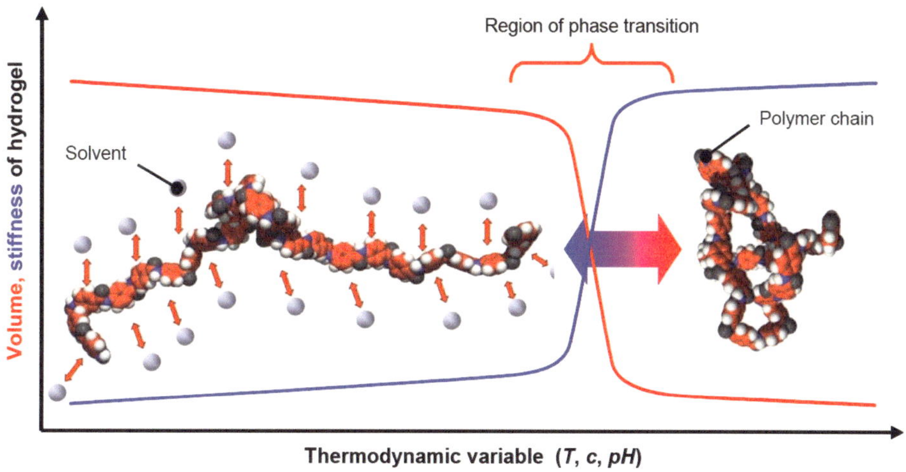

Figure 12. Phase transition behavior of stimuli-responsive hydrogels.

Different pH-sensitive behaviors and degrees of swelling can be achieved by using different monomers. The most commonly studied ionic polymers for pH-responsive behavior include poly(acrylamide) (PAAm), PAA, poly(methacrylic acid) (PMAA), poly(diethylaminoethyl methacrylate) (PDEAEMA), and poly(dimethylaminoethyl methacrylate) (PDMAEMA). Polymers containing phosphoric acid derivatives have also been reported.

4.2. Temperature-sensitive hydrogels

Temperature-sensitive hydrogels (thermogels) are aqueous monomer/polymer solutions, which have the ability to form a gel upon temperature change and have a slightly hydrophobic characteristic due to the presence of groups such as methyl, ethyl, and propyl, which preferably interact with water molecules by hydrogen bonds that cause the hydrogel to swell. These hydrogen bonds are correlated to the temperature. The structures of some of the temperature-sensitive hydrogels are shown in **Figure 13**.

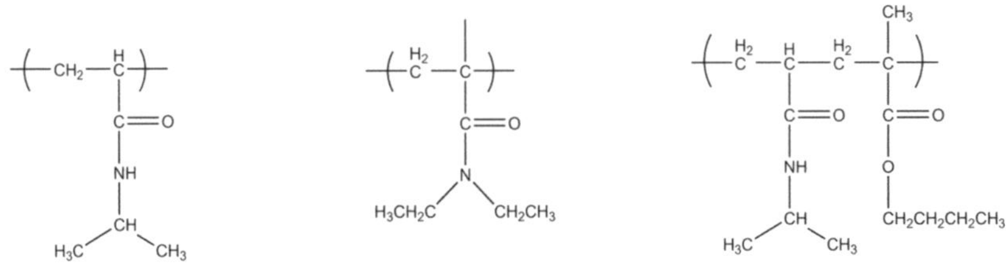

Poly(N-isopropylacrylamide) (PNIAAm) — Poly(N,N-diethylacrylamide) (PDEAAm) — P(NIAAm-co-BMA)

Figure 13. Structures of some temperature-sensitive polymers.

As can be seen, the common characteristic of temperature-sensitive polymers is the presence of hydrophobic groups. Most polymers increase their water solubility as the temperature increases. However, in some cases water solubility decreases with an increase in temperature (inverse or negative temperature dependence) [9]. This unusual behavior produces a phenomenon of polymer phase transition as the temperature is raised to a critical value called the "lower critical solution temperature" or LCST, which is an entropy-driven process. In the case of hydrogels with negative thermosensitivity, right below the LCST, water is a good solvent for the polymer, and hydrogen bonding interactions between the polymer and water molecules lead to enhanced dissolution in water. However, when the temperature exceeds the LCST, these interactions are broken, and the polymer chains collapse and then precipitate in the media [10, 11]. These types of hydrogels comprise polymer chains that either possess moderately hydrophobic segments (if too hydrophobic, the polymer chains will not dissolve in water at all) or contain a mixture of hydrophilic and hydrophobic groups.

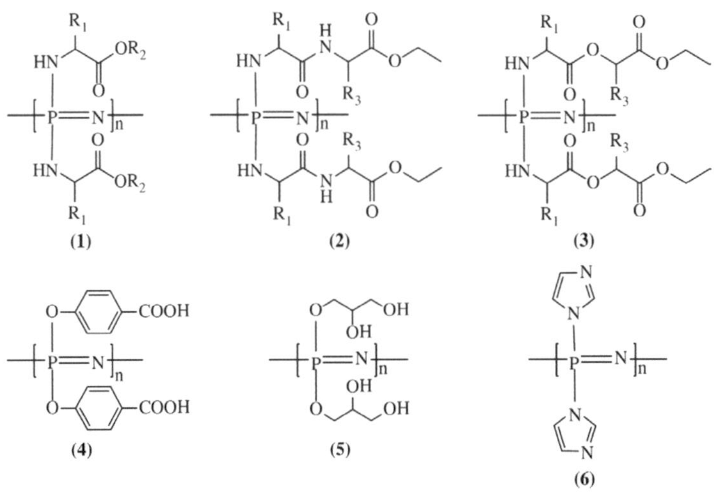

Figure 14. Some examples of poly(organophosphazene) thermogels.

As the temperature increases, positive thermosensitive hydrogels exhibit just the opposite behavior of negative thermosensitive hydrogels. The LCST of hydrogels can be modulated to increase by adding a hydrophilic component, or to decrease with a hydrophobic one. Due to this property, temperature-sensitive hydrogels swell below the LCST and collapse in an aqueous environment above this temperature, being thus suitable for controlled drug delivery. Among others, poly(*N*-isopropylacrylamide) (PNIPAM) is the most studied thermosensitive hydrogel in tissue engineering investigations. This is due to the ability of PNIPAM to squeeze out the absorbed drug when temperature is near that of the human body [19].

Other examples of thermosensitive hydrogels are collagen, agarose, hyaluronic acid, poly(organophosphazenes), and chitosan [58, 59] (**Figure 14**).

4.3. Electro-sensitive hydrogels

Electro-sensitive hydrogels, as the name indicates, undergo shrinking or swelling in the presence of an applied electric field. Like pH-sensitive hydrogels, they are usually composed of polyelectrolytes. Under the influence of an electric field, a force on counterions and immobile charged groups is produced in the network, which attracts mobile ions to the electrodes. As a result, the hydrogel can swell and shrink regionally at the cathode and anode, respectively. This phenomenon leads to bending of the hydrogel, which is caused by ion concentration difference inside the hydrogel network and culture medium and can be explained by Flory's theory of osmotic pressure [12–17]. The extent of bending depends on hydrogel structure and electrical field characteristics including strength, direction, and duration of the electrical stimulus. Electro-sensitive hydrogels can selectively be permeable for a specific molecular size and adjust the water permeability by expanding and contracting in micropore size under electrical stimulation [18]. Because electro-responsive hydrogels can transform electrical energy into mechanical energy and have promising applications in biomechanics, sensing, energy transduction, sound dampening, chemical separations, controlled drug delivery [33], and tissue engineering [20, 21], these polymers are an increasingly important class of smart materials. Hydrogels of acrylamide and carboxylic acid derivatives like PAA have been utilized as electro-sensitive and biocompatible smart muscle-based devices [22, 23].

4.4. Light-responsive hydrogels

Photo-responsive hydrogels undergo a change in their properties when irradiated with light of the appropriate wavelength. Typically, these changes are the result of light-induced structural transformations of specific functional groups along the polymer backbone or side chains. Light-sensitive hydrogels can expand and contract upon exposure to ultraviolet (UV) or visible light. Visible light offers many advantages over UV light including wide availability, low cost, ease of manipulation, and clean operation. The mechanism of visible light-induced volume change of hydrogels is based on the induction of temperature changes by incorporating a photo-responsive functional group (chromophore) (e.g. trisodium salt of copper chlorophyllin) into the polymer network. Under exposure to a specific wavelength, the chromophore absorbs light which is then dissipated locally as heat, increasing the "local" temperature of the hydrogel [26]. The resulting temperature change alters the swelling behavior of the thermosensitive hydrogel [9]. Because of the thermal nature of the infrared radiation, it can also be used to elicit a hydrogel response in the absence of chromophores. If an additional functional group, such as an ionizable moiety of PAA, is incorporated into the hydrogel network, the light-responsive hydrogels become sensitive to pH changes also. This type of hydrogel can be induced to shrink by visible light and can be induced to swell by increasing the pH. The UV-sensitive hydrogels can be synthesized by introducing a leuco derivative molecule into the polymer network. Leuco derivatives are normally neutral but dissociate into ion pairs upon UV exposure. At a fixed temperature, the hydrogels discontinuously swelled in response to UV irradiation but shrank when the UV light was removed [24]. The potential applications of light-responsive hydrogels in the development of artificial muscles [25, 64], reversible valves in microfluidic devices [65], and temporal drug delivery were proposed.

5. Applications of hydrogels

Hydrogels are used in many fields. This is due to their specific structures and compatibility with different conditions of use. Flexibility of hydrogels, which is because of their water content, makes it possible to use them in different conditions ranging from industrial to biological, and the biocompatibility of the materials used to produce them and also their chemical behavior in biological environments, which can be nontoxic, extends their applications to the medical sciences.

Major applications and some examples of hydrogel usages are listed below. Note that it is not a complete listing but considers the most practical applications of hydrogels in medicine and industry.

5.1. Drug delivery

Controlled drug delivery systems (DDS), which are used to deliver drugs at certain rates for predefined periods of time, have been used to overcome the limitations of regular drug formulations. The marvelous properties of hydrogels make them a great choice in drug delivery applications. The hydrogel structures with high porosity can be obtained by controlling two factors: the degree of cross-linking in the matrix and the affinity of hydrogel to the aqueous environment in which swelling occurs. Due to the porous structures, hydrogels are highly permeable to different kinds of drugs and thus drugs can be loaded and, in proper conditions, released [27]. The possibility of releasing pharmaceuticals for long periods of time (sustained release) is the main advantage obtained from hydrogels in drug delivery investigations, which results in supplying a high concentration of an active pharmaceutical substance to a specific location over a long period of time.

Both physical (electrostatic interactions) and chemical (covalent bonding) strategies can be employed to enhance the binding between a loaded drug and the hydrogel matrix to extend the duration of drug release. Hydrogels can store and protect various drugs from hostile environments, and release them at a desired kinetics of the release. Drug release can be activated on demand by local changes in pH, temperature, the presence of specific enzymes, or by remote physical stimuli.

5.1.1. pH-sensitive hydrogels in DDS

Since the pH change occurs at many specific or pathological body sites, it is one of the important environmental parameters for DDS. The human body exhibits variations of pH along the gastrointestinal tract and also in some specific areas such as certain tissues (and tumoral areas) and subcellular compartments. Both acidic and basic polymers are employed in pH-sensitive DDS. PAA, PMAA, poly(L-glutamic acid), and polymers containing sulfonamide are the most commonly used acidic polymers for drug delivery. Typical examples of the basic polyelectrolytes include poly(2-(dimethylamino) ethyl methacrylate) and poly(2-(diethylamino) ethyl methacrylate), poly(2-vinylpyridine), and biodegradable poly(β-amino ester).

pH-sensitive hydrogels were also used for extraction and determination purposes by different methodologies [28–31].

5.1.2. Temperature-sensitive hydrogels in DDS

Thermosensitive polymers, like pH-responsive systems, offer many possibilities in biomedicine.

Among many temperature-sensitive polymers, poly(*N*-isopropylacrylamide) (PNIPAAm) and poly(*N*,*N*-diethylacrylamide) (PDEAAm) find many applications. PDEAAm has a low value of LCST (a critical temperature below which the components of a solution with any composition are miscible) in the range of 25–32°C, which is near to normal body temperature.

5.2. Dyes and heavy metal ions removal

Heavy metal pollution is commonly found in wastewater of many industrial processes and has been known to cause severe threats to the public health and ecological systems. The removal of heavy metal ions from various water resources is of great scientific and practical interest.

Synthetic cross-linked polyacrylate hydrogels have been used to remove heavy metal toxicity from aqueous media [27]. However, application of these synthetic materials on large scales may not be a practical solution because they are very costly.

The pollution caused by heavy metal ions can be removed by well-known adsorption processes which, alongside flexibility in design and operation, offer the advantage of reusing the treated effluent. Also because of general reversibility of adsorption process, it is usually possible to regenerate the adsorbent to make the process most cost-effective.

The use of hydrogels as adsorbents for the removal of heavy metals, recovery of dyes, and removal of toxic components from various effluents has been studied. Adsorbents with carboxyl, sulfonic, phosphonic, and nitrogen groups on their surface favor metal ion adsorption [77].

The hydrogels were proven to be excellent dye adsorbent materials with extremely high amounts of methylene blue adsorption.

Among hydrogel-forming materials, polyelectrolytes have a special significance in heavy metal ions' removal. Many applications of polyelectrolytes in this area are due to their ability to bind oppositely charged metal ions to form complexes.

In fact, having both cationic and anionic charges on the micro- or nano-gel provides additional advantages for the removal of two distinct species simultaneously. Hydrogels are versatile and viable materials that show potential for environmental applications.

Chitosan, alginate, starch, and cellulose derivatives are biopolymer-based hydrogels, which were used to remove metal ions from aqueous media. It has been shown that the sorption mechanism and sorption capacity of heavy metal ions were influenced by the functional groups of the hydrogel. This is because of the participation of other processes like chelating and ion exchange rather than simple sorption in removal of metal ions [78, 79].

Chitosan-based hydrogels are applicable in the removal of heavy metal ions due to the presence of multiple amino (NH_2) and hydroxyl (OH) groups in their structure. This applicability originates from the tendency of metal ions to form chelates with the so-called amino groups. But after reaction of chitosan with cross-linkers, its alkalescence which is related to adsorption capacity is decreased. Chemical modification of these functional groups can improve not only the adsorption capacity but also the physicochemical properties of chitosan [79, 80]. Different approaches were employed by researchers to modify chitosan including the use of amino acid esters, oxo-2-glutaric acid, pyridyl, ethylenediamine, carbodiimide, aromatic polyimides, amine-functionalized magnetic nanoparticles, and glycine [81–83]. It is shown in these studies that both adsorption capacity and mechanical resistance of chitosan-based hydrogels will improve after modification of functional groups.

5.3. Scaffolds in tissue engineering

Tissue engineering is defined as a combination of materials, engineering, and cells to improve or replace biological organs. This needs finding proper types of cells and culturing them in a suitable scaffold under appropriate conditions. Hydrogels are an appealing scaffold material because their structures are similar to the extracellular matrix of many tissues, they can often be processed under relatively mild conditions, and they may be delivered in a minimally invasive manner [32]. Adequate scaffold design and material selection for each specific application depends on several variables, including physical properties, mass transport properties, and biological properties and is specified by the intended scaffold application and environment into which the scaffold will be placed. For example, the type of scaffold used to produce artificial skin must be different from that used for artificial bone and thus different structures for materials are needed.

Both synthetic and naturally derived materials can be used to form hydrogels for tissue engineering scaffolds.

http://dx.doi.org/10.5772/64301

Synthetic hydrogels are interesting because it is easy to control their chemistry and structure and thus alter their properties. Examples of polymeric synthetic materials which can be used to form hydrogels are poly(ethylene oxide) (PEO), poly(vinyl alcohol) (PVA) and poly(propylene fumarate) (PPF) [33] (**Figure 15**).

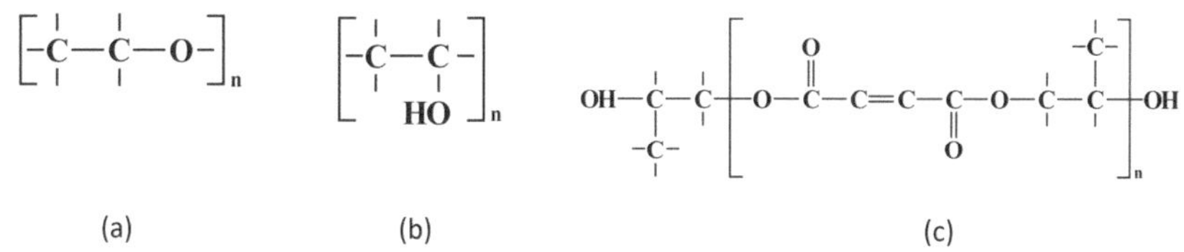

Figure 15. Structure of synthetic hydrogel-forming polymers: (a) PEO, (b) PVA (100% hydrolyzed), and (c) PPF.

Naturally derived hydrogel-forming polymers are other candidates for use in tissue engineering scaffolds because they either are natural extracellular matrix components or have properties similar to these matrices and they interact in a favorable manner in vivo. Examples are alginate and chitosan [32–34] (**Figure 16**).

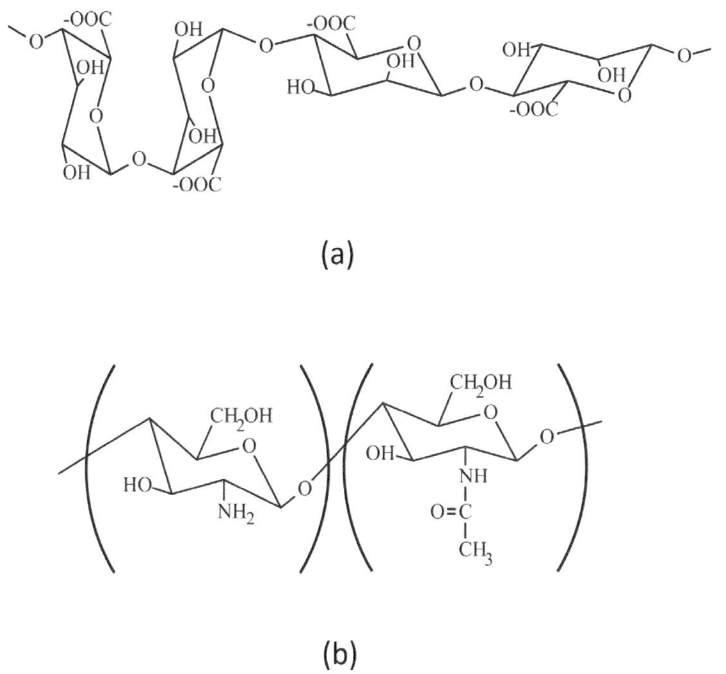

Figure 16. Structure of naturally derived hydrogel-forming polymers: (a) alginate and (b) chitosan.

Hydrogels are used for three purposes in tissue engineering applications. They may be used as agents for filling vacant spaces, carriers for delivery of bioactive molecules, and 3D structures that act as a support for cells and help the formation of an ideal tissue.

Agents for filling vacant spaces (space-filling agents) include scaffolds that provide bulking, prevent adhesions, or act as bioadhesives [33]. To reach this aim, the most basic design requirements for a hydrogel are the abilities to keep a desired volume and structural integrity for the required time.

Hydrogel scaffolds based on alginate, chitosan, and collagen show potential for use as general bulking agents. Synthetic hydrogels are often used as anti-adhesive materials because cells lack adhesion receptors to them and proteins often do not readily absorb to them if designed appropriately. Polyethylene glycol (PEG) has been used to prevent post-operative adhesions [32, 33].

Hydrogels composed of chitosan and chitin derivatives are now used as biological adhesives in surgical procedures to seal small wounds out of which air and body fluids could leak, and to improve the effectiveness of wound dressings [37, 38].

Another application of scaffold hydrogels that is quite different includes using them as vehicles to stabilize and deliver bioactive molecules to the target tissues and to encapsulate secretory cells. Currently, most drugs are delivered into patients systemically without the use of a scaffold, so large doses are usually required for a desired local effect because of enzymatic degradation of the drug and nonspecific uptake by other tissues. This is a costly process and can cause serious side effects. In addition, many factors, which are necessary or beneficial to the target tissue, may be toxic to other tissues. Thus, a vehicle or scaffold allowing for local and specific delivery to the desired tissue site is highly desirable in many situations. Ionically cross-linked alginate hydrogels and glutaraldehyde cross-linked collagen sponges are some of the examples to fulfill this requirement [39, 40].

As hydrogels are highly hydrated 3D networks of polymers, they can provide chemical and mechanical signals and also an environment for cells to adhere, proliferate, and differentiate; thus, they are suitable for cell delivery and tissue development goals. Nowadays, hydrogel scaffolds are being used to produce a wide range of tissues, including cartilage, bone, muscle, fat, liver, and neurons. Based on the type of the desired tissue, different kinds of hydrogels can be utilized. For example, alginate has been used more widely than other hydrogels to assess the in vivo potential of hydrogel scaffolds for cartilage engineering and also as Schwann cell matrices in the area of nerve grafting, and collagen has been used for engineering large blood vessels [33].

5.4. Contact lenses

A key area in the use of synthetic hydrogels for bioapplications is ophthalmology, especially contact lenses. A contact lens is a small optical device placed directly on the cornea to alter the corneal power. The first concept of using contact lenses was described by Leonardo da Vinci in 1508; this consisted of immersing the eye in a bowl of water. At the end of 1960, poly(2-hydroxyethyl methacrylate) (PHEMA) lenses were developed by Professor Otto Wichterle; this invention represents the most important step in contact lens development and the start of soft lenses' era [41].

http://dx.doi.org/10.5772/64301

Direct placing of contact lenses on the surface of cornea prevents the exchange of atmospheric oxygen and thus disturbs the natural physiological metabolism of the cornea known as hypoxic stress, so a good contact lens must have maximum oxygen permeability. Mechanical stress to the cornea produces the same problems as the hypoxic stress, such as mitosis of the epithelial cells, elevated activity in protease and glycosidase, corneal sensitivity, and changes in corneal hydration and transparency. To reduce these stresses, the proper choice of contact materials and their shape are necessary.

Hydrogels used for production of contact lenses can cover most of the requirements needed when using in different physiological conditions. For a hydrogel material that is used as a contact lens, there are some necessities to make it comfortable during usage. These necessities include amount of water content, good mechanical properties, permeability toward oxygen, wettability of surface, good optical facilities, stability toward hydrolysis and sterilization, having nontoxic nature, and having enough biological tolerance for living cells.

In order to increase the water content of hydrogel and achieve an enhanced swelling effect, different types of monomers can be used. These include dihydroxy methacrylates, methacrylic acid, acrylamides, and many other monomers.

Silicone hydrogels represent an independent group of contact lens materials. The evolution of basic hydrogels gave rise to the production of this class, and they have good swelling properties and high permeability toward oxygen, which make them suitable for use in lenses. These properties are owing to their structure in which hydrophobic silicones are connected with hydrophilic chains in such a way that makes the resulting composite suitable both mechanically and optically.

High oxygen permeability is achieved with the siloxymethacrylate monomer commonly referred to as "TRIS." The methylene groups in the structure of TRIS represent the sites for hydrophilic modification (**Figure 17**).

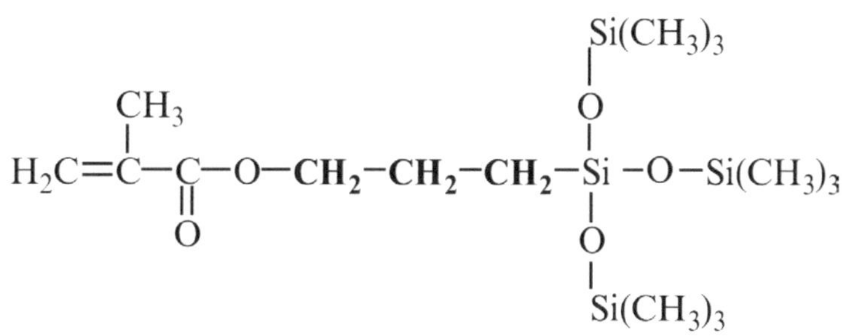

Figure 17. Structure of siloxymethacrylate monomer (TRIS).

It is possible to incorporate linear or branched hydrophilic polymer chains into the structure of the polymer to form an interpenetrated network to reduce the drying by the lenses when using them normally. This means that the "wetting chains" are fixed only by physical bonds without any covalent attachments to the patterned hydrogels' network [43].

5.5. pH-sensors

Stimuli-responsive polymers or hydrogels can change their volume significantly in response to small alterations of certain environmental parameters. Cationic polyelectrolytes dissolve (swell) more at low pH and anionic polyelectrolytes vice versa and this is due to ionization [44].

Two types of transducers are used in pH-sensitive hydrogel sensors: transducers based on mechanical work performed by hydrogel swelling and shrinking and those observing changes in properties of free swelling gels [45].

The ability of hydrogels to deform or to strain mechanically a transduction element resulting in a change of a special property of that element or in a change of a detectable distance is the basis of operation of transducers based on mechanical work of the hydrogel. They are classified as optical transducers, including reflective diaphragms and fiber Bragg grating sensors, and mechanical transducers, including microcantilevers and bending plate transducers.

Transducers of free swelling gels have to directly observe changes in one or more hydrogel properties and include optical, conductometric, and oscillating transducers. Optical transducers can directly measure changes in optical properties of hydrogels. A different approach is based on the observation of special fillings or surface coatings, which are changed or moved due to hydrogel swelling. Oscillating transducers are devices that keep changing their resonance frequency. Changes in the properties of a load result in a shift of this resonance frequency. This can be accompanied by a change of the signal amplitude. Conductometric transducers are based on measuring the conductivity of hydrogel as the degree of swelling changes [44].

5.6. Biosensors

Combining physical and chemical sensors results in a biosensor. There are two definitions for what a biosensor can do: it can be thought as a device that can sense and report a biophysical property of the system under study or a device that can deliver useful analytical information from transforming biochemical data. A common aspect in all biosensors is the presence of a biological recognition part that makes it possible to analyze biological information. Biosensors are becoming increasingly important as practical tools to cover a wide variety of application areas including point-of-care testing, home diagnostics, and environmental monitoring. Biological recognition part known as bioelement consists of different structures like enzymes, antibodies, living cells, or tissues but the point is its specificity toward one analyte and zero response to other interferents. There are various methods for coupling biomolecules with sensors including entrapment into membranes, physical adsorption, entrapment into a matrix, or covalent bonding [42, 44].

The high water content and hydrophilic nature of hydrogels are similar to the void-filling component of the extracellular matrix and render them intrinsically biocompatible. Hence, an apparent application of hydrogels in biosensors is the protection and coating function of sensor parts to prevent undesired interaction with biological molecules or cells. Hydrogels can be used as immobilization matrices for the biosensing elements and provide excellent environments for enzymes and other biomolecules to preserve their active and functional structure.

The interaction between analyte and sensing element results in a volume change in response to target component and this volume change is the basis of recognition in hydrogel-based sensors (**Figure 18**).

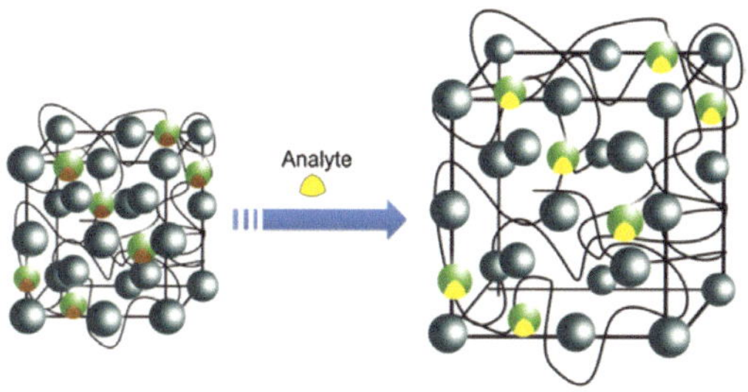

Figure 18. Volume increase of a hydrogel biosensor in response to an analyte.

Several types of sensing elements are used based on the nature of analyte but these elements can be categorized in two distinct groups: molecular interactions and living sensors.

Molecular interactions used for sensing analytes encompass different mechanisms. One of them is enzyme-substrate interactions. As enzymes are highly specific and efficient in their reactions with substrates, they may be used for precise determination of desired analytes' concentrations. There are many examples for different sensors based on enzyme-substrate interactions in hydrogel matrices including detection of organic-phase alcohols, amino acids, ammonia, urea, glucose, hydrogen peroxide, etc. Glucose-responsive hydrogels that are capable of acting as long-term insulin depots in response to increased blood glucose levels and automatically release doses of insulin at appropriate times are a promising development and could obviate the need for frequent injection and therefore provide a more convenient treatment option that would improve treatment efficacy and quality of life for hundreds of millions of people. The swelling of a hydrogel in the presence of glucose molecules makes it possible to release insulin in a controlled manner [46] (**Figure 19**).

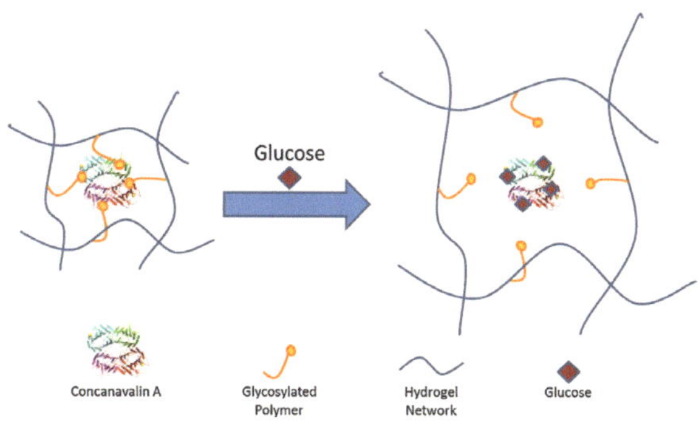

Figure 19. Concanavalin A-based glucose-responsive hydrogel swelling mechanism.

Antibody-antigen-based sensors that are affinity-based devices with a coupling of immunochemical reactions are another class of molecular interactions group. The general working principle of these sensors is based on the specific immunochemical recognition of antibodies (or antigens) immobilized on a transducer to antigen (or antibodies) that produce signals which depend on the concentration of the analyte. It is possible to use quartz crystal microgravimetry (QCM), surface plasmon resonance (SPR), or electrochemical methods for detection of the analyte.

There are several other examples of molecular interaction-based sensing of analytes like nucleotide, oligonucleotide, DNA, etc.

Another group of sensing elements is living sensors. They are combinations of hydrogels with living cells and microorganisms to form living cell-polymer composites for biosensing application. Microorganisms can detect a wide range of chemical substances, they are amenable to genetic modification and have a broad operating pH and temperature range, making them ideal as biological sensing materials. 3D structures, high water content, and biocompatibility are the main advantages of hydrogels that provide the ability to entrap cells or bacteria inside their networks enabling them to exchange gases in high rates and nourish the entrapped cells and in this way provide the possibility of usage of the cell-polymer composites in a biosensor. An instance is the use of *Arxula adeninivorans* LS3 as a biological recognition element for the rapid determination of the concentration of biodegradable pollutants in wastewater on a Clark-type oxygen electrode [47, 48].

There are two different methods to use hydrogels in biosensors: they can be coated on the surface of a sensing device like an electrode or be used as a 3D matrix or support to maintain bioelements such as cells. Preservation of cells for certain time periods in a hydrogel matrix and pathogen sensing are other examples of applications in this group (**Figure 20**).

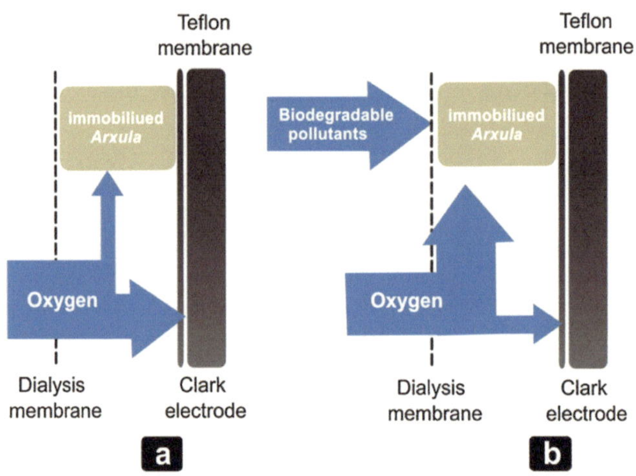

Figure 20. Schematic diagram of the A. *adeninivorans* LS3 microbial sensor illustrates the microbial consumption of dissolved oxygen (a) before and (b) after the addition of biodegradable pollutant.

http://dx.doi.org/10.5772/64301

5.7. Injectable hydrogel for spinal cord regeneration

Spinal cord injury (SCI) is a complex regenerative problem because of the multiple facets of growth inhibition that occur following trauma to the cord tissue. Many of these injuries do not hurt the dura mater and some of the axons are yet alive in the injury site and can be recovered. In such conditions, inserting a preformed frame or DDS into the damaged spinal cord by surgical operations may cause subsequent lesion. One alternative for this method is the use of in situ-forming scaffolds. What happens after injection into the injured cord area is the fast conversion of viscoelastic hydrogel from liquid form to gel and adaptation to the tissue of injury site [49].

The small spaces between spinal cord tissue and even transected parts formed after SCI will be filled by in vivo conversion of liquid hydrogel to the gel form. The gel, which now serves as a scaffold, will eliminate vacant spaces and forms a template for regeneration of the injured cord tissue by helping cellular penetration and matrix. In this way, it is not necessary to create preformed scaffolds for each patient individually and disconnecting viable tissue at the injury site to implant the preformed scaffold, which can cause further damage and loss of functionality, will be avoided [50, 51] (**Figure 21**).

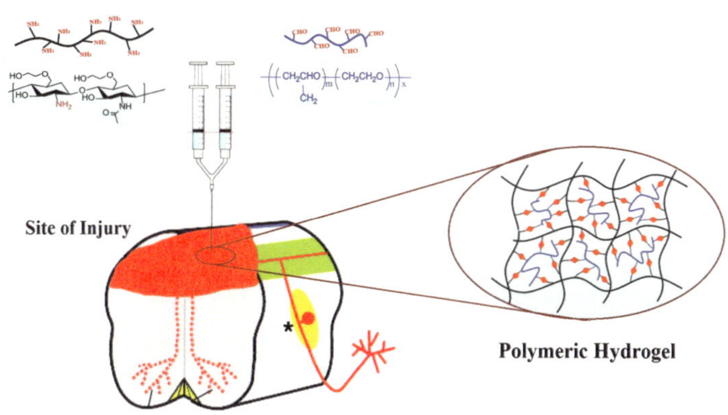

Figure 21. Injection of liquid hydrogel into the site of injury.

Injectable materials, in their liquid state, can be uniformly mixed with cells and other therapeutics prior to delivery into the injury site. The mechanical properties of gel scaffolds can more closely match the properties of the spinal cord tissue, compared to most preformed biomaterial matrices [49, 51].

There are some requirements for the design and use of injectable systems based on their functions and design parameters.

Their functions include [49, 52–54]:

1. Create a scaffold for cellular infiltration and axonal ingrowth. The gel material itself will serve to bridge the lesion site.

2. Encapsulation of drugs and maintenance of bioactivity throughout gelation and release. Injectable systems can provide a sustained and tunable delivery of these agents locally to the lesion site.

3. Support of suspended cell populations prior to injection, throughout the solidification process, and within the lesion site. Cellular therapies are more effective when delivered and maintained locally in the injured area as opposed to being delivered systemically (**Figure 22**).

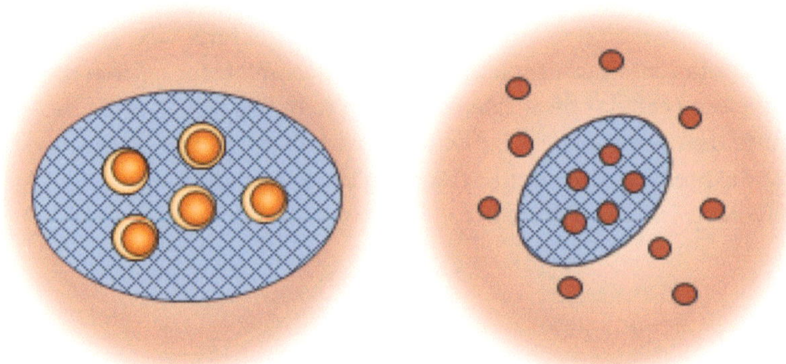

Figure 22. Functions of injectable hydrogels.

The importance of design parameters is originating from the difficulty in isolating the effects of cross-linking and macromer concentration-dependent material properties such as mechanical stiffness, mesh or pore size, degradation rate, and bioactive ligand density.

Design parameters include biocompatibility of used materials with the tissue of injured site, mild solidification conditions, suitable porosity and mesh size of the designed scaffold, mechanical properties of the gel material, degradation rate, and bioactivity [55–57].

Injectable hydrogels can be natural or synthetic with their own benefits and disadvantages. They can also be classified as physical and chemical gels [49, 62].

Generally, physical hydrogels have the limitation of weak mechanical properties; thus, a combination of chemical and physical cross-linking has been used to overcome this weakness. For example, PNIPAAm-co-glycidyl methacrylate (GMA) and polyamidoamine (PAMAM) macromers undergo a dual-hardening physical and chemical gelation precess and form PNIPAAm-co-glycidyl methacrylate (GMA)/polyamidoamine (PAMAM) injectable hydrogel [58].

Injectable hydrogel systems are minimally invasive and patient friendly. Cells or bioactive molecules are easy to mix with polymer solutions and these mixtures are in situ and easily form the 3D microenvironments into any desired defect shapes.

5.8. Supercapacitor hydrogels

The fast developments in portable electronic equipments industry such as wearable devices, arbitrarily curved displays and even transparent mobile phones, require the fabrication of flexible, transparent, lightweight and efficient storage options [73]. To this end, the key issue is the simultaneous incorporation of mechanical robustness, optical transmittance, and electronic conductivity [74]. Because of the importance of high-performance flexible supercapacitors, the technique of the supercapacitors is still making rapid progresses. Recently, several strategies for flexible supercapacitors have been demonstrated, including coating active materials, such as RuO2 [66], MnO2 [67], V2O5 [68], NiOH [69], and graphene nanosheets [70] onto conductive fibers by electrochemical deposition or casting and fabrication of hydrogel or aerogel films based on graphene [73]. However, these methods suffer from several disadvantages that hinder their large-scale commercialization, such as high cost of noble metals or expensive carbon support materials, limited ionic/electronic conductivity, poor mechanical flexibility, and scalable electrochemical synthesis conditions.

Electrically conducting polymer hydrogels show great potential for the expected integration due to their excellent solid-liquid interface, good electric characteristics, and mechanical flexibility, and represent a promising material platform for emerging flexible energy storage devices [71, 72]. Conducting polymers such as polyaniline, polypyrrole, and their derivatives provide the unique electrical properties of metals or semiconductors, as well as attractive properties associated with conventional polymers, such as ease of synthesis and flexibility in processing; therefore, supercapacitor hydrogels are attracting much attention as new power sources [73]. Flexible solid-state supercapacitors provide high power density, long cycle life, and the potential to achieve relatively high energy density.

Shi et al. have recently synthesized a 3D nanostructured conductive polypyrrole hydrogel via an interfacial polymerization method [73]. The high-performance flexible solid-state supercapacitor demonstrated promising capacitive properties and good electrochemical stability during long-term cycling. So far, many aspects such as conductivity and morphology of conductive polymer hydrogels have been extensively studied. However, the combination of stretchability and transparency is unique, and particularly long cycle stability has not been achieved before. In this regard, Hao et al. demonstrated a facile and smart strategy for the preparation of structurally stretchable, electrically conductive, and optically semitransparent α-cyclodextrin polyacrylamide-polyaniline hybrid hydrogel networks as electrodes, which show a high performance in supercapacitor application [74].

6. Conclusions

This chapter aims to introduce briefly the hydrogels: a class of natural or synthetic polymeric materials that have the ability to hold huge amounts of water because of their specific structures and subsequent swelling properties. Based on this ability, they found a wide variety of applications, and because of the possibility to modify the polymeric structure to obtain desired functionality, the areas of applications are rapidly expanding. They can be designed in such a way that they can respond to a specific stimulus including pH, temperature, light, etc. at a predefined level and thus be stimuli responsive. Among their amazing characteristics, the biocompatibility and biodegradability make them a powerful candidate to use in biological and environmental applications as implants or materials for removal of toxic pollutants. In addition, conducting hydrogels are often a good choice in designing and fabrication of supercapacitors, which promise the most rapid developments in electronics.

Author details

Morteza Bahram*, Naimeh Mohseni and Mehdi Moghtader

*Address all correspondence to: morteza.bahram@gmail.com

Department of Analytical Chemistry, Faculty of Chemistry, Urmia University, Urmia, Iran

References

[1] O. Wichterle, D. Lím, Hydrophilic Gels for Biological Use, Nature. 1960; 185: 117-118, DOI: 10.1038/185117a0

[2] A.K.A. Silva, C. Richard, M. Bessodes, D. Scherman, O.W. Merten, Growth Factor Delivery Approaches in Hydrogels, Biomacromolecules. 2009; 10(1): 9-18, DOI: 10.1021/bm801103c

[3] Faheem Ullah, Muhammad Bisyrul Hafi Othman, Fatima Javed, Zulkifli Ahmad, Hazizan Md. Akil, Classification, Processing and Application of Hydrogels: A Review, Materials Science & Engineering C. 2015, DOI: 10.1016/j.msec.2015.07.053

[4] N.Das, Preparation Methods and Properties of Hydrogel: A Review, Int. J. Pharm. Sci. 2013; 5(3): 112-117

[5] A. Cretu, R. Gattin, L. Brachais, D. Barbier-Baudry, Synthesis and Degradation of Poly (2-Hydroxyethyl Methacrylate)-Graft-Poly (ε-caprolactone) Copolymers, Polym. Degrad. Stab. 2004; 83(3): 399-403, DOI: 10.1016/j.polymdegradstab.2003.09.001

[6] B. Kim, N.A. Peppas, Poly (ethylene glycol)-Containing Hydrogel for Oral Protein Delivery Applications, Biomed. Microdevices. 2003; 5(4): 333-341, DOI: 10.1023/A:1027313931273

[7] A. Richter, G. Paschew, S. Klatt, J. Lienig, K.F. Arndt, H.J.P. Adler, Review on Hydrogel-based pH Sensors and microsensors, Sensors. 2008; 8(1): 561-581, DOI: 10.3390/s8010561

[8] K. Na, K.H. Lee, Y.H. Bae, pH-sensitivity and pH-dependent Interior Structural Change of Self-assembled Hydrogel Nanoparticles of Pullulan Acatate/Oligo-sulfonamide Conjugate, J. Controlled Release. 2004; 97(3): 513-525, DOI: 10.1016/j.jconrel.2004.04.005

[9] Y. Qiu, K. Park, Environment-sensitive Hydrogels for Drug Delivery, Adv. Drug Deliv. Rev. 2001; 53(13): 321-339, DOI: 10.1016/S0169-409X(01)00203-4

[10] A.K. Bajpai, J. Bajpai, R. Saini, R. Gupta, Responsive Polymers in Biology and Technology, Polym. Rev. 2001; 51(1): 53-97, DOI: 10.1080/15583724.2010.537798

[11] M.A. Ward, T.K. Georgiou, Thermoresponsive Polymers for Biomedical Applications, Polymers. 2011; 3(3): 1215-1242; DOI: 10.3390/polym3031215

[12] T. Tanaka, I. Nishio, S. T. Sun, S. Ueno-Nishio, Collapse of Gels in an Electric Field, Science. 1982; 218(4571): 467-469, DOI: 10.1126/science.218.4571.467

[13] M. Doi, M. Matsumoto, Y. Hirose, Deformation of Ionic Gels by Electric Field, Macromolecules. 1992; 25(20): 5504-5511, DOI: 10.1021/ma00046a058

[14] T. Shiga, T. Kurauchi, Deformation of Polyelectrolyte Gels under the Influence of Electric Field, J. Appl. Polym. Sci. 1990; 39(1-2): 2305-2320, DOI: 10.1002/app1990.070391110

[15] S. Murdan, Electro-responsive Drug Delivery from Hydrogels, J. Controlled Release. 2003; 92(1-2): 1-17, DOI: 10.1016/S0168-3659(03)00303-1

[16] J. Shang, Z. Shao, X. Chen, Electrical Behavior of a Natural Polyelectrolyte Hydrogel: Chitosan/Carboxymethylcellulose Hydrogel, Biomacromolecules. 2008; 9(4): 1208-1213, DOI: 10.1021/bm701204j

[17] M.J. Garland, T.R.R. Singh, A.D. Woolfson, R.F. Donnelly, Electrically Enhanced Solute Permeation across Poly(ethyleneglycol)-crosslinked Poly(methyl vinyl ether-co-maleic acid) Hydrogels: Effect of Hydrogel Crosslink Density and Ionic Conductivity, Int. J. Pharm. 2011; 406(1-2): 91-98, DOI: 10.1016/j.ijpharm.2011.01.002

[18] S. Indermun, Y. E. Choonara, P. Kumar, L. C. du Toit, G. Modi, R. Luttge, V. Pillay, An Interfacially Plasticized Electro-responsive Hydrogel for Transdermal Electro-activated and Modulated (TEAM) Drug Delivery, Int. J. Pharm. 2014; 462(1-2): 52-65, DOI: 10.1016/j.ijpharm.2013.11.014

[19] J.E. Chung, M. Yokoyama, M. Yamato, T. Aoyagi, Y. Sakurai, T. Okano, Thermoresponsive Drug Delivery from Polymeric Micelles Constructed Using Block Copolymers of Poly(N-isopropylacrylamide) and Poly(butylmethacrylate), J. Control. Release. 1999; 62: 115-127, DOI: 10.1016/S0168-3659(99)00029-2

[20] A. Khademhosseini, R. Langer, Microengineered Hydrogels for Tissue Engineering, Biomaterials. 2007; 28(34): 5087-5092, DOI: 10.1016/j.biomaterials.2007.07.021

[21] N. Rahimi, D.G. Molin, T.J. Cleij, M.A. van Zandvoort, M.J. Post, Electrosensitive Polyacrylic Acid/Fibrin Hydrogel Facilitates Cell Seeding and Alignment, Biomacromolecules. 2012; 13(5): 1448-1457, DOI: 10.1021/bm300161r

[22] X.L. Wang, I. K. Oh, S. Lee, Electroactive Artificial Muscle Based on Crosslinked PVA/SPTES, Sens. Actuators B. 2012; 150(1): 57-64, DOI: 10.1016/j.snb.2010.07.042

[23] M.M. Fitzgerald, K. Bootsma, J.A. Berberich, J.L. Sparks, Tunable Stress Relaxation Behavior of an Alginate-Polyacrylamide Hydrogel: Comparison with Muscle Tissue, Biomacromolecules. 2015; 16(5): 1497-1505, DOI: 10.1021/bm501845j

[24] A. Mamada, T. Tanaka, D. Kungwachakun, M. Irie, Photoinduced Phase Transition of Gels, Macromolecules. 1990; 23(5): 1517-1519, DOI: 10.1021/ma00207a046

[25] A. Suzuki, T. Tanaka, Phase Transition in Polymer Gels Induced by Visible Light, Nature 1990; 346: 345-347, DOI: 10.1038/346345a0

[26] A. Suzuki, T. Ishii, Y. Maruyama, Optical Switching in Polymer Gels, J. Appl. Phys. 1996; 80(1): 131-136, DOI: 10.1063/1.362768

[27] M. Bahram, N. Nurallahzadeh, N. Mohseni, pH-sensitive Hydrogel for Coacervative Cloud Point Extraction and Spectrophotometric Determination of Cu(II): Optimization by Central Composite Design, J. Iran. Chem. Soc. 2015; 12(10): 1781-1787, DOI: 10.1007/s13738-015-0653-5

[28] M. Bahram, F. Keshvari, P. Najafi-Moghaddam, Development of Cloud Point Extraction Using pH-sensitive Hydrogel for Preconcentration and Determination of Malachite Green, Talanta. 2011; 85(2): 891-896, DOI: 10.1016/j.talanta.2011.04.074

[29] M. Bahram, F. Keshvari, N. Mohseni, A Novel Hydrogel Based Microextraction of Analytes, J. Saudi Chem. Soc. 2013, DOI: 10.1016/j.jscs.2013.05.002

[30] N. Mohseni, M. Bahram, Kh. Farhadi, P. Najafi-Moghaddam, F. Keshvari, Spectrophotometric Determination of Paracetamol Using Hydrogel Based Semi Solid-liquid Dispersive Microextraction, Sci. Iran. C. 2014; 21(3): 693-702

[31] M. Bahram, F. Hoseinzadeha, Kh. Farhadi, M. Saadat, P. Najafi-Moghaddam, A. Afkhami, Synthesis of Gold Nanoparticles Using pH-sensitive Hydrogel and Its Application for Colorimetric Determination of Acetaminophen, Ascorbic Acid and Folic Acid, Colloids Surf. A: Physicochem. Eng. Asp. 2014; 441: 517-524, DOI: 10.1016/j.colsurfa.2013.09.024

[32] K.Y. Lee, D.J. Mooney, Hydrogels for Tissue Engineering, Chem. Rev. 2001; 101(7): 1869-1877, DOI: 10.1021/cr000108x

[33] J.L. Drury, D.J. Mooney, Hydrogels for Tissue Engineering: Scaffold Design Variables and Applications, Biomaterials. 2003; 24(24): 4337-4351, DOI: 10.1016/S0142-9612(03)00340-5

[34] J.K.F. Suh, H.W.T. Matthew, Application of Chitosan-based Polysaccharide Biomaterials in Cartilage Tissue Engineering: A Review, Biomaterials. 2000; 21(24): 2589-2598, DOI: 10.1016/S0142-9612(00)00126-5

[35] J.L. West, S.M. Chowdhury, A.S. Sawhney, C.P. Pathak, R.C. Dunn, J.A. Hubbell, Efficacy of Adhesion Barriers. Resorbable Hydrogel, Oxidized Regenerated Cellulose and Hyaluronic Acid, J. Reprod. Med. 1996; 41(3): 149-154

[36] J.L. Hill-West, S.M. Chowdhury, A.S. Sawhney, C.P. Pathak, R.C. Dunn, J.A. Hubbell, Prevention of Postoperative Adhesions in the Rat by In Situ Photopolymerization of Bioresorbable Hydrogel Barriers, Obstet. Gynecol. 1994; 83(1): 59-64

[37] K. Ono, Y. Saito, H. Yura, K. Ishikawa, A. Kurita, T. Akaike, M. Ishihara, Photocrosslinkable Chitosan as a Biological Adhesive, J. Biomed. Mater. Res. 2000; 49(2): 289-295, DOI: 10.1002/(SICI)1097-4636(200002)49:2<289::AID-JBM18>3.0.CO;2-M

[38] X. Zhao, K. Kato, Y. Fukumoto, K. Nakamae, Synthesis of Bioadhesive Hydrogels from Chitin Derivatives, Int. J. Adhesion. Adhesives. 2001; 21(3): 227-232, DOI: 10.1016/S0143-7496(01)00003-3

[39] Y. Tabata, M. Miyao, M. Ozeki, Y. Ikada, Controlled Release of Vascular Endothelial Growth Factor by Use of Collagen Hydrogels, J. Biomater. Sci. Polymer Edn. 2000; 11(9): 915-930, DOI: 10.1163/156856200744101

[40] F. Lim, Microencapsulation of Living Cells and Tissues-Theory and Practice. In: Lim F, editor. Biomedical Applications of Microencapsulation. Boca Raton, FL: CRC Press (1984). 137-154

[41] Michalek, R. Hobzova, M. Pradny, M. Duskova, Hydrogels Contact Lenses in: Biomedical Applications of Hydrogels Handbook, Springer (2010). 303-315

[42] A. Mateescu, Y. Wang, J. Dostalek, U. Jonas, Membranes. 2012; 2: 40-69, DOI: 10.3390/membranes2010040

[43] B. Tighe, Silicone hydrogel materials-how do they work? In: Sweeney D. (ed) Silicone Hydrogels: The Rebirth of Continuous Wear Contact Lenses. Butterworth-Heinemann, Oxford (2000). 1-21

[44] M.C. Koetting, J.T. Peters, S.D. Steichen, N.A. Peppas, Stimulus-responsive Hydrogels: Theory, Modern Advances, and Applications, Mater. Sci. Eng. R. 2015; 93: 1-49, DOI: 10.1016/j.mser.2015.04.001

[45] P. Bawa, V. Pillay, Y.E. Choonara, L.C. du Toit, Stimuli-responsive Polymers and Their Applications in Drug Delivery, Biomed. Mater. 2009; 4(2): 1-15, DOI: 10.1088/1748-6041/4/2/022001

[46] G.G. Adams, Y. Cui, J.H. Mitchell, M.J. Taylor, Rheological and Diffusion Properties of a Dextran-con A Polymer in the Presence of Insulin and Magnesium, Rheol. Acta. 2006; 45(5): 611-620, DOI: 10.1007/s00397-005-0013-y

[47] D. Buenger, F. Topuz, J. Groll, Hydrogels in Sensing Applications, Prog. Polym. Sci. 2012; 37(12): 1678-1719, DOI: 10.1016/j.progpolymsci.2012.09.001

[48] T. Renneberg, R.C.H. Kwan, C. Chan, G. Kunze, R. Renneberg, A Salt-tolerant Yeast-based Microbial Sensor for 24 Hour Community Wastewater Monitoring in Coastal Regions, Microchim. Acta. 2004; 148(3): 235-240, DOI: 10.1007/s00604-004-0266-7

[49] D. Macaya, M. Spector, Injectable Hydrogel Materials for Spinal Cord Regeneration: A Review, Biomed. Mater. 2012; 7(1): 012001, DOI: 10.1088/1748-6041/7/1/012001

[50] B.M. Gillette, J.A. Jensen, B.T. Tang, G.J. Yang, A. Bazargan-Lari, M. Zhong, S.K.Sia, In Situ Collagen Assembly for Integrating Microfabricated Three-Dimensional Cell-seeded Matrices, Nat. Mater. 2008; 7(8): 636-640, DOI: 10.1038/nmat2203

[51] H. Nomura, Y. Katayama, M.S. Shoichet, C.H. Tator, Complete Spinal Cord Transection Treated by Implantation of a Reinforced Synthetic Hydrogel Channel Results in Syringomyelia and Caudal Migration of the Rostral Stump, Neurosurgery. 2006; 59(1): 183- 192, DOI: 10.1227/01.NEU.0000219859.35349.EF

[52] Y. Zhong, R.V. Bellamkonda, Biomaterials for the Central Nervous System, J.R. Soc. Interface. 2008; 5: 957-975, DOI: 10.1098/rsif.2008.0071

[53] E.J. Bradbury, S.B. McMahon, Spinal Cord Repair Strategies: Why Do They Work?, Nat. Rev. Neurosci. 2006; 7(8): 644-653, DOI: 10.1038/nrn1964

[54] E. Eftekharpour, S. Karimi-Abdolrezaee, M.G. Fehlings, Current Status of Experimental Cell Replacement Approaches to Spinal Cord Injury, Neurosurg. Focus. 2008; 24(3-4): E19., DOI: 10.3171/FOC/2008/24/3-4/E18

[55] M.J. Mahoney, K.S. Anseth, Three-dimensional Growth and Function of Neural Tissue in Degradable Polyethylene Glycol Hydrogel, Biomaterials. 2006; 27(10): 2265-2274, DOI: 10.1016/j.biomaterials.2005.11.007

[56] K.S. Straley, S.C. Heilshorn, Independent Tuning of Multiple Biomaterial Properties Using Protein Engineering, Soft. Matter. 2009; 5(1): 114-124, DOI: 10.1039/B808504H

[57] L.A. Flanagan, Yo-El Ju, B. Marg, M. Osterfield, P.A. Janmey, Neurite Branching on Deformable Substrates, Neuroreport. 2002; 13(18): 2411-2415, DOI: 10.1097/01.wnr. 0000048003.96487.97

[58] X.J. Loh, In-Situ Gelling Polymers for Biomedical Applications, Springer, Singapore (2015).

[59] T. Führmann, J. Obermeyer, C.H. Tator, M.S. Shoichet, Click-crosslinked Injectable Hyaluronic Acid Hydrogel Is Safe and Biocompatible in the Intrathecal Space for Ultimate Use in Regenerative Strategies of the Injured Spinal Cord, Methods. 2015; 84: 60-69, DOI: 10.1016/j.ymeth.2015.03.023

[60] J.Y. Sun, X. Zhao, W.R.K. Illeperuma, O. Chaudhuri, K.H. Oh, D.J. Mooney, J.J. Vlassak, Z. Suo, Highly Stretchable and Tough Hydrogels, Nature. 2012; 489: 133-136, DOI: 10.1038/nature11409

[61] P.A. Janmey, J.P. Winer, J.W. Weisel, Fibrin Gels and Their Clinical and Bioengineering Applications, J. R. Soc. Interface. 2006; 6(30): 1-10, DOI: 10.1098/rsif.2008.0327

[62] J. Yang, J. Yeom, B.W. Hwang, A.S. Hoffman, S.K. Hahn, In Situ-forming Injectable Hydrogels for Regenerative Medicine, Prog. Polym. Sci. 2014; DOI: 10.1016/j.progpolymsci.2014.07.006

[63] P.B. Welzel, S. Prokoph, A. Zieris, M. Grimmer, S. Zschoche, U. Freudenberg, C. Werner, Modulating Biofunctional starPEG Heparin Hydrogels by Varying Size and Ratio of the Constituents, Polymers. 2011; 3(1): 602-620, DOI: 10.3390/polym3010602

[64] Y. Takashima, S. Hatanaka, M. Otsubo, M. Nakahata, T. Kakuta, A. Hashidzume, H. Yamaguchi, A. Harada, Expansion-contraction of Photoresponsive Artificial Muscle Regulated by Host-guest Interactions, Nat. Commun. 2012; 1270, DOI: 10.1038/ncomms2280

[65] J. ter Schiphorst, S. Coleman, J.E. Stumpel, A.B. Azouz, D. Diamond, A.P.H.J. Schenning, Molecular Design of Light-responsive Hydrogels, for In Situ Generation of Fast and Reversible Valves for Microfluidic Applications, Chem. Mater. 2015; 27 (17): 5925-5931, DOI: 10.1021/acs.chemmater.5b01860

[66] C.C. Hu, W.C. Chen, K.H. Chang, How to Achieve Maximum Utilization of Hydrous Ruthenium Oxide for Supercapacitors, J. Electrochem. Soc. 2004; 151(2): A281-A290, DOI: 10.1149/1.1639020

[67] L. Yu, G. Zhang, C. Yuan, X.W. Lou, Hierarchical $NiCo_2O_4$@MnO_2 Core-shell Heterostructured Nanowire Arrays on Ni Foam as High-performance Supercapacitor Electrodes, Chem. Commun. 2013; 49(2): 137-139, DOI: 10.1039/C2CC37117K

[68] W. Zhou, C. Cheng, J. Liu, Y.Y. Tay, J. Jiang, X. Jia, J. Zhang, H. Gong, H.H. Hng, T. Yu, H.J. Fan, Epitaxial Growth of Branched α-Fe_2O_3/SnO_2 Nano-Heterostructures with Improved Lithium-ion Battery Performance, Adv. Funct. Mater. 2011; 21(13): 2439-2445, DOI: 10.1002/adfm.201100088

[69] B. Wang, J.S. Chen, Z. Wang, S. Madhavi, X.W. Lou, Green Synthesis of NiO Nanobelts with Exceptional Pseudo-capacitive Properties, Adv. Energy Mater. 2012; 2(10): 1188-1192, DOI: 10.1002/aenm.201200008

[70] M. Pumera, Graphene-based Nanomaterials for Energy Storage, Energy Environ. Sci. 2011; 4(3): 668-674, DOI: 10.1039/C0EE00295J

[71] G.A. Snook, P. Kao, A.S. Best, Conductive-polymer-based Supercapacitor Devices and Electrodes, J. Power Sources. 2011; 196(1): 1-12, DOI: 10.1016/j.jpowsour.2010.06.084

[72] Y. Zhao, B. Liu, L. Pan, G. Yu, 3D Nanostructured Conductive Polymer Hydrogels for High-performance Electrochemical Devices, Energy Environ. Sci. 2013; 6(10): 2856-2870, DOI: 10.1039/C3EE40997J

[73] Y. Shi, L. Pan, B. Liu, Y. Wang, Y.Cui, Z. Bao, G. Yu, Nanostructured Conductive Polypyrrole Hydrogels as High-performance Flexible Supercapacitor Electrodes,, J. Mater. Chem. A 2014; 2(17): 6086-6091, DOI: 10.1039/C4TA00484A

[74] G.P. Hao, F. Hippauf, M. Oschatz, F.M. Wisser, A. Leifert, W. Nickel, N.M. Noriega, Z. Zheng, S. Kaskel, Stretchable and Semi-transparent Conductive Hybrid Hydrogels for Flexible Supercapacitors, ACS Nano. 2014; 8(7): 7138-7146, DOI: 10.1021/nn502065u

[75] Y.X. Zhang, F.P. Wu, M.Z. Li, E.J. Wang, pH Switching On-off Semi-IPN Hydrogel based on Cross-linked Poly(acrylamide-co-acrylic acid) and Linear Poluallyamine, Polymer. 2005; 46: 7695-7700, DOI: 10.1016/j.polymer.2005.05.121

[76] C.A. Lorenzo, A. Concheiro, A.S. Dubovik, N.V. Grinberg, T.V. Burova, V.Y. Grinberg, Temperature-sensitive Chitosan-poly(N-isopropylacrylamide) Interpenetrated Network with Enhanced Loading Capacity and Controlled Release Properties, J. Controlled Release. 2005; 102: 629-641, DOI: 10.1016/j.jconrel.2004.10.021

[77] E. Ramírez, S.G. Burillo, C.B. Díaz, G. Roa, B. Bilyeu, Use of pH-sensitive Polymer Hydrogels in Lead Removal from Aqueous Solution, J. Hazard. Mater. 2011; 192: 432-439, DOI: 10.1016/j.jhazmat.2011.04.109

[78] E.S. Dragan, Design and Applications of Interpenetrating Polymer Network Hydrogels, Chem. Eng. J. 2014; 243: 572-590, DOI: 10.1016/j.cej.2014.01.065

[79] A. Ramesh, H. Hasegawa, W. Sugimoto, T. Maki, K. Ueda, Adsorption of Gold(III), Platinum(IV) and Palladium(II) onto Glycine Modified Crosslinked Chitosan Resin, Bioresource Technol. 2008; 99: 3801-3809, DOI: 10.1016/j.biortech.2007.07.008

[80] N.G. Kandile, A.S. Nasr, Environment Friendly Modified Chitosan Hydrogels as a Matrix for Adsorption of Metal Ions, Synthesis and Characterization, Carbohydr. Polym. 2009; 78: 753-759, DOI: 10.1016/j.carbpol.2009.06.008

[81] C. Jeon, W.H. Holl, Chemical Modification of Chitosan and Equilibrium Study for Mercury Ion Removal, Water Res. 2003; 37: 4770-4780, DOI: 10.1016/S0043-1354(03)00431-7

[82] I. Kavianinia, P.G. Plieger, N.G. Kandile, D.R.K. Harding, Fixed-bed Column Studies on a Modified Chitosan Hydrogel for Detoxification of Aqueous Solutions from Copper(II), Carbohydr. Polym. 2012; 90: 875-886, DOI: 10.1016/j.carbpol.2012.06.014

[83] X. Liu, Q. Hu, Z. Fang, X. Zhang, B. Zhang, Magnetic Chitosan Nanocomposites: A Useful Recyclable Tool for Heavy Metal Ion Removal, Langmuir. 2009; 25: 3-8, DOI: 10.1021/la802754t

Chapter 3

Characterization and Tailoring the Properties of Hydrogels Using Spectroscopic Methods

Gabriela Ionita

Additional information is available at the end of the chapter

http://dx.doi.org/10.5772/62900

Abstract

Hydrogels represent heterogeneous systems that consist of a large amount of water retained by a three-dimensional network. The hydrogel network is the result of assembly through physical interactions or chemical cross-linking of polymers or small molecules. The applications of hydrogels (water purification, tissue regeneration, therapeutic delivery, bio-detection or bio-imaging, etc.) depend on their physicochemical properties and structural features. Although electron microscopy and viscoelastic measurements provide general information about a gel material, the spectroscopic methods complement these methods and also afford a deep insight into the gel structure. In this chapter, the applications of several spectroscopic methods for characterizing polymeric or supramolecular hydrogels are discussed. Thus, this review highlights the particular application of vibrational spectroscopy, circular dichroism, fluorescence (these providing information on assembly in the network), interactions that occur between network and solvent (water), pulsed-field gradient NMR (determination of mesh size) and EPR spectroscopy (a method that can provide extensive information regarding the assembly process, diffusion and release).

Keywords: hydrogels, IR, Raman, fluorescence, circular dichroism, PFG-NMR, EPR

1. Introduction

Gels represent a class of soft materials consisting of a 'solid-like' network that holds a large volume of solvent via surface tension or capillary effect [1]. 'Hydrogels' refer to the case in which the solvent retained in the frame of a gel network is water. The term hydrogel was first mentioned in the literature at the end of the nineteenth century describing a colloidal gel of inorganic salts

[2], but the current meaning referring to a water-swollen gel network became prevalent six decades ago with the developing research area of polymeric hydrogels.

Hydrogels are classified using different criteria, such as the nature of the gelators, the type of interactions that contribute to the building of the 'solid network' and physical properties of the gel network (**Figure 1**). The gel networks result by chemical or physical cross-linking of gelators, which can be either polymers or low molecular weight compounds able to generate fibrillary networks [1–3]. The best-known hydrogels are those obtained by physical or chemical assembly of natural and synthetic polymers.

Chemical hydrogels are the result of interconnection of polymer chains or molecular building blocks through the formation of covalent bonds, which obviously are non-reversible. Once the gel network is broken, the chemical hydrogels are unable to self-heal. The mechanical properties of these gels usually are easier to tune by varying the reaction conditions that finely influence the number of cross-links. Another particularity of chemical hydrogels is that their volume changes significantly during the transition from solution to gel state [1, 4, 5].

Physical hydrogels refer mainly to self-assembly of synthetic or natural polymers, but in recent years, there are more and more studies dedicated to designing new low molecular weight compounds able to hydrogelate [3]. The knowledge gained after studying numerous low molecular weight gelators (LMWG) is still not enough for predicting the ability of a new molecule to generate gel systems through non-covalent supramolecular interactions. Frequently, the LMWGs are characterized by amphiphilicity and the capacity to generate non-covalent interactions (π–π interactions, hydrogen-bonding and charge interactions among the molecules, host–guest interactions) allowing the formation of a three-dimensional fibrillary network.

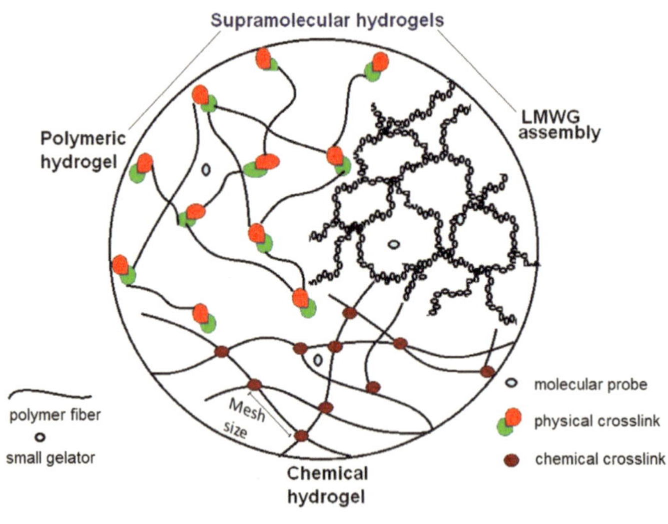

Figure 1. Classification of hydrogels function of the nature of interactions building the network.

It can be accepted that hydrogels formed through non-covalent interactions leading to assembly either of LMWG or polymers form a large family of supramolecular hydrogels [1, 3].

Research in the gels field is closely bound with their various applications. Gels find applications in daily life, being frequently used in cosmetics and food chemistry [3], but there are specific or high-interest domains that correlate structural gel characteristics with applications. Thus, whether their building blocks are biocompatible or are structural biomaterials, hydrogels are oriented towards regenerative medicine and tissue engineering, obtaining of enzyme-responsive hydrogels, enzyme–hydrogel hybrid materials, drug delivery and therapeutic agents with release control, separation processes or water purification [1, 2, 6]. As smart materials, gels may find applications in optoelectronic, responsive systems to pH, ions and light harvesting systems [7]. Supramolecular hydrogels resulting from assembly of LMWG can find application in catalysis generating either self-construction of the catalytic system or by generation of new catalytic properties arising from association modes [8, 9]. Metallic nanoparticles can be generated *in situ* in the chemical hydrogel matrices for further use as catalysts [8].

2. Spectroscopic methods used in hydrogels studies

Unravelling the molecular organization inside a gel represents a complex issue and involves a multitude of physicochemical methods. The majority of gel studies use electron microscopy to demonstrate the existence of gel fibrils, while mechanical properties expressed by rheological parameters reflect the density of cross-linking, thickness of gel fibril, all of these aspects being the result of molecular packing of gelators [10]. Hydrogels represent systems with a dynamic nature given the interactions between building blocks, the encapsulated species and interaction between solvent entrapped and gel fibres. Therefore, gel properties need to be investigated through a large variety of complementary physicochemical methods. In this chapter, the applications of several spectroscopic methods for characterizing polymeric or supramolecular hydrogels are discussed. **Figure 1** schematically represents different networks that can be investigated by the methods discussed in this review: vibrational spectroscopy, circular dichroism, fluorescence, pulsed-field gradient NMR and electron paramagnetic resonance (EPR) spectroscopy. Some methods can be involved in characterization of hydrogels irrespective of the driving forces that build the network, such as vibrational spectroscopy (IR and Raman). Circular dichroism can be relevant for assembly of LMWG, especially as preponderantly these molecules are chiral or in the particular case of polypeptides assembled into gel networks. Pulsed-field gradient NMR is a method applied to estimate the mesh size of the gel network, while other methods can be used to determine diffusion coefficients of solutes in gels or to monitor the release of encapsulated species (e.g. vibrational spectroscopy, fluorescence spectroscopy). Circular dichroism can also be used to estimate the transition from sol to gel in the case of supramolecular hydrogels following changes in ellipticity. Fluorescence and EPR spectroscopies are suitable methods to test the environment around the sensing groups (fluorescent and paramagnetic, respectively), and thus can be used to demonstrate the heterogeneity of the hydrogel systems, but also provide information on the dynamics of encapsulated species or the gel network.

2.1. IR spectroscopy

Vibrational spectroscopy or infrared (IR) spectroscopy is a suitable method to obtain substantial information on the self-assembly process leading to formation of supramolecular gels. IR spectroscopy deals with the infrared region of the electromagnetic spectrum. The photon energies (1–15 kcal/mol) corresponding to the IR region are not large enough to excite the electrons but may induce vibrational excitation. The active vibrations in IR cause a change in the dipole moment. Vibrations of bonds or a group of bonds include stretching, bending, scissoring, rocking and twisting [11–14]. The analysis of intensities or shifts of wavenumbers can probe involvement of specific bonds in gel network building.

Comparison of the IR spectrum in the gel state with the spectra of solution or solid states of gelators can provide information on the interactions driving the formation of a gel network. The formation of a supramolecular gel network involves different types of non-covalent interactions: hydrogen bonding, van der Waals and π–π stacking. These interactions can be identified in IR spectra by the appearance and disappearance of vibrational bands characteristic of associated and free groups. In many cases, the gelators have in their structure functional groups with characteristic vibration bands (carboxy, hydroxy, amino and amide groups) that are sensitive to formation of weak physical interactions. The hydrogen bonding involving, for example, the carbonyl or amide groups, which often contribute to the building of hydrogel network, is frequently hidden by the water hydrogen bonding. In such cases, the replacement of water with deuterated water is required [13, 14].

The following examples are selected from studies reported in the literature highlighting the applicability of IR spectroscopy in attempting to determine the assembly mode of gelators (polymers or LMWG) in a supramolecular network. Suzuki and co-workers [13] synthesized a lysine derivative, *N*α-hexanoyl-*N*ζ-lauroyl-l-lysine and studied the organogelation and hydrogelation of the gelator in combination with corresponding alkali salts (Li^+, Na^+ and K^+). Deconvolution of the broadbands arising in the interval 1750–1500 cm^{-1} (which summarize the bands corresponding to the stretching vibration of the carboxylic acid, amide I and II and carboxylate groups) and analysis of the CH_2 stretching vibrations arising in the region 2850–2950 cm^{-1} demonstrated that the gel network is the result of hydrogen-bonding interactions between amide groups, interactions between carboxylic acid and alkyl carboxylate and hydrophobic interactions involving CH_2 groups.

Urea-based gelators represent a class of molecules suitable to be investigated by IR spectroscopy due to the presence of the ureido group, which is likely to be involved in hydrogen bonding. In general, a condition for obtaining supramolecular hydrogels resulting by assembly of urea-based gelators is the presence of other moieties able to establish hydrophobic bonds like π–π stacking or van der Waals interaction. The lack of a moiety able to interact through hydrophobic forces from the structure of a potential gelator can be compensated by other interactions. For instance, the study of Kleinsmann and co-workers [15] on the hydrogelation capability of *N*-[(uracil-5-yl)methyl]urea in different buffers revealed that phosphate ions can trigger the assembly in a gel network. IR spectra show that the characteristic N–H stretching band shifts from 3314 to 3146 cm^{-1} due to strong hydrogen bonding with C=O groups during the self-assembly, while the carbonyl stretching vibration band observed in solution at 1632 cm^{-1} shifts to 1675 cm^{-1} due to the hydrogen-bonding interactions with the uracil N–H protons. Moreover, the phosphate bands arising at 1070 and 982 cm^{-1} show intensity increase as a result of immobilization of the phosphate ion into the supramolecular assembly.

In some cases, the formation of hydrogels through hydrogen-bonding interactions can be demonstrated by FT-IR analyses of xerogels. This was the case of the supramolecular hydrogel formed between 1,4-bi(phenylalanine-diglycol)-benzene (PDB) and sodium alginate (SA), a polysaccharide composed of (1–4)-linked β-D-mannuronic acid (M) and α-L-guluronic acid (G) residues [16]. In this hydrogel, nanofibres of PDB alternate with SA chains. The network represents a semi-interpenetrating polymer network (semi-IPN) through hydrogen bonds, which has been proven by analysis of the IR bands specific to carboxylate groups from SA and the amide group from the PDB structure.

There is a large literature dedicated to polymeric hydrogels with medical applicability that contain biocompatible polymers such as poly(*N*-isopropyl acrylamide) (PNIPAM) or natural polymers such as dextran. PNIPAM has the amido group in its structure, whose vibrational modes are sensitive to association through hydrogen bonding. Inclusion of PNIPAM polymer in the structure of a hydrogel network can be thus demonstrated by FT-IR [17].

Dextran is a polysaccharide consisting of chains of 1–6 linked α-D-glucopyranosyl units with 1–2, 1–3 and 1–4 ramifications. The presence of hydroxyl groups allows chemical modifications by grafting various functional groups. One of the most reported dextran derivatives refers to introduction of methacrylic groups, which further can be involved in realisation of gel networks that have been tested in drug delivery experiments. These systems are pH sensitive, and changes in IR spectra corresponding to vibration motion associated with the bonds formed between dextran groups and other reactive groups can be used to monitor synthesis or degradation of dextran-based hydrogels [18–20]. In recent years, there has been an increasing interest in studying self-healing polymer hydrogels. These hydrogels are characterized by their ability to autonomously repair themselves without external stimuli utilising a multitude of co-operative non-covalent interactions such as hydrogen bonds, hydrophobic interactions and host–guest interactions [21, 22]. IR spectroscopy shows that both hydrogen-bonding and hydrophobic interactions make significant contributions to obtaining self-healing materials [23–26].

IR investigations on hydrogels usually demonstrate the interactions involving gelator molecules in building the 'solid network'. The non-homogeneous character of a hydrogel system is not only given by the presence of two phases—the network and the solvent—but also by the different water organization in the proximity of the fibres and in the solvent pools. Okazaki and Satoh [27] have shown that FT-IR measurements can probe the hydration and water properties in a hydrogel formed by cross-linking of poly(4-vinyl phenol) with different amounts of ethylene glycol diglycidyl ether. The O–H stretching band around 3300 cm^{-1} was deconvoluted into four sub-bands. On the basis of the relative band area and the peak wavenumber, it was suggested that hydrogen bonding of water in the gel is the most stabilized when the acidic proton of the phenol residue is not involved in chemical cross-linking. The authors also studied the influence of salts in hydrophobic hydration of the polymer.

Hydrogels can incorporate inorganic materials (e.g. graphene or graphene oxide (GO), oxidic and metallic nanoparticles) resulting in hybrid materials with improved properties and potential applications in biomedical fields. IR spectroscopy can be a method for the investigation of these hybrid materials. In a systematic study on the poly(vinyl alcohol) (PVA)/graphene oxide composite hydrogels, Xue and co-workers [28] noticed that an increase of the PVA molecular weight favours the formation of hydrogels containing GO. GO contains various oxygen functional groups (e.g. hydroxyl, epoxide and carbonyl groups) on their basal planes and edges that assure the possibility to be dispersed in water and at the same time to be involved in supramolecular architectures. In such systems, both polar non-covalent interactions and π–π interactions are possible [28, 29]. For PVA–GO systems, the FT-IR spectra suggested that oxygen-containing groups on the GO surface possibly interact with the PVA molecules through hydrogen bonding involving the C=O group from the surface of GO and H–O bonds from PVA and water molecules.

Another example of hybrid material is represented by the composite chitosan/ZnO hydrogel, which can be used as a component for production of bandages for wound dressing due to the antibacterial action of ZnO and adhesion properties of chitosan [30]. FT-IR spectroscopy was among the various techniques involved in characterization of such material. IR spectra proved the interaction involving ZnO nanoparticles and chitosan through hydrogen bonding, which determined the broadening of O–H at 3400 cm^{-1}.

2.2. Raman spectroscopy

Raman spectroscopy also studies the vibrational energy of molecules, being a suitable method to investigate the interactions that generate the hydrogel network and also hybrid hydrogel/nanoparticles materials. Both IR and Raman methods provide fingerprints of a system, although the information is complementary and selection rules are different. In the case of Raman spectroscopy, a vibration is active if this causes a change in the polarizability of the molecule [31]. An advantage of using Raman spectroscopy compared with IR spectroscopy in studying hydrogels is the fact that Raman spectroscopy is insensitive to aqueous absorption bands, while IR spectra are dominated by the broad water absorbance band. Among various Raman techniques, surface-enhanced Raman scattering (SERS) is characterized by high sensitivity [32]. Herein will be illustrated the suitability of Raman investigation on structural organization or characterization of diffusion properties of solutes in hydrogels.

Raman measurements are useful in characterization of carbon materials like graphene or GO that are incorporated in hydrogels. The oxidation state of graphene sheet incorporated in hydrogel can be determined by the ratio of the Raman peaks of GO that appear at 1606 and 1341 cm^{-1}. Zu and Han [33] obtained stabilized solutions of GO in the presence of Pluronic block copolymers. These solutions in the presence of α-cyclodextrin led to formation of hydrogels. The authors of this study exploited simultaneously the property of poly(ethylene oxide) (PEO) chains to form pseudo-polyrotaxanes and the ability of Pluronic to disperse graphene in aqueous solution to obtain a hybrid supramolecular gel. They proved that the presence of GO favours the gelation process of poly(ethylene glycol) (PEG)/α-cyclodextrin systems.

The SERS method provides enhancement of a Raman signal arising from molecules adsorbed on a metal surface in an aqueous medium. This method has a high sensitivity, permitting detection of molecules in very dilute solutions [14, 16] and demonstrating the interactions between different species. Thus, SERS is a method of choice for studying gel formation or hybrid inorganic nanoparticles/hydrogel composites.

Miljanic and co-workers [14] investigated the self-assembly process of bis-(S-phenylalanine) in the presence of silver or gold nanoparticles. SERS measurements probe the existence of adsorbed gelator molecules on the nanoparticles entrapped in a gel. The authors found that the presence of nanoparticles perturbs the common fibrous gel structure, but their presence allows identification of intermolecular interactions responsible for gelation.

Ou-Yang and co-workers [32] obtained a PVA hydrogel decorated with silver nanoparticles (Ag NPs), which were obtained in situ using β-cyclodextrin (β-CD) acting as a reducing agent and a shape-control agent. This material was then tested as a facile SERS sensor for determining sulphonamides. Uniform dispersion of Ag NPs in hydrogels assures reproducibility of detection of this type of antibiotic, which has a weak affinity for nanoparticles in the absence of β-CD.

The study of IR and Raman vibrations in the spectra of hydrogels resulting from cross-linking of β-CD with a pyromellitic anhydride (PMA) in different ratios gave the possibility to explore the structural changes in the polymer network during the hydration process. Rossi and co-workers have studied intensively this type of nanosponge (NS) material, which can swell in water. First, a gel network was obtained by a cross-linking process between β-CD and PMA in anhydrous dimethyl sulfoxide (DMSO), followed by washing and drying to obtain a NS. The hydrogel was obtained by adding the NS to water [34–37]. Deconvolution of the spectral band arising from active vibrations either in IR or Raman demonstrated the type of interactions between water confined in gel and the functional groups from the network. The formation of hydrogen bonds was proved by comparing the spectra of hydrated networks with the corresponding dry networks. The analysis of IR and Raman spectra allowed estimation of the hydration level of the polymer network. Such analysis has an impact on the understanding of the contribution of the interactions determining the formation and stabilization of a hydrogel, which further allow a rational design of such systems.

Later, the same group of researchers used a combination of UV–Raman spectroscopy with IR spectroscopy and offered an insight into the molecular mechanism related to the thermal response of cyclodextrin-based hydrogels [34]. The results provided information on the degree of H-bond association of water molecules entrapped in the gel network and the extent of intermolecular interactions involving the hydrophobic/hydrophilic moieties of the polymer matrix, which determine the pH-dependent thermal activation of hydrogels. The authors [24] analysed the shift of the maximum of bands arising in the 1500–1800 cm^{-1} region corresponding to stretching motions of the C=C bonds of the aromatic moiety of PMA, towards lower wavenumber values. This effect was observed in the presence of Na_2CO_3, and it was explained by the formation of intermolecular hydrogen bonds C–H•••O–H involving C–H from the aromatic rings of PMA and O–H from water molecules.

Equally, Raman spectroscopy can be a valuable tool in investigation of structural changes of water and polymer networks during the dehydration process of a hydrogel. In a number of papers, Ikeda-Fukazawa and co-workers [38] investigated the dehydration of poly(*N,N*-dimethylacrylamide) (PDMAA), PVA [39] and polyacrylamide (PAA)-based hydrogels [40]. It is accepted that the water in hydrogels exists in three states, namely bound water, intermediate water and free water [41] and dehydration occurs in two stages, which implies structural changes in the polymer network and water. In the case of dehydration of PDMAA, the free water is evaporated in the first stage, while in the second stage, the intermediate and bound water to gel fibres evaporates [38]. The evolution of the dehydration process can be followed by changes in the vibrational bands corresponding to H–O and C=O bonds of the polymer, but also following the peaks corresponding to water molecules involved in various types of hydrogen bonding. The authors found that the polymer network shrinks with water evaporation and at the end in the gel the residual water forms a tetragonal structure [28]. Similar behaviour on dehydration has been noticed for all polymeric systems [38–40].

Diffusion characteristics in polymer hydrogels are relevant for possible applications in drug delivery and controlled release because this can predict the release rates and the transport properties. Raman spectroscopy is a method of choice to estimate such parameters. For instance, Kwak and Lafleur [42] measured the mutual diffusion coefficients of poly(ethylene glycol)s with different sizes in Ca-alginate gels, analysing the intensity of the C–H stretching band as a parameter to estimate the concentration of solute in gel.

The number of studies regarding the applications of hydrogels in regenerative medicine is continuing to grow due to various synthetic routes and the possibility to investigate the final materials through various physicochemical methods. Among them, Raman spectroscopy is a suitable method for investigating the cross-link type and cross-linking density as well as conformation changes of polymers under various stimuli in a gel network. Moreover, this method has the advantage of hydrogels monitoring in cell cultures [31].

2.3. Fluorescence spectroscopy

Fluorescence spectroscopy can be used in characterizing gel systems that contain fluorophore species or moieties capable of absorbing energy allowing excitation from its ground electronic state to one of the vibrational states in an excited electronic state. Then, the excited state of the fluorophore reaches the lowest vibrational level of the excited state (singlet). This process involves conformational changes and interaction with the microenvironment. The third step involves transition from an excited electronic state to the ground electronic state and is accompanied by a photon emission characterized by a lower energy than that corresponding to the excitation photon.

To monitor the properties of gel formation using fluorescence methods, the presence of fluorescent moieties in the system is required. An impressive number of studies regarding investigation on gel systems that involve fluorescent methods have been reported. In most cases, fluorescence measurements are performed to demonstrate the application of gel materials as sorbents or as drug delivery agents, but can highlight as well the dynamics of molecules inside the gel. To demonstrate the suitability of fluorescence spectroscopy in characterizing different aspects of gel systems, several examples are mentioned in this review.

Attachment of a fluorophore to polymeric hydrogels can find potential application in the detection and removal of toxic cations such as Cd^{2+}, Cu^{2+}, Pb^{2+} and Hg^{2+}. Chitosan has been proved to be a good absorbent for Hg^{2+}, and at the same time, this natural polysaccharide can be easily functionalized. Geng et al. [6] described the synthesis of a fluorescent three-dimensional chitosan-based hydrogel via cross-linking with glutaric aldehyde. This new hydrogel has been tested as a solid-phase fluorescent probe for the detection of Hg^{2+} in water. This cation determines the fluorescence quenching, and in the case of the chitosan-based hydrogel a remarkable selectivity and sensitivity determination of Hg^{2+} compared with other systems was noticed. The detection limit achieved with this material was 0.9 nM, a value at least one order of magnitude lower than with other fluorescent dye molecules.

An interesting example is represented by the study by Fraix and co-workers [43] reporting the formation of a supramolecular hydrogel based on assembly of four components: two polymeric structures—poly(β-CD) polymer and a hydrophobically modified dextran by attaching lauryl chains—a commercial zinc phthalocyanine and a tailored nitric oxide photodonor. The gel network resulted by inclusion complexes formed between cyclodextrin units and lauryl chains. The photoactive components were also held in the gel through host–guest interaction with the cyclodextrin cavity, this type of interaction avoiding their aggregation, which usually results in inactivation of photodynamic properties. Steady-state and time-resolved fluorescence methods showed that the photoactive components do not interfere when they are enclosed in a hydrogel assuring their simultaneous operation under the light stimuli. The Zn phthalocyanine complex is a well-known red photo emitter and an effective 1O2 photogenerator [44]. The structure of the other component, which generates NO under the light action, is a derivative of nitroaniline and a 4-amino-7-nitrobenzofurazan (known for its emission in the green region) [45]. The system described assures simultaneous generation of active species with therapeutic properties.

A similar example of hydrogel incorporating photoporphyrin units formed in this case by chemical cross-linking between a PEG-derivative porphyrin and sodium alginate has been reported by Dong et al. [46]. The system has been characterized by various methods: rheological analysis, thermogravimetric analysis and UV–visible and fluorescence spectrophotometry. This hydrogel with porphyrin as a sensing unit has been prepared giving biocompatibility of alginate and recognized bio-applications of porphyrin and tested for fluorescent imaging. The alginate acted as a spacer for the porphyrin units, preventing aggregation and preserving the fluorescent properties, which assure the tracking of hydrogel within organisms. Multispectral fluorescence imaging for drug delivery from the hydrogel shows high yields and excellent biocompatibilities have been demonstrated using doxorubicin, a fluorescent drug. The photoporphyrin and the drug used have different spectra, this technique offering the possibility of simultaneous tracking of two (or, in extension, more) probes.

Montalti and co-workers [47] demonstrated the organization of gel fibres in a supramolecular gel formed by assembly of 1,3,5-cyclohexyltricarboxamide-based gelator comprising two hydrophilic moieties and one hydrophobic substituent containing a naphthalene group, which represents the fluorophore. This has been possible by analysing the energy transfer from the fluorescent supramolecular hydrogel to a hosted fluorophore. The organization of gelator molecules through naphthalene π–π stacking interactions has been demonstrated by introduction in the gel system of a fluorescent probe with similar structure, propyldansylamide, characterized by sensitivity to the polarity of the microenvironment. In general, the formation of one-dimensional assemblies of gelator represents a recognition process, but the possibility to insert probes with similar structures cannot be excluded. In this case, the dansyl derivative could be inferred in the gel fibres. Analysis of the fluorescence spectrum of dansyl derivative, excited state life-time measurements, steady-state and time-resolved fluorescence anisotropy measurements demonstrated the partition of the dansyl probe between the water environment and gel fibres. The measurements revealed a strong immobilization of the dansyl moiety in gel fibres. The authors also demonstrated that the percentage of photon transfer absorbed by the gelator through the dansyl probe depends on the ratio between the species; a higher concentration of probe increases the percentage. This system can be of particular interest in the field of technology and solar energy conversion.

2.4. Circular dichroism

Circular dichroism spectroscopy represents a valuable tool for investigating the assembly of gelators that include chiral centres in their structures [48, 49]. Circular dichroism (CD) refers to the differential absorption of left and right circularly polarized light by chiral molecules and is usually expressed as the ellipticity of the transmitted light [48]. In the case of achiral molecules, both polarized rays are equally absorbed and the result is a 'zero' spectrum. Formation of gel networks, in general, and hydrogel in particular, can be explored by CD spectroscopy in the cases of assemblies resulting from organization of chiral biomacromolecules, such as proteins or incorporating DNA, polypeptides or chiral LMWGs. A CD signal is observed where the molecules exhibit absorbing UV bands. In many cases, the CD band of chiral molecules is weak, but assembly in nanofibers often is translated into an increase of CD ellipticity [11, 50]. The possibility of recording CD spectra at variable temperature is useful to demonstrate the transition from sol phase to assemblies at the nanoscale level. CD measurements can be extended to the IR region [49], but in this case, there are only a limited number of studies compared with UV-CD. This new technique can provide information on the structural fragments of molecules involved in realization of gel networks, as in the case of self-assembled peptide [50], chiral bis-urea gelators [7] or guanosine-5′-hydrazide [51].

The vast number of studies on formation of gels through self-assembly of LMWGs stressed the observation that the presence of a chiral centre favours assembly into a gel network. In many cases, the urea unit appears in the structure of potential LMWG due to unidirectionality of the non-covalent interactions assuring formation of gel fibres and the ability to form hydrogen bonding. Rodriguez-Llansola et al. [7] reported studies on a self-assembly process of bis-urea molecules bearing aromatic hydrophobic moieties that allow π–π stacking arrangements and chiral centres. They found that similar compounds without the chiral centres do not form gels. The molecular solutions of a chiral molecule either in the R or S enantiomer did not exhibit the dichroic band. Once the gel was formed, the CD spectra showed the positive and negative dichroic bands for the R and S enantiomers, respectively. Surprisingly, the presence of an achiral molecule with similar structure added to a solution of a chiral gelator results in a gradual increase of chiral bands attributed to assemblies of R enantiomer up to a certain concentration, while a further increase produces precipitation of the mixture and disappearance of the CD bands.

CD measurements can demonstrate the preservation of the natural state of proteins encapsulated in some hydrogels. For instance, Song et al. [52] reported that the formation of hydrogelator resulting from self-assembly of amylopectin grafted with lauryl chains. Bovine serum albumin (BSA) can be entrapped in this hydrogel without affecting its secondary structure. Moreover, recording the CD spectra at different temperatures, it was observed that the BSA is protected in the gel against thermal denaturation.

Polypeptides are often used as building blocks in hydrogelator structure. These fragments exhibit CD bands depending on the chirality of the amino acid used and analyses of the CD spectra can monitor changes in the secondary structure of polypeptide induced by the formation of various self-assemblies. For example, poly(L-alanine) grafted to PEG with different chain lengths changes its β-sheet structure observed in water to α-helical structure in gel fibres. The change depends gradually on the PEG chain length. This effect of PEG in peptide assembly results in a steric hindrance of polypeptide packing side-by-side characteristic of the β-sheet, favouring α-helix assembly [53].

CD measurements offer useful information on the assembly of hydrogelators with a common backbone highlighting the differences induced by specific groups. Vilaca et al. [54] characterized by various methods the dipeptide-type hydrogelators containing tryptophan N-capped with the non-steroidal anti-inflammatory drug naproxen and C-terminal dehydroamino acids, dehydrophenylalanine (ΔPhe), dehydroaminobutyric acid and dehydroalanine (ΔAla). The assembly into gel fibres of these gelators is assured by the capacity to generate hydrogen bonding, hydrophobic and aromatic π–π interactions. CD measurements revealed that assembly is similar for these gelators, all spectra exhibiting a broad positive cotton effect, assigned to the π–π^* short-axis polarized transitions of naphthalene (with maximum around 287 nm). The presence of positive and negative bands at 219 and 233 nm, respectively, suggests left-handed helical naphthalene arrangements through chiral stacking.

There is a trend to use chirality to tune the self-assembly, which determines later the applications of the resulting gels in the biomedical area [55, 56]. Many chiral hydrogels result by self-assembly of biomolecules, such as amino acids, peptides, cholesterol and saccharides [57–61], and they have potential applications, which include chiral adsorption and release and chiral catalysis [57, 62–64]. For example, hydrogels obtained by free-radical co-polymerization using N-acryloyl-L-alanine as chiral hydrophilic monomer and octadecyl acrylate as hydrophobic monomer are pH sensitive and exhibit enantio-differentiating release ability to chiral drugs such as ibuprofen [57]. This aspect is logical knowing that many biological processes are defined as chiral-recognition types. Marchesan et al. [55] obtained a series of eight tripeptides with the sequence Phe–Phe–Val having a combination of D and L enantiomers of amino acids. The authors investigated in what manner the self-assembly is influenced by the chirality of gelator and found that chirality of the CD signal is determined by the chirality of the central amino acid. The peptides exhibit either a positive maximum or a negative minimum in the region 225–235 nm due to π–π stacking of phenylalanine with either L or D chirality, respectively.

Ma and co-workers [65] obtained new pentapeptides bearing aromatic fluorophore units (fluorenyl, pyrenyl or naphthyl groups) as potential hydrogelators and characterized the systems using CD and fluorescence spectroscopy combined with electron microscopy and rheological measurements. In the absence of these fluorescent aromatic units, they did not observe formation of a gel. Introducing proline in the peptide sequence, it was possible to control the supramolecular assembly avoiding β-sheet arrangements [66]. This has been proven by CD measurements, which demonstrated formation of α-helix assemblies.

CD measurements can also monitor the transition that occurs in the assembly of a peptide amphiphile solution induced by the presence of salt additive on supramolecular assembly. It was considered that the salt facilitated the α-helix to β-sheet transition and induced gelation in amphiphilic peptide due to the electrostatic screening effect [67].

Huang et al. [68] investigated the thermohydrogelation of PEG and an oligo(tyrosine) block copolymers with various PEG and tyrosine (Tyr) chain lengths. These copolymers self-assemble due to β-sheet organization of Tyr blocks and gradual dehydration of PEG by increasing temperature favouring entanglement of the PEG chains. The β-sheet conformation adopted in the hydrogel state has been proven by CD spectra, which exhibit an increased negative band at temperatures corresponding to the gel state [68].

A library of naphthalene–dipeptide hydrogelators has been investigated by CD, X-ray and electron microscopy to demonstrate structural aspects that govern the assembly in gel fibres [69]. The CD spectra demonstrated that fibres are the result of the π–π stacking of the naphthalene moieties as they exhibit bands corresponding to exciton couplets of the naphthalene groups. At the same time, the sign of the naphthalene exciton is determined by the dipeptide chiral properties, indicating that the handedness of the chiral arrangements in the fibres is directed by the dipeptide chain.

These few examples chosen from the very broad types of studies demonstrate that CD spectroscopy complements the electron microscopy methods, offering valuable information on the gel organization.

2.5. Estimation of mesh-size distribution by pulsed-field gradient NMR spectroscopy

Mesh size (**Figure 1**) of a hydrogel or correlation length, ξ, represents the average distance between consecutive cross-links [70]. Mesh-size distribution of a hydrogel is a characteristic relevant for the ability of the entrapped species (ions, drugs and biomolecules) to diffuse through the gel network [71, 72].

This parameter indicates the maximum size of solutes that can pass through the gel network. To characterize the mesh size of hydrogels, different methods can be used that rationally offer an average value [70, 73]. Methods such as solute exclusion, mercury porosimetry, nitrogen adsorption/capillary condensation and microscopy were used in the beginning [74]. Solute exclusion is a method suitable to estimate the mesh size of polymeric networks that was introduced more than 50 years ago to investigate the penetration of polymer molecules in cellulose fibres [75]. The methods can be applied for wet samples. The other three methods mentioned above require dry samples of polymeric structures. The drying process can induce changes in network organization and distances between fibres. Other methods that can be applied to determine the mesh size for both polymeric and supramolecular hydrogels are based on determination of diffusivity of probes that have no specific interaction with gel fibres. Pulsed-field gradient NMR (PFG-NMR) is one of the methods based on investigation of the chaotic motion of molecules determined by molecular collisions. This method allows estimation of the average pore size in a hydrogel network from the ratio of the diffusion coefficient of the probe molecules in the gel network (D), to their diffusion coefficient in the solvent (D0) [75, 76].

Matsukawa [77] analysed in a review article the diffusional behaviour of molecular probes in polymer networks. Two parameters, the spin–lattice relaxation time and spin–spin relaxation time, reflect the solvent and polymeric probes' behaviours in the gel network. The strength of the field gradient has to be chosen as a function of diffusion coefficient. Thus, for molecules with a diffusion coefficient of the order of 10^{-5} cm^2 s^{-1} the strength of the field gradient is about 100 G cm^{-1}. A larger magnetic field gradient, increased up to 2000 G cm^{-1}, is necessary for probes with small diffusional coefficients. The scheme for pulsed-field gradient pulse sequence for measuring the diffusion coefficient D is represented in **Figure 2**:

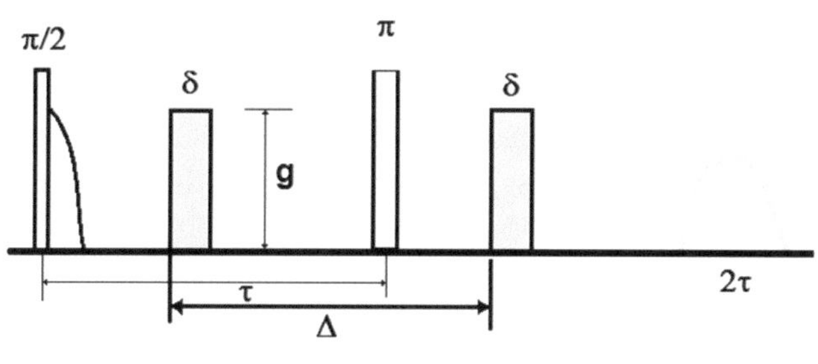

Figure 2. Pulsed field gradient NMR sequence for determination of diffusion coefficient.

The diffusion coefficient is extracted from Eq. (1):

$$ln\left[A(\delta)/A(0)\right]) = -\gamma 2g2D\delta 2(\Delta - \delta/3), \tag{1}$$

in which A(δ) and A(0) are echo signal intensities at 2τ with and without the magnetic field gradient pulse of length δ, respectively, τ is the pulse interval, γ the gyromagnetic ratio of the proton, g the field gradient strength, D the self-diffusion coefficient and Δ the gradient pulse interval [77, 78]. The diffusion coefficient is related to the hydrodynamic radius of the probe. Usually, NMR gives values of the hydrodynamic size, which is larger than the pore size of the network estimated from thermoporometry. Scherer estimated a relation between the values provided by NMR and thermoporometry [79]. Estimation of the differences provided by different methods is relevant for studies aiming to predict whether the network is capable of encapsulation of large molecules.

Using this method, Matsukawa et al. [77, 78] analysed the diffusion behaviour in various polymeric networks such as polypeptide gels, starch gel, gellan gum gels, polyacrylamide gel or swollen polyisoprene.

Pescosolido and co-workers [72] investigated the properties of interpenetrating polymer network hydrogels based on calcium alginate and a dextran methacrylate derivative, due attention being given to determination of mesh size. The authors found significantly different values for pore size provided by the two methods used in their study: 44.5 nm from cryoporometry and 22.5 nm from low-field NMR (using a pulse sequence). The differences have been explained by the effect of water solidification, which occurs in the cryoporometry measurement. Release experiments of a model protein (myoglobin) from the interpenetrating polymer alginate/dextran methacrylate network hydrogel confirmed the value for pore size provided by NMR experiments.

Another example of application of the PFG-NMR technique refers to the microstructural analysis of the calcium alginate gels resulting from introducing $CaCO_3$ as an insoluble salt, which is dissolved gradually upon acidification [80]. In the presence of glucono-δ-lactone (GDL) as a proton donor, the calcium salt is dissolved. To tailor the microstructural properties of the gel, the ratio between alginate, Ca^{2+} and the co-solvent methanol were varied. Hydrophilic dendrimers with PEG surface groups have been used as probes. Image analysis of transmission electron microscopy (TEM) micrographs has revealed that the strand radii are in a narrow range of 2–2.3 nm and are not influenced by Ca^{2+} concentration or the presence of methanol. Dendrimer diffusion into these gels studied by PFG-NMR provided, as expected, larger values for hydrodynamic radii, and demonstrated that the presence of methanol affects the diffusion of the dendrimers into the gel networks. Walther et al. [81] studied the self-diffusion of PEG into *K*-carrageenan gels with different microstructures tailored by adding potassium or sodium chloride using nuclear magnetic resonance (NMR) diffusometry and TEM. As expected, small voids observed in the case of potassium-induced gels reduced diffusion coefficients and gave a strong dependence of the self-diffusion coefficient, while in the case of sodium-induced gelation, the voids increase and the ratio D/D0 is close to 1.

Using PFG-NMR spectroscopy, Wallace et al. [82] determined the network's mesh size in a supramolecular hydrogel formed upon addition of Ca^{2+} to solutions of naphthalene diphenylalanine (2FF). A series of dextran probes with molecular weights in the range 6–2000 kDa and hydrodynamic diameters in the range 2–60 nm was used. These probes are soluble in water and have no specific interaction with naphthalene diphenylalanine.

The models applied in analysing the variation of the decrease in the diffusion coefficients relative to those measured in dilute solutions suggested a value of mesh size for supramolecular hydrogel of approximately 40 nm. It was found that probes with hydrodynamic diameter <40 nm move freely through the network, while a dextran probe with molecular weight 2000 kDa has a restricted diffusion.

Jowkarderis and Van de Ven [83] prepared a hydrogel by cross-linking of cellulose nanofibrils with diamines, and mesh-size analysis was performed by solute exclusion and PFG-NMR spectroscopy using dextran-type probes. Diffusion coefficients have been determined both in hydrogels (D) and in dilute solutions of dextrans (D0). The pore sizes of the hydrogel network were determined by analyses of the ratio D/D0. It was found in this case that the average mesh size was around 15 nm. They also found a group of large pores accessible to all dextran probes.

The spectroscopic methods referred to above are well recognized in studying the complex and diverse hydrogel systems. Compared with these, EPR spectroscopy is a less-used method. Therefore, the next part of this review is dedicated to general aspects of EPR spectroscopy and will summarize the applicability of this method in studying formation and properties of hydrogels.

2.6. EPR spectroscopy

Electron paramagnetic resonance (EPR) is a spectroscopic technique that detects transitions induced by electromagnetic radiation between the energy levels of electron spins, in the presence of a static magnetic field. EPR spectral features (e.g. resonance frequencies, splitting, line shapes and line widths) are sensitive to the electron distribution, molecular orientation, molecular motion and the local environment. Numerous books and review articles cover the theoretical and experimental EPR methods, targeting various fields of applications: materials and polymer sciences, physical chemistry, biochemistry and medicine, catalysis, environmental sciences, radiation dosimetry and geological dating [84–90]. In this section, the application of X-band spectroscopy in providing structural and dynamics (in the time range of 10^{-10}–10^{-7} s) information on the hydrogels will be mentioned.

Hydrogels are usually EPR-silent systems, therefore, to have access to the type of information that EPR spectroscopy can provide, it is necessary to introduce paramagnetic reporters, spin labels and probes. Spin probes are stable free radicals or in some cases paramagnetic transition metal ions introduced in the system of interest, while spin labels are stable radicals covalently bound to a constituent of the studied system. Spin labelling involves modification of the system or material studied by introducing a new moiety (with a paramagnetic property) [84–88].

The most widely used spin probes or labels are nitroxides, which are stable paramagnetic molecules with a paramagnetic N–O• fragment bearing an unpaired electron. This fragment is surrounded by shielding substituents (usually methyl groups) [84]. Nitroxides have the advantage that by varying their chemical structure, the paramagnetic properties do not suffer significant changes. TEMPO- and PROXYL-type nitroxides (**Figure 3**) are commercially available, offering the possibility to be covalently attached to other molecules of interest through reactive groups R (carboxy, amino, oxo and hydroxy). Depending on the system analysed and the information needed to be obtained, they can be used directly as spin probes or labels.

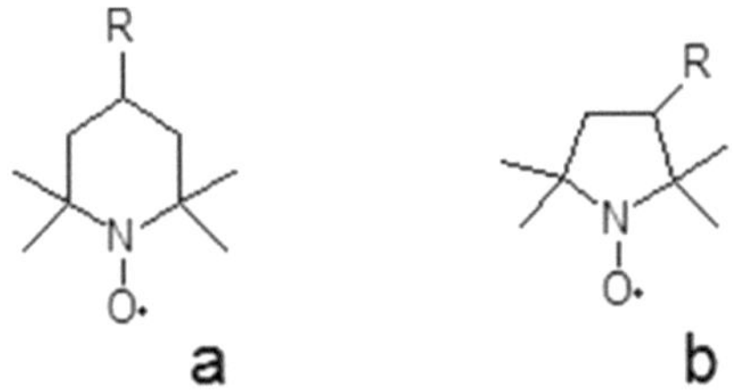

Figure 3. Structures of the most common classes of nitroxide spin probes and labels. (a) TEMPO-type nitroxides, (b) PROXYL-type nitroxides.

The EPR parameters of nitroxides, the rotational correlation time and nitrogen hyperfine splitting constant (a_N), are usually correlated with the properties of the microenvironment sensed by the paramagnetic probes, like local viscosity or polarity. Although the information provided by EPR spectroscopy is local and not global as in the case of CD, IR or Raman spectroscopy, these can be further correlated with macroscopic properties of a system, in particular, a hydrogel. Sol-to-gel transition can be easily demonstrated involving methods providing global information on the system (vide supra), but the macroscopic observed changes are the result of reorganization at microscopic or nanoscopic levels. Thus, a method-like EPR spectroscopy can be used to demonstrate such a transformation.

Compared with other spectroscopic techniques, EPR spectroscopy has been very rarely used for studying gel properties and dynamic aspects of these systems. The sol-to-gel transition can be demonstrated in particular cases by other spectroscopic methods, such as CD if the gelators are chiral or assembly induces a chiral conformation, vibrational spectroscopy, following vibrational band intensities of groups involved in the assembly process generating the gel network. For dynamic studies, EPR spectroscopy can be used only if such changes take place in the range of 10^{-6}–10^{-9} s. However, a hydrogel represents a complex system and a good strategy for such investigation needs to involve different techniques. The literature indicates hundreds or thousands of studies involving the spectroscopic methods described above, while the number of EPR spectroscopy studies is very small. Most of the EPR studies reported in the literature regard polymeric gels [89, 91, 92]. For instance, the EPR spectroscopy with its spin probe method had been proved a valuable method for studying thermoresponsive systems and in particular thermoresponsive hydrogels. Using an appropriate spin probe, it is possible to demonstrate the inhomogeneities inside a gel system [89, 93–95]. EPR spectroscopy and EPR imaging were used to study the translational diffusion process of spin probes in PVA hydrogels, with vast applications in biotechnological field, to get information on the polymer dynamics. The diffusion of 4-*N*-butylamino-TEMPO and 4-amino-TEMPO (4-NH_2T) in PVA hydrogel and in polymeric blends of PVA with other polymers has been determined as a function of various parameters: molecular weight and concentration of polymer [91].

Natural polysaccharides are often part of hydrogel networks and the literature indicates a number of EPR studies in studying the hydrogels resulting from assembly of polysaccharides. A series of EPR studies has been reported on investigation of polysaccharide hydrogels resulting from complexation of polymeric chains with various cations or dynamics of synthetic spin probes in polysaccharide hydrogels [92, 96–101].

Alginate or pectin form hydrogels in the presence of numerous divalent cations, most of them being paramagnetic. Kajsheva [92] analysed the EPR spectra of metals in alginates and pectinates to get information about the configuration of the corresponding complexes with Cr(III), Cu(II), Ni(II), Mn(II) and Co(II).

EPR spectroscopy has been applied to study the binding of paramagnetic metal ions to extracellular polysaccharide surfaces produced by cyanobacteria *Anabaena spiroides* [98]. The results underline the potential application of this complexation process in water purification. Kanesaka and co-workers [100] investigated the effect of Cu^{2+} on gelation of gellan solution using EPR, CD and viscoelasticity measurements. The CD and EPR experiments revealed valuable information regarding the sol-to-gel transition of gellan in the presence of this cation, which can be considered as a two-step process. In the first phase Cu^{2+} binds to the gellan chain determining a coil-to-helix transition. In the second stage, the hydrogen bonds formed between gellan chains determined a further coil–helix transition.

Kempe [96] used spin-labelled insulin as a spin probe to analyse its behaviour in chitosan/glycerol-2-phosphate (beta-GP) gel in terms of dynamics and pH sensitivity. In addition, the release of this drug from the chitosan gel has been monitored. Bertholon and co-workers [97] prepared a series of spin-labelled polysaccharides (dextran, dextran sulfate or chitosan), which were incorporated into poly(isobutylcyanacrylate) nanoparticles aiming to study mobility of the nanoparticle surface groups. The EPR spectra showed a two-component feature, the ratio between them depending on the ability of polysaccharide to fold on the nanoparticle surface. Numerous cations of transitional metals are paramagnetic and form complexes with various starches. Such complexes were investigated by EPR [99].

The formation of alginate gel in the presence of different divalent cations has been investigated by EPR spectroscopy using spin-labelled alginate (ALG-L) obtained by reaction of alginate

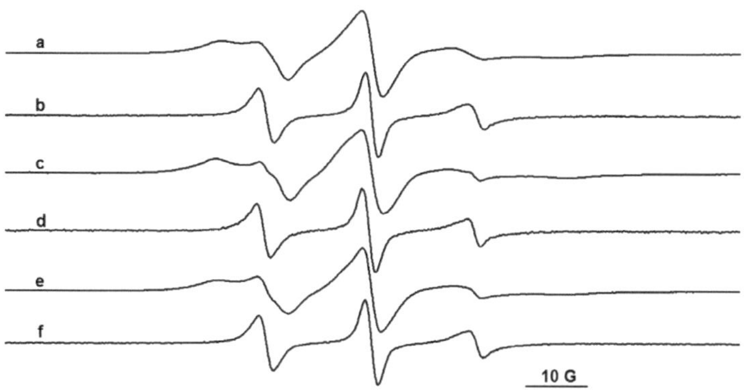

Figure 4. The EPR spectra of 4-NH_2-T-ALG-L in: Zn^{2+}/alginic acid gel (a), alginic acid/Zn^{2+}/EDTA mixture (b), Ba^{2+}/alginic acid gel (c), alginic acid/Ba^{2+}/EDTA mixture (d), Ca^{2+}/alginic acid gel (e), alginic acid/Ca^{2+}/EDTA mixture (reproduced from [90]).

with amino-TEMPO [101]. Sol-to-gel transition induced by the presence of dications was demonstrated by change of EPR features of spin-labelled alginate reflecting different dynamics. Thus, in the presence of dications (Ca^{2+}, Zn^{2+} and Ba^{2+}) spin-labelled alginate became immobilized (**Figure 4**, spectra a, c and e). It was noticed that the distance between the outer lines in the EPR spectra of complexes can be correlated with the strength of the gel.

The complexes of divalent cations are dissolved in the presence of a stronger ligand for cations, like EDTA. This is expressed in the EPR spectra by the change in the dynamics of spin-labelled alginate, which became more mobile (**Figure 4**, spectra b, d and f).

Diffusion experiments revealed that both the cation and alginate polyanion in the gel fibres can exchange with cations and molecules, respectively, in solution [101].

As EPR spectroscopy is sensitive to environmental changes around a paramagnetic probe and its dynamics, Ionita and co-workers investigated by EPR spectroscopy the synthesis [102] and properties [103] of covalent hydrogels that resulted via cross-linking of diisocyanate-terminated PEG with β-CD [104, 105]. In fact, these hydrogels were obtained by replacing the reaction solvent (anhydrous DMF) with water. The synthesis has been monitored following the dynamics of various spin probes (like commercially available TEMPO derivatives or synthesized spin probes: spin-labelled cyclodextrins [106, 107], spin-labelled oligo(ethylene glycol)s [108]). Spin-labelled cyclodextrins (SL-CD) introduced in the reactions showed progressive immobilization as the gel network was formed (**Figure 5**).

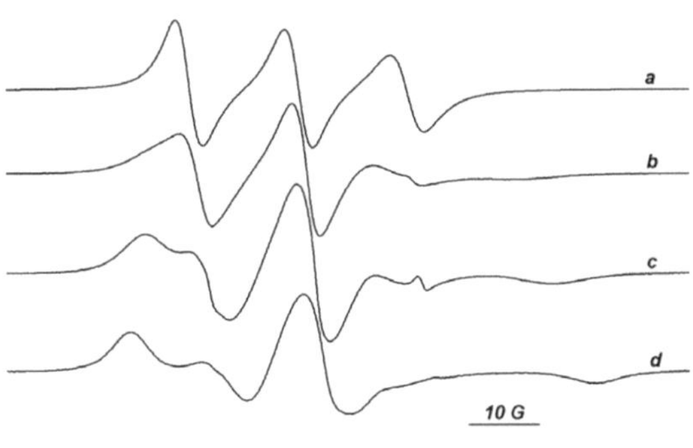

Figure 5. EPR spectra of SL-CD: (a) in DMF, initial solution, (b) during gel formation in DMF, (c) after gel formation in DMF and (d) after replacing the DMF form gel network with water (reproduced from [102]).

Using both spin probes and spin-labelling methods, it was possible to explore by EPR spectroscopy the changes in the gel network as a function of temperature, solvent, polymer chain length and initial ratio between reactants [103].

The EPR spectra of spin-labelled PEG/β-CD gels are dominated by two components, showing strongly immobilized and relatively mobile spin labels (**Figure 6**). These components were assigned to two different positions occupied by the β-CD units in the network. The immobilized component is associated with the β-CD units located at the cross-linking points of the gel network, while the mobile component corresponds to the spin labels at the ends of the PEG chains. **Figure 6** shows EPR spectra of spin-labelled gels with different PEG/β-CD ratios and different PEG chain lengths at 293 K. These demonstrate that mobility of spin labels attached to cyclodextrins representing the cross-links in the hydrogel network is due to the flexibility of PEG chains connecting the spin label to the β-CD. In this sense, longer PEG chains determine higher mobility.

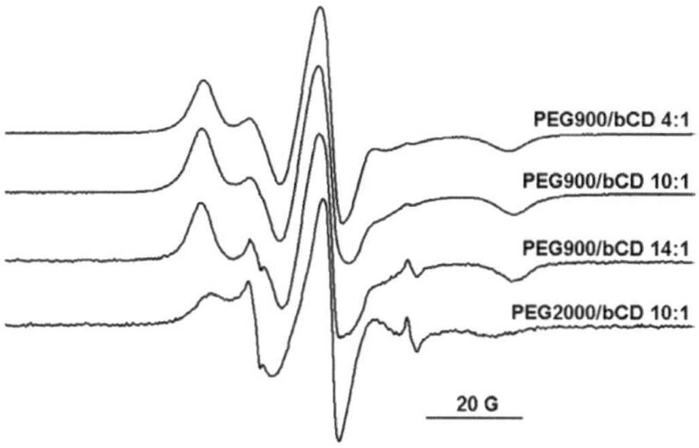

Figure 6. EPR spectra of spin-labelled gels PEG900/β-CD and PEG2000/β-CD recorded at 293 K in water (reproduced from [92]).

Hydrogels can encapsulate in the solvent pools various species as long as their size is smaller than the mesh size of the network. In this particular case of chemical hydrogel resulting from reaction of β-CD with PEG, the ability of cyclodextrin to act as a host for hydrophobic compounds enhanced the encapsulation properties of the hydrogel. To probe the host properties of the β-CD groups in the gel, a spin probe with high affinity for β-CD was used adamantane–TEMPO (AT) 92. Diffusion into PEG/glycerol (glycerol replacing β-CD) and PEG900/β-CD gels of this spin probe was followed using water or dichloromethane as solvents (**Figure 7**). In the case of PEG900/glycerol gel, the EPR spectra of AT (**Figure 7**, spectrum A) showed only one rapidly tumbling component consistent with the spin label dissolved in the solvent pool. The same feature of EPR spectra was observed for the PEG900/β-CD gel, when DCM was used as a solvent. This behaviour is normal as host–guest interactions of cyclodextrin are demonstrated in water solutions. However, when AT was uploaded into the PEG900/β-

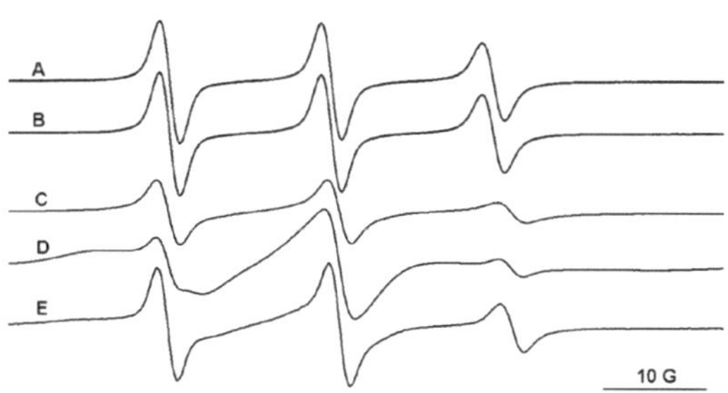

Figure 7. EPR spectra of AT in PEG900/β-CD gel in DCM (A); PEG/glycerol gel in DCM (B); PEG/glycerol gel in water (C); PEG900/β-CD gel in water (D), and PEG900/β-CD gel in water in the presence of a large excess of amino-adamantane (E) (reproduced from [92]).

CD hydrogel (**Figure 7**, spectrum D), the EPR spectrum showed two components, a rapidly tumbling AT situated in water pools of the gel network and a slowly tumbling AT complexed with the β-CD units at the cross-linking points of the gel network.

Complexation of AT to the β-CD cavity was further confirmed by a competition experiment with 1-adamantylamine. The sample of PEG900/β-CD gel loaded with AT was placed in a 10^{-1} M solution of 1-adamantylamine. The EPR solution spectra recorded after equilibrium was reached indicated that AT had partially diffused from the gel to the solution. The EPR spectra of the gel (**Figure 7**, spectrum E) showed that the rapidly tumbling component became dominant. This is a consequence of the replacement of AT from the β-CD cavities inside the gel structure by 1-adamantylamine. This result proves that the β-CD cavities in the gel retain their host–guest properties. Analyses of the EPR spectra of the TEMPO spin probe corresponding to frozen hydrogel samples showed that gel fibres prevent ice formation.

Few EPR studies reported in the literature regard the supramolecular gels resulting from assembly of LMWG [109–112]. One explanation is probably that the self-assembly process represents a reorganization and thus selective process and insertion of a different molecule in the fibrillar assembly is, in general, less probable. Some studies involving EPR measurements in studying formation of supramolecular hydrogels through assembly LMWG have appeared recently.

Takemoto et al. [110] reported the synthesis of an amphiphilic optically active compound that contains a paramagnetic (2S,5S)-2,5-dimethyl-2,5-diphenylpyrrolidine-N-oxyl radical fixed in the inner position with a hydrophobic long chain and a hydrophilic R-alanine residue in the opposite terminal position. They found that this compound acted as a LMWG in water, resulting in a spin-labelled hydrogel. From the variable temperature EPR spectra recorded, it was possible to determine the gel-to-sol transition temperature by analysing the EPR spectral line-width of the central line from peak-to-peak (Hpp). This parameter increased at the sol-to-gel transition due to spin–spin interactions and decreased in the heating run. This study demonstrated the suitability of EPR measurements in evaluation of the temperature for the sol-to-gel transition, which depends on various factors such as gelator structure or solvent nature.

Hydrogelation of three isomers of carboxy-dibenzylidene sorbitol induced by lowering the pH from 13 to 4 has been investigated by rheological measurements, SEM and EPR spectroscopy. Introducing in the system the corresponding spin-labelled carboxy-dibenzylidene sorbitol, it was concluded that these are not included in the gel fibres, as gelation was not accompanied by a change in the probe's dynamics. Surprisingly, in the case of the spin probe 5-doxyl stearic acid, its aggregation was noticed once the gel was formed from water solution. The behaviour of spin probes in gel systems depends on the solvent used. Thus, it was observed that the spin probe 5-doxyl stearic acid is entrapped in the gel fibres when the solvent composition water/DMSO was 1:1 (v/v) [111].

These few examples presented above demonstrate that EPR spectroscopy is a valuable tool for characterizing hydrogel properties, which can accompany other 'classical' well-established physicochemical methods for investigating complex hydrogel systems.

http://dx.doi.org/10.5772/62900

3. Conclusion

In summary, the aim of the review was to demonstrate the usefulness of different spectroscopic techniques in physical characterization of the complex systems represented by hydrogels. The results of these investigations orient each material towards a specific application. Nowadays, there is a trend in development of multifunctional nanocomposites based on different types of hydrogels acting as a matrix for various nanomaterials. Various physicochemical methods, including those presented in the review, can be used to characterize new materials. Only a limited number of spectroscopic techniques have been considered, most of them being classically considered for characterization of hydrogels. Some of them provide global information on the hydrogel system (vibrational spectroscopy, circular dichroism and PFG-NMR spectroscopy), others provide local information, like fluorescence or EPR spectroscopy. Although less involved in characterization of hydrogels, EPR spectroscopy can bring an important contribution not only in demonstrating the heterogeneity of hydrogel-based materials by using spin-labelling and spin-probe methods mentioned here, but also to monitor radical processes taking place in hydrogels. The choice of spectroscopic methods that can be combined with microscopic and rheological methods depends on structural characteristics of the investigated hydrogels.

Acknowledgements

This work was supported by a grant from the Romanian **National Authority for Scientific Research, CNCS – UEFISCDI, Project Number PN-II-ID-**PCE-2011-3-0328.

Author details

Gabriela Ionita

Address all correspondence to: gabi2ionita@yahoo.com

Institute of Physical Chemistry 'Ilie Murgulescu' of the Romanian Academy, Splaiul Independentei, Bucharest, Romania

References

[1] Appel EA, del Bario J, Loh XJ, Scherman OA. Supramolecular polymeric hydrogels. Chem Soc Rev. 2012;41:6195–6214.

[2] Buwalda SJ, Boere KW, Dijkstra PJ, Feijen J, Vermonden T, Hennink WE. Hydrogels in a historical perspective: from simple networks to smart materials. J Control Release. 2014;190:254–273.

[3] Du X, Zhou J, Shi J, Xu B. Supramolecular hydrogelators and hydrogels: from soft matter to molecular biomaterials. Chem Rev. 2015;115(24):13165–13307.

[4] Peppas NA, Huang Y, Torres-Lugo M, Ward JH, Zhang J. Physicochemical foundations and structural design of hydrogels in medicine and biology. Annu Rev Biomed Eng. 2000;2:9–29.

[5] Jonker AM, Löwik DWPM, van Hest JCM. Peptide- and protein-based hydrogels. Chem Mater. 2012;24(5):759–777.

[6] Geng Z, Zhang H, Xiong Q, Zhang Y, Zhao H, Wang G. A fluorescent chitosan hydrogel detection platform for sensitive and selective determination of trace mercury (II) in water. J Mater Chem A. 2015;3:19205–19217.

[7] Rodriguez-Llansola F, Hermida-Merino D, Nieto-Ortega B, Ramirez FJ, Navarrete JTL, Juan Casado J, Hamley IW, Escuder B, Hayes W, Miravet JF. Self-assembly studies of a chiral bisurea-based superhydrogelator. Chem Eur J. 2012;18(46):14725–14731.

[8] Escuder B, Rodrıguez-Llansola F, Miravet JF. Supramolecular gels as active media for organic reactions and catalysis. New J Chem. 2010;34:1044–1054.

[9] Hapiot F, Menuel S, Monflier E. Thermoresponsive hydrogels in catalysis. ACS Catal. 2013;3(5):1006–1010.

[10] Raeburn J, Cardoso AZ, Adams DJ. The importance of self-assembly process to control mechanical properties of low molecular weight hydrogels. Chem Soc Rev. 2013;42(12): 5143–5156.

[11] Yu G, Yan X, Hana C, Huang F. Characterization of supramolecular gels. Chem Soc Rev. 2013;42:6697–6722.

[12] Silverstein RM, Webste FX, Kiemle DJ. Infrared spectroscopy in: spectrometric identification of organic compounds. John Wiley & Sons, Ltd: New Jersey. 2005.

[13] Suzuki M, Yumoto M, Shirai H, Hanabusa K. Supramolecular gels formed by amphiphilic low-molecular-weight gelators of Nα,N$\circledcirc$-diacyl-l-lysine derivatives. Chem Eur J. 2008;14:2133–2144.

[14] Miljanic S, Frkanec L, Biljan T, Meic B, Zinic M. Surface-enhanced Raman scattering on colloid gels originated from low molecular weight gelator. J Raman Spectrosc. 2008;39:1799–1804.

[15] Kleinsmann AJ, Weckenmann NM, Nachtsheim BJ. Phosphate-triggered self-sssembly of N-[(Uracil-5-yl)methyl]urea: a minimalistic urea-derived hydrogelator. Chem Eur J. 2014;20:9753–9761.

[16] Li P, Dou XQ, Tang YT, Zhu S, Gu J, Feng CL, Zhang D. Gelator-polysaccharide hybrid hydrogel for selective and controllable dye release. J Colloid Interface Sci. 2012;387:115–122.

[17] Sun T, Zhu J, Wang M, Lu M, Ding J, Lv Z, Huab P, Zhang Y. A glucosyl triblock copolymer: synthesis and its injectable thermo-and pH-responsive behaviours. RSC Adv. 2015;5:24231–24238.

[18] Zhao Z, Zhang Z, Li Chen, Cao Y, He C, Chen X. Biodegradable stereocomplex micelles based on dextran-blockpolylactide as efficient drug deliveries. Langmuir. 2013;29:13072–13080.

[19] Vlugt-Wensink KDF, Jiang X, Schotman G, Kruijtzer G, Vredenberg A, Chung JT, Zhang Z, Versluis C, Ramos D, Verrijk R, Jiskoot W, Crommelin DJA, Hennink WE. In vitro degradation behavior of microspheres based on cross-linked dextran. Biomacromolecules. 2006;7:2983–2990.

[20] Giannuzzo M, Corrente F, Feeney M, Paoletti L, Paolicelli P, Tita B, Vitali F, Casadei MA. pH-Sensitive hydrogels of dextran: synthesis, characterization and in vivo studies. J Drug Target. 2008;16(9):649–659.

[21] Ding F, Wu S, Wang S, Xiong Y, Yan Y, Li B, Deng H, Du Y, Xiao L, Shi X. A dynamic and self-crosslinked polysaccharide hydrogel with autonomous self-healing ability. Soft Matter. 2015;11:3971–3976.

[22] Yang X, Yu H, Wang L, Tong R, Akram M, Chen Y, Zhai X. Self-healing polymer materials constructed by macrocycle-based host–guest interactions. Soft Matter. 2015;11:1242–1252.

[23] Cui Y, Tan M, Zhu A, Guo M. Non-covalent interaction cooperatively induced stretchy, tough and stimuli-responsive polyurethane–urea supramolecular (PUUS) hydrogels. J Mater Chem B. 2015;3:2834–2841.

[24] Maiti B, Ruidas B, De P. Dynamic covalent cross-linked polymer gels through the reaction between side-chain β-keto ester and primary amine groups. React Funct Polym. 2015;93:148–155.

[25] Vivek B, Prasad E. Reusable self-healing hydrogels realized via in situ polymerization. J Phys Chem B. 2015;119:4881–4887.

[26] Sharma M, Mondal D, Mukesh C, Prasad K. Self-healing guar gum and guar gum-multiwalled carbon nanotubes nanocomposite gels prepared in an ionic liquid. Carbohydr Polym. 2013;98(1):1025–1030.

[27] Okazaki E, Satoh M. Hydrations and water properties in the super salt-resistive gel probed by FT-IR. Polymer. 2010;51:4362–4367.

[28] Xue R, Xin X, Wang L, Shen J, Ji F, Li W, Jia C, Xu G. Systematic study of the effect of molecular weights of polyvinyl alcohol on polyvinyl alcohol/graphene oxide composite hydrogels. Phys Chem Chem Phys. 2015;17:5431–5440.

[29] Xu YX, Wu Q, Sun YQ, Bai H, Shi GQ. Three-dimensional self-assembly of graphene oxide and DNA into multifunctional hydrogels. ACS Nano. 2010;4:7358–7362.

[30] Kumar PTS, Lakshmanan VK, Anilkumar TV, Ramya C, Reshmi P, Unnikrishnan AG, Nair SV, Jayakumar R. Flexible and microporous chitosan hydrogel/nano ZnO composite bandages for wound dressing: in vitro and in vivo evaluation. ACS Appl Mater Interfaces. 2012, 4, 2618-–2629.

[31] Mather ML, Tomlins PE. Hydrogels in regenerative medicine: towards understanding structure–function relationships. Regen Med. 2010;5(5):809–821.

[32] Ou-Yang L, Zhu L, Ruan Y, Tang H. Preparation of native β-cyclodextrin modified plasmonic hydrogel substrate and its use as a surface-enhanced Raman scattering scaffold for antibiotics identification. J Mater Chem C. 2015;3:7575–7582.

[33] Zu SZ, Han BH. Aqueous dispersion of graphene sheets stabilized by pluronic copolymers: formation of supramolecular sydrogel. J Phys Chem C. 2009;113:13651–13657.

[34] Rossi B, Venutic V, D'Amicoa F, Gessinia A, Mele A, Puntad C, Meloned L, Crupic V, Majolinoc D, Trotta F, Masciovecchioa C. Toward an understanding of the thermosensitive behaviour of pH responsivehydrogel based on cyclodextrins. Soft Matter. 2015;11:5862–5871.

[35] Crupi V, Majolino D, Mele A, Rossi B, Trotta F, Venutia V. Modelling the interplay between covalent and physical interactions in cyclodextrin-based hydrogel: effect of water confinement. Soft Matter. 2013;9:6457–6464.

[36] Rossi B, Venuti V, D'Amico F, Gessini A, Castiglione F, Mele A, Punta C, Melone L, Crupi V, Majolino D, Trotta F, Masciovecchio C. Water and polymer dynamics in a model polysaccharide hydrogel: the role of hydrophobic/hydrophilic balance. Phys Chem Chem Phys. 2015;17:963—971.

[37] Rossi B, Venuti V, Mele A, Punta C, Melone L, Crupi V, Majolino D, Trotta F, D'Amico F, Gessini A, Masciovecchio C. Probing the molecular connectivity of water confined in polymer hydrogels. J Chem Phys. 2015;142:014901.

[38] Sekine Y, Ikeda-Fukazawa T. Structural changes of water in a hydrogel during dehydration. J Chem Phys. 2009;130:034501.

[39] Kudo K, Ishida J, Syuu G, Sekine Y, Ikeda-Fukazawa T. Structural changes of water in poly(vinyl alcohol) hydrogel during dehydration. J Chem Phys. 2014;140:044909.

[40] Sekine Y, Takagi H, Sudo S, Kajiwara Y, Fukazawa H, Ikeda-Fukazaw T. Dependence of structure of polymer side chain on water structure in hydrogels. Polymer. 2014;55:6320–6324.

[41] John MS, Andrade JD. Water and hydrogels. J Biomed Mater Res. 1973;7:509–522.

[42] Kwak S, Lafleur M. Raman spectroscopy as a tool for measuring mutual diffusion coefficients in hydrogels. Appl Spectrosc. 2003:57(7);768–773.

[43] Fraix A, Gref R, Sortino S. A multi-photoresponsive supramolecular hydrogel with dual-color fluorescence and dual-modal photodynamic action. J Mater Chem B. 2014;2:3443–3449.

[44] Snow AW. Phthalocyanine aggregation. In: The Porphyrin Handbook. Phthalocyanines: Properties and Materials. Kadish KM, Smith KM, Guilard R, editors. Academic Press: San Diego, CA, USA 2003, vol. 17, 129-176.

[45] Uchiyama S, Santa T, Imai K. Fluorescence characteristics of six 4,7-disubstituted benzofurazan compounds: an experimental and semi-empirical MO study. J Chem Soc Perkin Trans. 2. 1999;11:2525–2532.

[46] Dong X, Wei C, Liu T, Lv F. Protoporphyrin Incorporated alginate hydrogel: preparation, characterization and fluorescence imaging in vivo. RSC Adv. 2015;5:96336–96344.

[47] Montalti M, Dolci LS, Prodi L, Zaccheroni N, Stuart MCA, Bommel KFC, Friggeri A. Energy transfer from a fluorescent hydrogel to a hosted fluorophore. Langmuir. 2006;22:2299–2303.

[48] Smith DK. Lost in translation? Chirality effects in the self-assembly of nanostructured gel-phase materials. Chem Soc Rev. 2009;38:684–694.

[49] Gotarrelli G, Lena S, Masiero S, Pierracini S, Spada GP. The use of circular dichroism spectroscopy for studying the chiral molecular self-assembly: an overview. Chirality. 2008;20:471–485.

[50] Schweitzer-Stenner R, Measey TJ. Simulation of IR, Raman and VCD amide I band profiles of self-assembled peptides. Spectroscopy. 2010;24:25–36.

[51] Setnicka V, Urbanova M, Volka K, Nampally S, Lehn JM. Investigation of guanosine-quartet assemblies by vibrational and electronic circular dichroism spectroscopy, a novel approach for studying supramolecular entities. Chem Eur J. 2006;12:8735–8743.

[52] Song F, Zhang LM, Li NN, Shi JF. In situ crosslinkable hydrogel formed from a polysaccharide-based hydrogelator. Biomacromolecules. 2009;10:959–965.

[53] Choi YY, Jang JH, Park MH, Choi BG, Chi B, Jeong B. Block length affects secondary structure, nanoassembly and thermosensitivity of poly(ethylene glycol)-poly(L-alanine) block copolymers. J Mater Chem. 2010;20:3416–3421.

[54] Vilaça H, Hortelão ACL, Castanheira EMS, Queiroz MJRP, Hilliou L, Hamley IW, Martins JA, Paula MT, Ferreira PMT. Dehydrodipeptide hydrogelators containing

naproxen N-capped tryptophan: self-assembly, hydrogel characterization, and evaluation as potential drug nanocarriers. Biomacromolecules. 2015;16:3562–3573.

[55] Marchesan S, Easton CD, Styan KE, Waddington LJ, Kushkaki F, Goodall L, McLean KM, Forsythec JS, Hartleya PG. Chirality effects at each amino acid position on tripeptide self-assembly into hydrogel biomaterials. Nanoscale. 2014;6:5172–5180.

[56] Zhang M, Qinga G, Sun T. Chiral biointerface materials. Chem Soc Rev. 2012;41:1972–1984.

[57] Cheng R, Liu J, Xie P, Wu Y, Deng J. Chiral, pH-sensitive polyacrylamide hydrogels: preparation and enantio-differentiating release ability. Polymer. 2015;68:246–252.

[58] Marchesan S, Easton CD, Styan KE, Waddington LJ, Kushkaki F, Goodall L, McLean KM, Forsythec JS, Hartley PJ. Chirality effects at each amino acid position on tripeptide self-assembly into hydrogel biomaterials. Nanoscale. 2014;6:5172–5180.

[59] Xu FM, Wang HN, Zhao J, Liu XS, Li DD, Chen CJ, Ji J. Synthesis of photoresponsive cholesterol-based azobenzene organogels: dependence on different spacer lengths. Macromolecules. 2013;46:4235–4246.

[60] Ogawa Y, Yoshiyama C, Kitaoka T. Helical assembly of azobenzene-conjugated carbohydrate hydrogelators with specific affinity for lectins. Langmuir. 2012;28:4404–4412.

[61] Clemente MJ, Tejedor RM, Romero P, Fitremann J, Oriol L. Maltose-based gelators having azobenzene as light-sensitive unit. RSC Adv. 2012;2:11419–11431.

[62] Shankar BV, Patnaik A. Chiral discrimination of a gemini-type surfactant with rigid spacer at the air–water interface. J Phys Chem B. 2007;111:11419–27.

[63] Bag DS, Alam S. Chiral chemical absorption property of a cross-linked poly(N-isopropyl acrylamide-co-sodium acrylate) thermoresponsive smart gel. Chirality. 2012;24:506–511.

[64] Roy S, Javid N, Sefcik J, Halling PJ, Ulijn RV. Salt-induced control of supramolecular order in biocatalytic hydrogelation. Langmuir. 2012;28:16664–16670.

[65] Ma M, Kuang Y, Gao Y, Zhang Y, Gao P, Xu B. Aromatic-aromatic interactions induce the self-assembly of pentapeptidic derivatives in water to form nanofibers and supramolecular hydrogels. J Am Chem Soc. 2010;132:2719–2728.

[66] Wood SJ, Wetzel R. Martin JD, Hurle MR. Prolines and amyloidogenicity in fragments of the Alzheimer's peptide β/A4. Biochemistry. 1995;34:724–730.

[67] Otsuka T, Maeda T, Hotta A. Solutions on the sol–gel transitions, the gelation speed, and the gel characteristics. J Phys Chem B. 2014;118:11537–11545.

[68] Huang J, Hastings CL, Duffy GP, Kelly HM, Raeburn J, Adams DJ, Heise A. Supramolecular hydrogels with reverse thermal gelation properties from (oligo)tyrosine containing block copolymers. Biomacromolecules. 2013;14 (1):200–206.

[69] Morris Kl, Chen L, Rodger A, Dave J, Adam DJ, Serpell LC. Structural determinants in a library of low molecular weight gelators†. Soft Matter. 2015;11:1174–1181.

[70] Kirchhof S, Abrami M, Messmann V, Hammer N, Achim M, Goepferich, Grassi M, Brandl FP. Diels–Alder hydrogels for controlled antibody release: correlation between mesh size and release rate. Mol Pharm. 2015;12 (9):3358–3368.

[71] Sutton S, Campbell NL, Cooper AI, Kirkland M, Frith WJ, Adams DJ. Controlled release from modified amino acid hydrogels governed by molecular size or network dynamics. Langmuir. 2009;25:10285–10291.

[72] Pescosolido L, Feruglio L, Farra R, Fiorentino S, Colombo I, Coviello T, Matricardi P, Hennink WE, Vermonden T, Grassi M. Mesh size distribution determination of interpenetrating polymer network hydrogels. Soft Matter. 2012;8:7708–7715.

[73] Canal T, Peppas NA. Correlation between mesh size and equilibrium degree of swelling of polymeric networks. J Biomed Mater Res. 1989;23:1183–1193.

[74] Gallegos DP, Munn K, Smith DS, Stermer DL. A NMR technique for the analysis of pore structure: application to materials with well-defined pore structure. J Colloid Interface Sci. 1987;119(1):127–140.

[75] Aggerbrandt LJ, Samuelson O. Penetration of water-soluble polymers into cellulose fibers. J Appl Polym Sci. 1964;8:2801–2812.

[76] Jowkarderis L, van de Ven TGM. Mesh size analysis of cellulose nanofibril hydrogels using solute exclusion and PFG-NMR spectroscopy. Soft Matter. 2015;11:9201–9210.

[77] Matsukawa S, Yasunaga H, Zhao C, Kuroki S, Kurosu H, Ando I. Diffusion processes in polymer gels as studied by pulsed field-gradient spin-echo NMR spectroscopy. Prog Polym Sci. 1999;24:995–1044.

[78] Matsukawa S, Ando I. A study of self-diffusion of molecules in polymer gel by pulsed-gradient spin-echo 1H NMR. Macromolecules. 1996;29:7136–7140.

[79] Scherer GW. Hydraulic radius and mesh size of gels. J Sol-Gel Sci Technol. 1994;1:285–291.

[80] Bernin D, Goudappel G-J, van Ruijven M, Altskar A, Strom A, Mats Rudemo M, Hermansson AM, Nyden M. Microstructure of polymer hydrogels studied by pulsed field gradient NMR diffusion and TEM methods. Soft Matter. 2011;7:5711–5716.

[81] Walther B, Loren N, Nyden M, Hermansson AM. Influence of K-carrageenan gel structures on the diffusion of probe molecules determined by transmission electron microscopy and NMR diffusometry. Langmuir. 2006;22:8221–8228.

[82] Wallace M, Adams DJ, Iggo JA. Analysis of the mesh size in a supramolecular hydrogel by PFG-NMR spectroscopy. Soft Matter. 2013;9:5483–5490.

[83] Jowkarderis L, van de Ven TGM. Mesh size analysis of cellulose nanofibril hydrogels using solute exclusion and PFG-NMR spectroscopy. Soft Matter. 2015;11:9201–9210.

[84] Lund A, Shiotani M, Shimada S. Principles and Applications of ESR Spectroscopy. Springer: Dordrecht Heidelberg London New York, 2011.

[85] Schlick S, Ed. Advanced ESR Methods in Polymer Research, Wiley Interscience: New York, 2006.

[86] Gerson F, Huber W. Electron Spin Resonance Spectroscopy of Organic Radicals. Wiley-VCH: Weinheim, 2003.

[87] Poole CP. Electron Spin Resonance, 2nd ed.; Dover Publications: New York, 1983.

[88] Eaton GR, Eaton SS, Salikhov KM., Eds. Foundations of Modern EPR, Singapore, World Scientific, 1998.

[89] Hinderberger D. EPR spectroscopy in polymer science. Top Curr Chem. 2011;321:67–90.

[90] Lucarini M, Mezzina E. EPR investigations of organic non-covalent assemblies with spin labels and spin probes. Electron Paramagn Reson. 2011;22:41–70.

[91] Pitt CG, Wang J, Shah SS, Sik R, Chignell CF. ESR Spectroscopy as a probe of the porphology of hydrogels and polymer-polymer blends. Macromolecules. 1993;26:2159–2164.

[92] Kajsheva NS. Configuration of central metals in polyuronates studied by EPR spectroscopy. Pharm Chem J. 2012;46(8):51–53.

[93] Kurzbach D, Reh MN, Hinderberger D. Nanoscale inhomogeneities in thermoresponsive triblock copolymers. ChemPhysChem. 2011;12:3566–3572.

[94] Junk MJN, Jonas U, Hinderberger D. EPR spectroscopy reveals nanoinhomogeneities in the structure and reactivity of thermoresponsive hydrogels. Small. 2008;4:1485–1493.

[95] Wasserman AM, Yasina LL, Motyakin MV, Aliev II, Churochkina NA, Rogovina LZ, Lysenko EA, Baranovsky VY. EPR spin probe study of polymer associative systems. Spectrochim Acta Part A. 2008;69:1344–1353.

[96] Kempe S, Metz H, Bastrop M, Hvilsom A, Contri RV, Mader K. Characterization of thermosensitive chitosan based hydrogels by rheology and electron paramagnetic resonance spectroscopy. Eur J Pharm Biopharm. 2008;68:26–33.

[97] Bertholon I, Hommel H, Labarre D, Vauthier C. Properties of polysaccharides grafted on nanoparticles investigated by EPR. Langmuir. 2006;22:5485–5490.

[98] Freire-Nordi CS, Vieira AAA, Nascimento OR. The metal binding capacity of Anabaena spiroides extracellular polysaccharide: an EPR study. Process Biochem. 2005;40:2215–2224.

[99] Ciesielski W, Lii CY, Yen MT, Tamasik P. Internation of starch with salt of metals from the transition groups. Carbohydr Polym. 2003;51:47–56.

[100] Kanesaka S, Watanabe T, Matsukawa S. Binding effect of Cu(2+) as a trigger on the sol-to-gel and the coil-to-helix transition processes of polysaccharide, gellan gum. Biomacromolecules. 2004;5:863–868.

[101] Ionita G, Ariciu AM, Smith DK, Chechik V. Ion exchange in alginate gels—dynamic behaviour revealed by electron paramagnetic resonance. Soft Matter. 2015;11(46):8968–8974.

[102] Ionita G, Chechik V. Exploring polyethylene glycol/cyclodextrin hydrogels with spin probes and EPR spectroscopy. Chem Commun. 2010;46:8255–8257.

[103] Ionita G, Ariciu AM, Turcu IM, Chechik V. Properties of polyethylene glycol/cyclodextrin hydrogels revealed by spin probes and spin labelling methods. Soft Matter. 2014;10:1778–1783.

[104] Cesteros LC, Ramirez CA, Pecina A, Katime I. Poly(ethylene glycol-β-cyclodextrin) gels: synthesis and properties. J Appl Polym Sci. 2006;102:1162–1166.

[105] Cesteros LC, Ramirez CA, Pecina A, Katime I. Synthesis and properties of hydrophilic networks based on poly(ethylene glycol) and β-cyclodextrin. Macromol Chem Phys. 2007;208:1764–1772.

[106] Ionita G, Chechik V. Spin-labelled cyclodextrins as hosts for large supramolecular assemblies. Org Biol Chem. 2005;3:3096–3098.

[107] Chechik V, Ionita G. Bis spin-labelled cyclodextrins. New J Chem. 2007;31:1726–1729.

[108] Ionita G, Meltzer V, Pincu E, Chechik V. Inclusion complexes of cyclodextrins with biradicals linked by a polyether chain—an EPR study. Org Biomol Chem. 2007;5:1910–1914.

[109] Caragheorgheopol A, Edwards W, Hardy JG, Smith DK, Chechik V. Using EPR spectroscopy as a unique probe of molecular-scale reorganization and solvation in self-assembled gel-phase materials. Langmuir. 2014;30:9210–9218.

[110] Takemoto Y, Yamamoto T, Ikuma N, Uchida Y, Suzuki K, Shimono S, Takahashi H, Sato N, Oba Y, Inoue R, Sugiyama M, Tsue H, Kato T, Yamauchi J, R. Tamura. Preparation, characterization and magnetic behavior of a spin-labelled physical hydrogel containing a chiral cyclic nitroxide radical unit fixed inside the gelator molecule. Soft Matter. 2015;11:5563–5570.

[111] Ariciu AM, Staicu T, Marin Micutz, Neacsu MV, Ionita P, Tecuceanu V, Munteanu C, Ionita G. Investigations on carboxy dibenzylidene sorbitol hydrogels using EPR apectroscopy. Appl Magn Reson. 2015;46:1395–1407.

[112] Kveder M, Andreis M, Makarević J, Jokić M, Rakvin B. The EPR study of low molecular weight organogels by means of a nitroxide spin probe. Chem Phys Lett. 2006:420;443–447.

Chapter 4

Hydrogels from Fundaments to Application

Dorota Kołodyńska, Alicja Skiba,
Bożena Górecka and Zbigniew Hubicki

Additional information is available at the end of the chapter

http://dx.doi.org/10.5772/63466

Abstract

Polymer superabsorbents commonly known as hydrogels are cross-linked highly molecular compounds able to absorb water from physicochemical fluids in the amounts from 10-fold to 100-fold larger than their dry mass. Numerous investigations have shown that they can help reduce irrigation water consumption, lower the death rate of plants, improve fertilizer retention in soil and increase plant growth rate. Besides water absorption and retention, the superabsorbent polymers have many advantages over conventional ones, such as a sustained supply of nutrition to plants for a longer time, thus increasing the phosphate fertilizer use efficiency and decreasing application frequency. The aim of this study is to investigate the influence of chemical conditions on hydrogels, kinetic and absorption behaviour towards metal ions in the presence of the chelating agent of a new generation. In this group, there are IDS, EDDS, GLDA, MGDA, etc. In the chapter, the research on the applicability of the effective absorption of metal complexes with a biodegradable complexing agent will be presented. The possibility of the preparation of slow-release fertilizers of controlled activity of a new generation in such system will also be discussed.

Keywords: superabsorbents, new complexing agents, IDS, EDDS, GLDA, MGDA, absorption, fertilizers

1. Introduction

Superabsorbents were first developed in the early 1960s in the United States Department of Agriculture by grafting acrylonitrile (AN) onto corn starch and saponifying the product. Over the past decades, significant progress has been made in the field of hydrogels as functional biomaterials. Since then in the group of the main global participants on the superabsorbent market, Dow Chemical, Hercules, General Mills Chemical, DuPont, National Starch & Chemical, Enka (Akzo), Sanyo Chemical, Sumitomo Chemical, Kao, Nihon Starch and Japan Exlan should be mentioned.

Polymer superabsorbents commonly known as hydrogels are three-dimensional (3D) cross-linked highly molecular compounds able to absorb liquids in the amounts from 10-fold to 100-fold larger than their dry mass. They are hydrophilic, cross-linked, and swelling in water polymer materials. They are sometimes called smart hydrogels as they can absorb up to 1000 g water. The literature reports a number of classifications of hydrogels. As for physicochemical properties, hydrogels can be divided into different modes. There are two main groups of superabsorbents. The first group is physical hydrogels called reversible and they have the 3D network in which the polymer chains are linked by electrostatic forces, hydrogen bonds and hydrophobic interactions. These gels are unstable and upon heating can be converted into a polymer mixture, e.g., gelatin and agar. The second group includes chemically stable gels with the 3D network, in which the polymer chains are linked by stable covalent bonds. Both starch and vinyl monomers such as acrylic acid, acrylamide, acrylonitrile and polyvinyl alcohol are of interest as they contain a number of hydrophilic functionalities in their structures such as hydroxyl and carboxyl groups. The ionic phase of hydrogels usually consists of both groups bound onto the polymer chains (which can be ionized) and a number of mobile ions—counter-ions and co-ions (due to the presence of solvent surrounding the hydrogel). As for the source they can be natural, synthetic and hybrid, as for the cross-linking they are physically and chemically linked and as for degradability they can be biodegradable and non-biodegradable. Homopolymeric, copolymeric and interpenetrating networks (IPN) can be recognized given the way of preparation [1, 2]. Homopolymeric gels contain the same monomers, which are basic structural units. Copolymeric gels comprise two or more different monomer species with at least one hydrophilic component whereas interpenetrating polymers or double networks are formed by swelling of the first network in a solvent containing monomers, which then forms the second intermeshing network structure. The double networks will either be hydrophobic or hydrophilic or the combination of both [3–5]. Depending on the charges of the bound groups, hydrogels can be ionic (cationic or anionic), neutral, amphoteric (ampholytic) containing both acidic and basic groups as well as zwitterionic (polybetaines) containing both anionic and cationic groups in each structural repeating unit. Response hydrogels are biochemically (antigens, enzymes, ligands), chemically (pH, solvent composition, ionic strength, molecular species) or physically (temperature, electric or magnetic field, light, pressure, sound) responsive.

The types of cross-linking agents can also be the criterion of classification. It is also possible to divide hydrogels into groups by their structure that is amorphous, semi-crystalline, crystalline and hydrocolloid aggregates [6].

The phenomenon of liquid absorption is a result of separation of polymer chain network that is manifested by swelling of polymer material assuming the gel form [1]. Hydrogels with characteristic properties such as desired functionality, reversibility, sterility and biocompatibility meet both material and biological requirements to treat or replace tissues and organs or the function of living tissues as well as to interact with the biological system. During years, the number of hydrogel formulations steadily grew and the problems such as low solubility, high crystallinity, non-biodegradability, unfavourable mechanical and thermal properties were solved. This was due to the combination of natural and synthetic polymers characterized by better biodegradation, solubility, crystallinity and biological activities. To avoid the disintegration (the hydrophilic linear polymer chain of hydrogel can dissolve in water due to the polymer and water thermodynamic compatibility) different cross-linked agents were proposed. Cross-linking can take place *in vitro*, during the preparation, or in vivo (*in situ*) after the application. The hydrophilicity of hydrogel network is due to the presence of hydrophilic groups such as $-NH_2$, $-COOH$, $-OH$, $-CONH_2$, $-CONH-$ and $-SO_3H$, capillary effect and osmotic pressure. It can be chemical or physical in nature. In the case of the chemical cross-linking, the polymer chains are covalently bonded by the cross-linking agent whereas in the case of the physical cross-linking hydrogen bonding, hydrophobic interaction, ionic complexation, post-process bulk modification, reshaping, biodegradation take place [7].

It should be remembered that swelling of hydrogels is a complex process comprising a number of steps. In the first step, the polar hydrophilic groups of the hydrogel matrix are hydrated by water, which appears in the form of primary bound water. In the second step, the water also interacts with the exposed hydrophobic groups and the secondary bound water layer is formed. Both the primary and secondary bound water forms the total bound water. In the third step, the osmotic driving force of the network towards infinite dilution is resisted by the physical or chemical cross-links, so the additional water is absorbed. The water absorbed into the equilibrium swelling is called bulk water or free water, which fills the spaces between the network or chains and the centre of larger pores. The amount of water absorbed by a hydrogel depends on the temperature and specific interactions between the water molecules and the polymer chains. **Table 1** summarizes the classification details for hydrogels.

Physical properties	Conventional		Smart
Source	Natural	Synthetic	Hybrid
Cross-linking	Physically linked		Chemically linked
Degradability	Biodegradable		Non-biodegradable
Preparation	Copolymeric	Homopolymeric	Interpenetrating network
Ionic charge	Cationic	Anionic	Non-ionic
Response	Physical agents: temperature, pressure, light, electric field, magnetic field		Chemically responsive: biochemical agents, antigens, enzymes, ligands

Table 1. Criterion of hydrogels classification.

2. Hydrogels preparation

Hydrogels can be produced in the form of films, membranes, rods, spheres and emulsions. They can be classed into three categories: starch-polyacrylonitrile graft polymers (starch copolymers), vinyl alcohol-acrylic acid co-polymers (polyvinyl alcohols) and acrylamide sodium acrylate co-polymers (cross-linked polyacrylamides) [8, 9]. They can be physical or chemical. Physical gels are obtained due to physical nature of the cross-linking process. This is achieved by physical processes such as hydrophobic association, chain aggregation, crystallization, polymer chain complexion and hydrogen bonding. On the other hand, a chemical process, i.e., chemical covalent cross-linking (simultaneously or post-polymerization) is utilized to prepare a chemical hydrogel. Physical hydrogels are reversible due to the conformational changes where chemical hydrogels are permanent and irreversible because of configurational changes. Another category is the double network hydrogels (interpenetrating networks), formed by the combination of physical and chemical cross-linked hydrogels due to an electrostatic interaction. Recently, they have been employed to overcome the disadvantages of solely using physical or chemical hydrogels with a high liquid uptake capacity over a wide range of pH and a higher sensitivity towards changes in the pH value as compared to chemical hydrogels.

The hydrogels are divided into two main classes, i.e., natural, containing two basic groups based on polypeptides (proteins) and polysaccharides, and another is artificial one [10, 11]. Natural hydrogels are usually prepared by the addition of some synthetic parts onto the natural substrates, e.g., graft copolymerization of vinyl monomers on polysaccharides. The synthetic route for the production of most synthetic hydrogel is the free radical multifunctional vinyl monomers. Each monomer contains a carbon double bond where an active centre can propagate to produce polymer chains. Active centres reaction also depends on the solvents, reaction conditions as well as particular monomers and can be initiated by different factors: heat (thermal initiators), light (photoinitiators), enzymes (bioinitiators) or electron beams. Generally, water-soluble natural or synthetic polymers are cross-linked to form hydrogels in a number of ways, such as (i) linking polymer chains by chemical reaction, (ii) using ionizing radiation to generate main-chain free radicals that can recombine as cross-link joints and (iii) interacting physically such as electrostatics, entanglements and crystallite interactions [12, 13]. Any of various polymerization techniques can be used to form hydrogels, including bulk, solution and suspension polymerizations. The three main components of hydrogels are monomers, initiators and cross-linkers which can be diluted in water or any solvent to control the polymerization heat. However, its disadvantage appears in the form of impurities left from the preparation process containing unreacted monomers, initiators, cross-linkers and side products. Hydrogels are commonly prepared from polar monomers of both natural and synthetic origins by graft polymerization, cross-linking polymerization, network formation in aqueous medium and by radiation cross-linking methods. Their short characteristic is presented below.

Solution polymerization (wherein the polymerization medium is a suitable solvent, which dissolves both monomers and initiator) and suspension polymerization (wherein monomer forming droplets, with the initiator in them, are completely insoluble and suspended in a solution) are the most common techniques for the production of a variety of hydrogel networks by reacting hydrophilic monomers with multifunctional cross-linkers and added initiators.

Polymerization reaction is frequently initiated with radiation, ultraviolet or chemical catalysts. Cross-linking of polymers is due to chemical reaction, ionizing radiation or physical interactions such as entanglements, electrostatics or crystallite formation. In the case of suspension polymerization, monomers and initiator are dispersed in the organic phase. The hydrogel properties depend on the viscosity of the monomer solution, agitation speed, rotor design and dispersant type [1, 14]. Recently radiation polymerization and grafting have been also used. In the case of the block polymerization to the liquid monomer, an initiator is added directly and the obtained hydrogels are characterized by inherent weak structure therefore they are frequently grafted. The suspension-gelation method (microcapsule templating method) was also described to prepare macroporous polymeric hydrogels [15]. An aqueous suspension is formed by dispersing the microcapsules as a pore template (porogen) into a pre-gel aqueous solution containing monomers and cross-linking monomers. Then, a composite hydrogel containing distributed microcapsules is prepared by copolymerizing the pre-gel aqueous solution. Finally, a macroporous hydrogel is obtained by breaking down the structure of the microcapsules dispersed in the composite hydrogel with a chemical treatment. The suspension-gelation method is conducted using only low toxic and inexpensive physically cross-linked microcapsules such as calcium alginate gel. Graft copolymerization resulting in development of physical and chemical properties of hydrogels makes them excellent smart materials [16–18].

Hydrogels originating from microbial poly(γ-glutamic acid) and poly(ε-lysine) can be prepared by γ-irradiation. These microbial poly(amino acid)s are water soluble, hydrodegradable and biodegradable [19]. Free radical polymerization of different vinyl monomers and aniline onto gum ghatti (Gg) under dissimilar reaction conditions was also described [18]. Polyaniline (PANI) is an attractive conductive polymer because of its simple methods of synthesis, high stability, variable structure as well as unique optical, magnetic, electrical, electrochemical and electromechanical properties. To optimize their properties, reaction parameters such as synthesis time, kind of solvent, pH, microwave power, cross-linker amount, aniline concentration, initiator concentration, monomer concentration and the percentage swelling can be varied. Both radiation and thermal cross-linking methods are inexpensive, safe, do not require a purification step and result in sterile hydrogels if a suitable combination of hydrophilic polymers is used. There are numerous papers and reviews focused on the synthesis of hydrogels [20].

3. Methods of hydrogels investigation

Fourier transform infrared analysis (FTIR) can be used for the analysis of the hydrogel structure, type of the functional groups, different stages of biodegradation, crystallization deformation of polymers, moisture uptake properties, to characterize the attached biomolecules, nature of interactions between components, etc. [21–23]. In FTIR-ATR (attenuated total reflection) spectroscopy technique, a beam of infrared light is passed through the ATR crystal in such a way that it reflects at least once off the internal surface in contact with the sample (a denser medium—crystal of diamond, include germanium, KRS-5 and zinc selenide or silicon, characterized by the high refractive index and a rarer medium in the form of the thin film of sample) must be in contact with each other. This reflection forms the evanescent wave which extends into the sample. This method is used for both solid and liquid samples analysis. For example, in the case of chitosan (CS), itaconic acid (IA) and methacrylic acid (MAA) hydrogels the characteristic bands are associated with the N–H and –OH stretching vibrations as well as those connected with the absorption process of Zn(II) were distinguished. It was also found that the stretching vibrations connected with the carboxylic groups (C=O) are weaker after absorption of Zn(II) and the strong absorption band at 1712 cm^{-1} shifted to the lower wave number 1705 cm^{-1}. The absorption band at 1540 cm^{-1}, assigned to the ionic interactions between CS and the acids, and the absorption band at 1165 cm^{-1}, assigned to the deformation absorption bands of –OH groups, were enhanced and shifted to higher wave numbers after Zn(II) absorption. On the basis of the FTIR spectra of unloaded and loaded hydrogels, it seems that $-NH_2$, –OH and –COOH groups are involved in Zn(II) ion bonding. This confirms the fact that the dominant type of bonding onto hydrogel is through chemical interactions because a signal corresponding to –COO anion is observed [24].

Atomic force microscopy (AFM) can be used for evaluation of the surface topography of hydrogels and changes in the surface chemical composition after polymerization, absorption and interaction with active agents. Grains of various sizes randomly aggregated can be seen. Another consequence of the agglomeration is the appearance of very large grains, displace ment of which leaves huge holes behind.

The X-ray diffraction (XRD) method allows to determine the plane of the network and to calculate the size of the crystallographic cell. Tests are also carried out by small-angle X-ray scattering (SAXS) and wide-angle X-ray scattering (WAXS) as well as differential scanning calorimetry (DSC). Differential scanning calorimetry is a technique by which the difference between the temperature of the sample and the reference samples is measured. It can be used to determine the glass-transition temperatures (T_g). The glass transition temperature is different for each polymer, but many polymers are above T_g at room temperature. The glass transition of a polymer is related to the thermal energy required to allow changes in the conformation of the molecules at a microscopic level, and above T_g there is sufficient thermal energy for these changes to occur. However, the transition is not a sharp one, nor it is thermodynamically well defined. Therefore, T_g depends on the polymer architecture and there are several factors influencing the transition: chain length, chain flexibility, side groups, branching, cross-linking, complexation. It was found that in the case of polymer metal complexes T_g and the decomposition temperature increase. For example, the decomposition temperature for the polymer complex of Cu(II) is 280°C and the T_g is about 140°C. This is the evidence that the thermal stability is enhanced by the incorporation of metal into the organic backbone. In addition, the T_g variation of metal incorporated polymer depends on the amount of metal ions and cross-linking effect of polymers. Based on the thermogravimetric analysis (TGA), the mechanism of decomposition of hydrogels can also be illustrated. The initial minor weight loss occurs due to the removal of moisture and volatile components; however, the maximum weight loss occurs due to the elimination of side chains, degradation of backbone polymer or breakdown of the cross-linked structure. Maximum weight loss observed after the first stage of decomposition of cross-linked samples indicates the chemical modification of hydrogels. In many cases, the results obtained from the TGA method and scanning electron microscopy (SEM) are consistent as for improvement of thermal stability after grafting and cross-linking.

In the case of scanning electron microscopy (SEM), the electron beam, being the collimated stream of accelerated electrons, is focused by magnetic lenses. In the case of the emission scanning electron microscope, the electron beam with a diameter of 10–50 nm is used. As for the tunnel scanning microscope, the current flow between the soda and the test material due to the tunnelling effect is recorded while the field scanning microscope records changes in the electric field. The radiation, passing through the material reflected by its surface or dispersed, contains information about the structure and after being synchronized with the incident beam is used to obtain an image. The images obtained from a scanning electron microscope make it possible to analyze the 3D structure of the test materials at a magnification from 20 to 100,000 times. Using the SEM method, change in surface morphology due to the formation of covalent bonds between different polymeric chains, structure of the material and its porosity as well as cross-linking can be observed. Interior morphology of synthesized hydrogels can be examined by the cross-sectional SEM analysis. For example, in the paper by Sato et al. [15], the surface of macroporous hydrogels in the form of less than 1 mm sliced was analyzed using a stereo-microscope. As for the SEM analysis, the swollen samples were freeze-dried to avoid shrinkage in the drying process and frozen using liquid nitrogen. Then, they were dried under reduced pressure and coated with platinum and palladium. The hydrogels have different morphologies depending on the synthesis process. Some of them have an interconnecting, 3D porous structure, whereas others have a rather irregular structure with very large completely empty spaces.

As for quasi-elastic light scattering (QELS), a monochromatic laser radiation is used. Beam of monochromatic laser radiation is reflected from the sample surface and forms a plurality of coherent elementary waves coming from different microscopic surface features. As a result of the interference of the waves, so-called interference fringes with different intensities are formed. Water microdroplets contained in the hydrogel material result in additional waves by reflection from the surface of the incident light.

Water retention behaviour and hydrodegradation of hydrogels is very important especially with regard to the difficulties of treating hydrogels waste by existing methods due to the inclusion of substantial amount of absorbed water [19]. Hydrolytic degradation studies of hydrogels can be determined at varying temperatures up to 100°C. After placing in a deionized water, the hydrogel samples were heated and filtered. The amount of gel dissolved in water can be measured using the total organic carbon analyser (TOC method). The degradation ratio can be calculated from the ratio of filtered TOC amount to the total TOC amount of gel. For example, it was that hydrogels were not degraded when held at 40°C for 3 h. When the medium temperature was raised to 100°C, the gel was degraded to 90% during 1 h. In the case of 60 and 80°C, the rates of degradation were increased (40 and 20% of hydrogel was degraded). This was due to the fact that the density of the polymer network decreased because of scission of the polymer main chain and the removal amount of degraded polymer increased.

3.1. Mechanical properties

Examination of cross-linking and mechanical strength (compressibility at break) is based on the measurement of the maximum clamping force, the change in solubility of the polymer in time using the compression testers. Typically, hydrogel samples are cut into cubes. The sample was placed on the sample plate of the compression tester. The compression plunger fell at a rate of 1 mm/5 s to compress the sample. For hydrogels, the compressive strength increased with the increasing concentration of monomer and the decrease in the γ-irradiation dose as well as with the decrease in the specific water content [19, 22, 23]. The hydrogels exhibited higher compressive strength in the case of higher monomer concentration or greater irradiation dose. This indicates that a higher cross-linked density causes a higher compressive strength of hydrogel. The cross-linking is connected with the swelling (P_s). It changes with temperature, pH and ionic strength.

4. Application

According to the Global Industry Analysts, Inc. report, the world demand for superabsorbent polymers will reach up to 1.9 million metric tons in the next years. The fast increase in demand will be seen on the developing markets and in new applications [25, 26]. Water-soluble polymers such as poly(acrylic acid), poly(vinyl alcohol), poly(vinylpyrrolidone), poly(ethylene glycol), polyacrylamide and some polysaccharides are the most common systems used to form hydrogels [27–31]. These water-soluble polymers are nontoxic. Their physicochemical properties such as good stability both in the swelling environment and during storage, re-wetting capability, good absorption ability (maximum equilibrium swelling), good rate of absorption, particle size and porosity, neutral pH, colourless, odourless, photo stability, low soluble content and residual monomer and non-toxicity, maximum biodegradability without formation of toxic groups and finally low price make them suitable for different applications [32, 33]. Among others, the most important are agriculture, drug delivery, sealing, coal dewatering, artificial snow production, food additives, pharmaceuticals, microfluidic control, biomedical applications, tissue engineering and regenerative medicines, diagnostics, biomimetic, biosensor/bioactuator, bioseparation and artificial skin and muscles, wound dressing, separation of biomolecules or cells and barrier materials to regulate biological adhesions and biosensor.

It should be mentioned that it is not possible to produce hydrogels with all these features, although some properties can be achieved such as porosity, response to either pH or temperature, for example, the lowest re-wetting, the highest absorption rate and the lowest residual monomer should characterize those used in drug delivery and production of personal hygiene agents [34–37]. In the case of pH-responsive hydrogels (anionic or cationic), they should contain groups that can accept or donate protons as response to pH change. Anionic hydrogels possess carboxylic or sulfonic groups and deprotonation occurs at pH greater than pK_a which causes their ionization. This, in turn, increases hydrogel swelling. Cationic hydrogels contain amine groups. In this case, ionization occurs at pH below the pK_b and increased swelling is observed. It was found that concentration, ionic charge, value of pK_a or pK_b, type of active groups, hydrophilicity or hydrophobicity, degree of ionization of both hydrogel and solution are very important. As for temperature-dependent hydrogels, their ability to swell and shrink when the temperature changes is detected. When the temperature is below the upper critical solution temperature, hydrogels contract and release solvents from the matrix and the dehydration process takes place. At temperatures higher than that the opposite process (swelling) is observed. In the group of positive temperature hydrogels those based on copolymer of polyacrylamide and *n*-butyl methacrylate (AAm-co-BMA) as well as derivatives of chitosan are found [27]. The group of negative temperature hydrogels contains those with two hydrophilic (–CONH–) and hydrophobic (R) parts.

4.1. Agriculture and horticulture industry

Recently, the research on the use of superabsorbents as water management materials for agricultural and horticultural applications has attracted great attention. In the case of agriculture, the main role of hydrogels is release of nutrients into soil. Hydrogels can be impregnated by fertilizer components (e.g., soluble phosphate, potassium ions, nitrogen compounds). Chemicals trapped in a polymer network cannot be immediately washed out by water, but gradually released into the soil and then absorbed by plants. The simplest way to use superabsorbents in agriculture is their mixing with the soil. Introduction of cross-linked hydrogels into the root zone of plants releases not only mineral fertilizers but also pesticides. In this way, they are used in the United States per year for 400,000 hectares of crops. For this aim, polyacrylamide, poly(acrylic acid) or polymethacrylic and their derivatives, much less frequently, cross-linked poly(vinyl alcohol) and chemically modified polymers based on polysaccharides, such as chitosan, pectin and carboxymethyl cellulose, have been used [5, 38–46]. However, the use of the latest is considerably limited due to their rapid biodegradability in soil [47]. Superabsorbents increase water capacity of the soil, at the same time counteracting the loss through seepage and evaporation [48–50]. They counteract rapid changes in soil moisture acting as buffer water. They also block channels for water loss from the soil surface. The effectiveness of the process depends on the soil properties such as aeration, temperature and nutrient transport, water uptake and transformation, which affect plant growth. Therefore, applicability of cellulose-based hydrogels can be recommended for the controlled release of water and nutrients in arid and desert areas. The main advantage is controlled release of water, long time maintaining soil humidity, increase of soil porosity and therefore better oxygenation of plant roots. It was confirmed that the amount of moisture retained in the soil is dependent on the concentration of the cellulose-based superabsorbent matrices. The advantages of this type of hydrogels are also eco-friendly sources, high holding capacity, low cost and biodegradability. Moreover, their application helps reduce irrigation water consumption, causes lower death rate of plants, improves fertilizer retention in soil and increases plant growth rate [51–54].

As for slow release fertilizers combined with superabsorbents to obtain both slow-release and water retention properties are well described in the literature. In general, fertilizers and superabsorbents are combined by two methods. In the first method, fertilizers are blended with superabsorbents. In the second approach, fertilizers are added to the reaction mixture and polymerized *in situ* whereby they are entrapped in the superabsorbents. These two methods always result in a higher release rate [55, 56].

Hydrogels are also used in the cultivation of nursery plants. Superabsorbents allow to reduce the costs associated with maintaining plants, increase the survival rate of planted crops and allow for non-invasive protect seedlings of various kinds of vaccines aimed at ensuring better development of plants [54].

Hydrogels as components of horticultural substrates increase the soli water capacity. During irrigation or rainfall they bind water in the soil, preventing it from seeping into deeper layers. As a result, water is continuously provided to plants limiting water stress in plants [55–57]. It should be mentioned that further research is needed because most of the superabsorbents are based on pure poly(methacrylate) and therefore they are too expensive and not suitable for saline containing water and soils. Non-ionic polymers have a much lower absorption, but they are also less sensitive to ions in the water. Due to that many authors propose introducing inorganic clays, such as kaolin, bentonite, montmorillonite, etc., into pure polymeric superabsorbents in order to improve swelling property, hydrogel strengths and reduce production costs [58–61].

Hydrogels can also be used as a carrier of pesticides. For example, chitosan is characterized by antimicrobial and insecticidal properties. In the mixture with other active substances, it can be used as a matrix to carry bioinsecticides designed to control, for example, proliferation of insect larvae. Similarly, microcapsules of alginate and chitosan were prepared, characterized and evaluated as a carrier system for imidacloprid, synthetic analogs of brassinosteroids and diosgenin derivatives, azadirachtin, etc. Bionanocomposite materials based on chitosan as well as chitosan and clay blends (with montmorillonite) can be used as absorbents for herbicides [57, 58].

4.2. Food production and packaging industry

According to the Food and Agricultural Organization of the United Nations (FAO), increase of the global population to 9.1 billion by 2050 will be connected with the increase of the world food production by 70% and double food production in the developing world [62–64]. The application of hydrogels provides both economic and environment benefits by avoiding contamination of the ground water. Due to using too much fertilizers, water supplies are polluted. It may also lead to eutrophication, a situation where there is not enough oxygen dissolved in the water for aquatic organisms to survive. The use of coated fertilizers is a promising alternative to improve many aspects of fertilization based on the concept of controlled release. The main advantages of using controlled release hydrogels are regular and continuous delivery of nutrients, low-frequency fertilizer application and elimination of damage to roots due to high salt concentrations [65–67].

Effective food production is an integral part of food storages. Development of the concept of long-term food storage is connected with using polymers of various chemical compositions and structures. Along with the required performance properties, such as chemical durability, physicomechanical and technological properties, polymer materials used in direct contact with foods should also meet high hygienic requirements. For this aim, typical synthetic polymers as well as biopolymers widespread in nature in the form of cellulose, starch, natural rubber, silk and various resins have been used. Besides a polymer binder, plasticizers, fillers, dyes, pore formers, lubricating agents and other components are made of these components. The most popular way to obtain degradable blends is mixing natural and synthetic polymers or incorporating inorganic fillers [68–70]. Hydrogels can also be applied to produce efficient biopolymer packaging materials with desirable properties. Stiff and rigid linear polysaccharides mixed with proteins tend to form gels in the form of sheets, membranes and coatings [71]. Several biopolymers such as starch, cellulose, chitosan, poly(lactic acid) (PLA), poly(caprolactone) (PCL), poly(hydroxybutyrate) (PHB) are used for packaging purposes. The current trend in food packaging is the use of blends of different biopolymers such as starch-PLA blends and starch-PCL blends. More globular and flexible polysaccharides form spherical structures [26]. Promising applications of hydrogels in food packaging industries include improved packaging (gas and moisture barriers), antibacterial packaging, product condition monitoring, nanoadditives, enhanced shelf life, protection from oxidation and task masking.

4.3. Biomedical applications

For the first time, poly-2-hydroxyethylmethacrylate (PHEMA) as a synthetic biocompatible hydrogel was used for contact lens applications in 1960. Hard contact lenses are primarily based on hydrophobic materials such as poly(methyl methacrylate) (PMMA) or poly(hexafluoroisopropyl methacrylate) (HFIM), whereas soft lenses are based on hydrogels. Nowadays many attempts are made to obtain lenses with good oxygen permeability. To this aim, the hydrophilic monomers: 4-t-butyl,2-hydroxycyclohexyl methacrylate, methacryloylamino-4-t-butyl-2-hydroxycyclohexane and 4-t-butyl,2-hydroxycyclopentyl methacrylate with HEMA and N-vinyl-2-pyrrolidone (NVP) are proposed [72].

Hydrogels are widely used in wounds treatment. Large wounds lead to the risk of infection and loss of large amounts of fluids. It was proved that chitosan hydrogels and their derivatives accelerate the division of fibroblasts and production of cytokines. The amount of collagen in wounds treated with chitosan materials is lower, therefore the growth of tissues does not occur so fast and the formation of small scars is observed. Chitosan is hypoallergenic and has natural antibacterial properties, which further support its use in field bandages [73, 74].

Bandages were introduced by HemCon, Inc., which develops market technologies to control severe bleeding for traumatic skin and organ injuries. Unlike the gauze, bandages made from chitosan hydrogels interact with the blood cells (chitosan molecules carry positive charge). All our cells, our blood cells, the outer membranes have negative charges. The negative charge of the blood cells is attracted by the positive charge of the chitosan. As soon as they touch each other, they fuse and form a tight, adherent clot on the surface of the wound. HemCon® bandages are conveniently packaged, stable at room temperature and simple to use. Being extremely robust and flexible, they can be used on irregular wound surfaces. They are also non-heat-producing and can easily be removed by the water. The other examples are Granugel® (ConvaTec), Intrasite Gel® (Smith & Nephew), Purilon Gel®(Coloplast), Aquaflo (Covidien) or Woundtab® (First Water) [14].

Applications involving cell immobilization in alginate hydrogels to obtain artificial organs are also very important. Hydrogels stimulate behaviour of human organs and they are sensitive to the changes of conditions such as pH, temperature, enzymes and electric field. The fact that they are hydrophilic in nature because of reactive groups such as –OH, –COOH, $–NH_2$, $–CONH_2$, $–SO_3H$ allows them to absorb or retain large quantities of water as well as to contain bioactive materials within their network. They could be used in controlled release systems without the need to remove them due to their biocompatibility. Therefore, development of pH-sensitive biodegradable and biocompatible hydrogels based on polysaccharides, polypeptides and proteins is observed. For example, polysaccharides/polyaniline based graft-copolymers have demonstrated their potential to be efficiently used for various biomedical applications as medical implants, prosthetic muscles or organs, diagnostic devices to artificial muscles, stabilization of bone implants and decreasing thrombosis [75, 76]. Used for production of urinary catheters, they prevent bacterial colonization on the surface and provide a smooth and slippery surface to improve its biocompatibility. One of the advanced applications of hydrogels was proposed by Caló and Khutoryanskiy [14]. Hydrogels can be used to convert electrochemical stimuli into mechanical work, i.e., reversible contraction and relaxation under physicochemical stimuli for the development of artificial muscles, which function like the human muscle and tissue but with an electrically driven muscle. Hydrogels obtained from natural polymers such as agarose due to its biocompatibility and the mild conditions of gelation make them suitable for applications in tissue engineering.

Hydrogels are also applied for insulin regulation in diabetes, treatment of hemophilia B, kidney failure (urea removal), liver failure, interferon production, treatment of Parkinson disease, hypocalcemia and chronic pain [77–79].

4.4. Hydrogels in cancer treatment

Advances in chemotherapy have resulted in development of many innovative drug delivery hydrogels which have significantly improved prognosis and quality of life in cancer patients. Peppas et al. [80] and Nguyen et al. [81] reported the PEG–PLGA graft copolymeric hydrogels application for cancer treatment and cancer imaging using light-sensitive polymers with stimuli sensitive properties for potential biomedical applications. Kanamala et al. [82] indicate that pH-sensitive hydrogels for tumour-targeted drug delivery have provided a new strategy for addressing the limitations of the conventional chemotherapy, i.e., HPMA polymer-coated liposomes to target tumour extracelluar pH_{ex}. Casolaro et al. [83] synthesized and studied two acrylic hydrogels as a biomaterial to load and release the chemotherapeutic agent cis-platin. Dual stimuli-responsive hydrogel has improved the release of platinum(II)-species for the chemotherapy of solid tumours.

The most interesting approach to treatment of pancreatic cancer has been proposed by David et al. [84]. Pancreatic cancer is a deadly form of neoplasm with the highest level of mortality among different cancers. They have developed a novel material by combination of quercetin and 5-fluororacil loaded on chitosan nanoparticles. The dual drug-loaded carrier system exhibited significant toxicity towards pancreatic cancer cells in both two-dimensional (2D) and 3D cultures.

In the case of congenital anomalies, traumatic injuries or tumour resections both biocompatible carriers and specific substances for bone reconstruction are applied. Many techniques applying tissue engineering are reported to promote osteogenesis with appropriate scaffolds for bone regeneration. Fujioka-Kobayashi et al. [85] created a drug delivery system that allows the controlled release of proteins—cholesterol-group- and acryloyl-group-bearing pullulan (CHPOA) nanogels. They were aggregated to form fast degradable hydrogels (CHPOA/hydrogels) by cross-linking with thiol-bearing polyethylene glycol. Two distinct growth factors, BMP2 and recombinant human fibroblast growth factor 18 (FGF18), were applied to a critical-size skull bone defect for bone repair by the CHPOA/hydrogel system. The CHPOA/hydrogel system can successfully deliver two different proteins to the bone defect to induce effective bone repair. The combination of the CHPOA/hydrogel system with the growth factors FGF18 and BMP2 might be a step towards efficient bone tissue engineering.

4.5. Heavy metal ions removal

In the case of heavy metal ions removal, different physicochemical properties of hydrogels can be taken into account [86–91]. Hydrogels (with and without magnetic properties) based on natural polysaccharides for Cd(II) removal from aqueous solutions were described in [92]. The chitosan-based hydrogels were synthesized with the addition of known amounts of chitosan, acrylic acid, and methylene-bisacrylamide in the presence of initiator in the form of ammonium persulfate and magnetite nanoparticles. It was found that the sorption process varies when magnetite is added during the synthesis of hydrogels. As for the contact time, pH, initial hydrogel mass and initial concentration of Cd(II) ions, it was found that the hydrogels without magnetic properties can be applied in the treatment of water and industrial effluents contaminated with Cd(II) more efficiently than those with magnetic properties. Application of new pH-sensitive chitosan-based hydrogel as a sorbent for Zn(II) ions from synthetic wastewater solutions was described in the paper by Milosavljević et al. [93]. The influence of different variables such as pH, temperature, contact time and initial concentration on the Zn(II) ions uptake was examined. The absorbate concentration (4.6, 9.2 and 14.1 mg/L), pH, absorption time (0.5–48 h) and temperature (25, 37 and 45°C) were varied. It was found that the maximum absorption of Zn(II) ions occurred at pH 5.5. The value obtained for maximum sorption capacity of 105.5 mg/g at 25°C. The hydrogel can be regenerated with 0.01 mol/L HNO_3.

The absorption of Cu(II) ions from aqueous solutions onto poly(acrylic acid-co-acrylamide) hydrogels prepared via free-radical solution polymerization was investigated by Orozco-Guareño et al. [94]. They found that hydrogel with a less cross-linking agent showed the highest capacity towards Cu(II) 121 mg/g. The effect of solution pH on the process effectiveness without dissolving of hydrogels was not observed. The Cu(II) absorption kinetics results showed that equilibrium was attained within 30 min. The absorption equilibrium data were better fitted by the Langmuir isotherm model. However, the effect of the amount of cross-linking agent was reflected in a poor fit of this isotherm equation attributed to the narrower interchain space. It was proposed that Cu(II) ions were coordinated through the carboxylic and amide functional groups to form a tetradentate surface copper complex [94].

4.6. Dyes removal

Hydrogels based on (meth)acrylates have been used in cationic dye removal, for example, (poly(methacrylic acid)-graft-cellulose/bentonite), poly(methacrylate-acrylic acid-vinyl acetate) and chitosan-graft-poly(acrylic acid) for methylene blue removal; poly(acrylic acid-acrylamide), semi-IPN of poly(acrylic acid-acrylamide-methacrylate) and amylase and poly(glycidylmethacrylate)-graft-cross-linked acrylate based resin for crystal violet, yellow 28 (BY28) removal [95–97]. Positive values of enthalpy, negative of entropy reveal favourable, spontaneous, endothermic processes. The Langmuir and Freundlich sorption isotherms are found to describe the experimental data best. Maximal sorption capacities corresponding to the complete monolayer coverage reached 102 mg/g and 157 mg/g for poly(methacrylic acid) based hydrogels with the neutralization degrees of monomer of 0 and 80%, respectively. Also the guar gum-based hydrogels such as guar gum-acrylic acid (Gg-AA) based cross-linked hydrogels were interpenetrated with PANI and evaluated for conductivity, antibacterial properties and dye absorption application [98].

5. Perspectives

Increasing knowledge about hybrid or composite hydrogel materials allows to design very useful shape-memory and self-healing, stimuli-responsive hydrogel materials. This is a consequence of the approach wherein minimally invasive treatments are needed. At the forefront are these which determine the immediate change from a low viscous solution before injection and quick formation of a strong network *in situ* requires careful selection of appropriate cross-linkers, possibilities to modulate release and degradation profiles after hydrogel administration, shielding pharmaceutical drugs in micro- and nanoparticles, etc. Such materials are also very interesting as sorbents because during the swelling process the network structure tends to be more extensive which allows access of various molecules to active centres. It should be mentioned that they can be readily and completely regenerated without no significant change of their properties.

5.1. Electroconductive hydrogels and biosensors

Electroconductive hydrogels are polymeric blends or conetworks that combine integrally conductive electroactive polymers (CEPs) with hydrated hydrogels. Electroconductive hydrogels belong to the class of multifunctional smart materials which combine the properties of constituent materials to give rise to technologically relevant properties for devices and systems as a biorecognition membrane layer in various biosensors. As an example, the biosensors of poly(HEMA)-based hydrogel and poly(aniline) (PANI) with the incorporated recombinant cytochrome P450-2D6 and poly(HEMA)] as well as polypyrrole (PPy) hydrogels with the incorporated analyte-specific enzyme can be mentioned. Such polymers have potential applications in bioactive electrode coatings and electrochemical devices.

5.2. Indicators

pH-sensitive hydrogels can be applied in modern analytical methods such as dispersive solid-phase extraction (DSPE) also known as QuEChERS (quick, easy, cheap, effective, rugged and safe), semisolid-liquid dispersive microextraction (SSLDM), dispersive liquid-liquid microex-

http://dx.doi.org/10.5772/63466

traction (DLLME) and solid-phase extraction (SPE) which are often applied for multi-residue analysis of active substances, plant extracts, pesticides etc. In many cases, some of them are free from organic solvents and therefore classified as a green analytical method (in this case, highly toxic, chlorinated solvent extractants are not applied). For example, pH-sensitive hydrogels were prepared from a catechol-conjugated alginate hydrogel and a pyrocatechol violet dye [98]. Chemical and mechanical stability in a wide range of pH values as well as a simple method of synthesis determine their application as visible sensors to withstand and monitor corrosive liquids (acids and bases) as well as radioactive compounds, hazardous wastes and infectious microorganisms.

6. Hydrogels in controlled release of fertilizers chelated by the biodegradable complexing agent IDS

Hydrogels found applications as slow-release fertilizers although the addition of fertilizers generally causes hydration reduction and affects their physical properties. Another aspect is the fact that most of the traditional hydrogels are polyacrylate-based products, thus not biodegradable and regarded as potential pollutants for the soil [99, 100]. Our research group has recently focused on the application of hydrogels in controlled release of fertilizes based on the biodegradable complexing agents. It was proved that the commercial hydrogels Luquasorb 1160 and Luquasorb 1280 can be used for the absorption of Cu(II), Zn(II), Mn(II), and Fe(III) complexes with the IDS [101].

It was found that the rate of absorption of the nutrients (macro- and micronutrients) by the plants depends not only on plant age but also on the form in which the element is supplied and the solution concentration. Micronutrients are applied in the form of simple complexes or chelates. Several compounds are used as the complexing agents. They can be both natural and synthetic. In this group, citric, formic, ascorbic, propionic, tartaric, succinic, lactic, gluconic and salicylic acids or their K, Na and NH_4^+ salts, lignosulfonates, natural and synthetic amino acids (glycine, cysteine, and glutamine), ethylenediaminetetraacetic can be listed. For example, citric acid is a natural biodegradable chelator. Because of its carboxyl groups, citric acid complexes are not stable in the acidic environment. EDTA forms strong water-soluble chelates with polyvalent metal ions over a fairly wide range of pH [102].

Liquid fertilizers applied for feeding of plants should be characterized by defined properties such as an appropriate content and ratio of macro- to micronutrients. In addition, the ligand of the sequestering agent should undergo a metabolic conversion or biodegradation. Therefore, in the present paper the new complexing agent for binding nutrient ions was proposed. ISD is characterized by excellent biodegradability.

This new, biodegradable complexing agent also known as Baypure CX 100 (Laxness, Germany) or tetrasodium salt of (*N*-1,2-dicarboxyethyl)-D,L-aspartate acid. Iminodisuccinic acid belongs to the group of biodegradable complexing agents of a new generation. It is environmentally friendly and non-toxic. It is prepared by thermal polymerization of aspartic acid and characterized by extremely rapid biodegradation, which equals approximately 80% after just

7 days [103–105]. It can solubilize in water at any ratio. Its ability to form soluble complexes with metal ions dissolved in water is in the range from 4 to 10. Formation of IDS complexes with metal ions can be described as follows (Eq. (1)):

$$M^{2+} + H_n ids^{n-4} \rightarrow \left[M\left(H_n ids\right)\right]^{n-2} \tag{1}$$

where *ids* is the ligand iminodisuccinic, *M* is the metal ion, n = 0–3. IDS is a pentadentate complexing agent. It forms chelates of octahedral structure with many metal ions. For example, for Cu(II) ion there are known the following complexes: $[Cu(Hids)]^{-}$, $[Cu(ids)]^{2-})]^{-}$, $[Cu(OH)(ids)]^{3-}$. The stability constants of metal complexes with IDS were presented in [106, 107]. In our previous paper [108], the influence of chemical conditions on hydrogel, kinetic and absorption behaviour towards Cu(II), Zn(II), Mn(II) and Fe(III) ions in the presence of the chelating agent of a new generation, i.e., IDS was studied. The studies were carried out to investigate the effect of the sorbent dose, pH of the solution, initial concentration as well as phase contact time and temperature on the absorption efficiency. The kinetic parameters based on the kinetic models, i.e., the pseudo first order (PFO), the pseudo second order (PSO) and the intraparticle diffusion (IPD) equations were also determined. From the linear dependence of the Langmuir and Freundlich isotherms, the maximal absorption capacities and the constants of the studied hydrogel in relation to the complexes of Cu(II), Zn(II), Mn(II) and Fe(III) with IDS were determined. Now the new hydrogels such as TerraHydrogel®Aqua, THA (Terra, Poland), Agro® hydrogel, AH (EverChem, Poland) and Zeba® hydrogel, ZH (Agrecol, Poland) were investigated. Their physicochemical properties are listed in **Table 2**.

Superabsorbent	TerraHydrogel®Aqua (THA)	Agro® hydrogel (AH)	Zeba® hydrogel (ZH)
Matrix	Cross-linked polyacrylate	Cross-linked acrylamide polyacrylate	Cross-linked polyacrylate
Form	Anionic	Anionic	Anionic
Commercial form	Na^+	K^+	n.a.
Appearance	White granules	White powder	White-yellow granules
The water absorbency	180–300 g H_2O/g	380 g H_2O/g	n.a.
Bead size (mm)	0.177–0.255	0.300–1.000	n.a.
Operating pH range	6–8	5–9	5–9

Abbreviation: n.a., not available.

Table 2. The physicochemical properties of THA, AH and ZH hydrogels.

In the first stage of investigations, the moisture retention capability (Q_{H2O}%) at time *t* was measured. To determine the water absorption capacity of the hydrogels, a gravimetric method was applied. It was found Q_{H2O}% values were changed from 34 to 89% for THA after the measured time intervals [104]. For AH and ZH hydrogels, they were as follows: from 25 to 73% and from 12 to 87%, respectively (**Figure 1**).

http://dx.doi.org/10.5772/63466

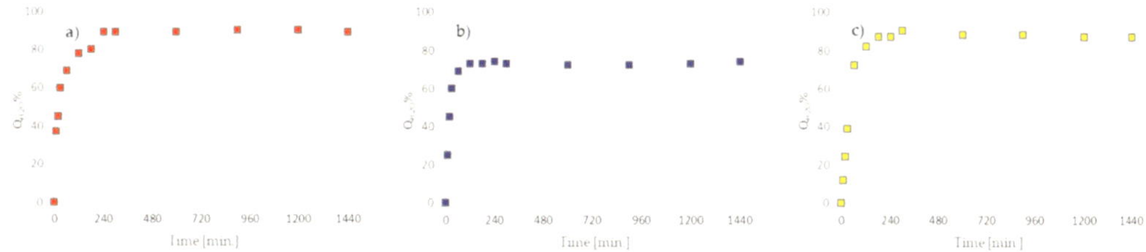

Figure 1. Comparison of water absorption capacity Q_{H2O}% of the (a) THA, (b) AH and (c) ZH hydrogels.

In the next stage THA, AH and ZH hydrogels were immersed in NaCl solution (initial concentrations in the range 0.2–1.5 M) at room temperature to reach the swelling equilibrium. After 24 h time, the samples were separated from the solution by filtration. The NaCl solution (Q_{NaCl}%) absorbency of THA, AH and ZH hydrogels were determined by weighing the swollen and dry samples. All the experiments were carried out three times to obtain the average values. It was found that Q_{NaCl}% values were equal to 10, 8 and 15% for 1.5 M NaCl solution, respectively. The obtained results were presented in **Figure 2**.

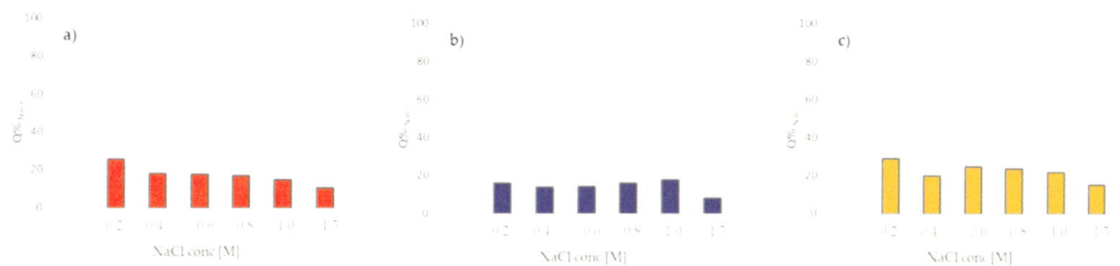

Figure 2. Comparison of NaCl absorption capacity Q_{NaCl}% of the (a) THA, (b) AH and (c) ZH hydrogels.

After immersing in water and NaCl solutions, the polymer network of negative charged due to the presence of carboxylic groups can absorb and retain a large volume of water or salt solutions. As follows from the literature data [25, 109] the rate of solution transport can be much lower than the relaxation of the polymer chains (Fickian diffusion) or diffusion is very rapid compared with the relaxation process (the rate of water movement is determined by relaxation) as well as in the intermediate case the diffusion and relaxation rates are comparable. Those estimations can be based on the equation (Eq. (2))

$$m_t / m_\infty = kt^n \tag{2}$$

where m_t is the mass of water absorbed at time t, m_∞ is the mass of water or NaCl solution absorbed at equilibrium, $k$is a characteristic constant of the polymer and n is a diffusional exponent. For $n < 0.5$, the rate of solution transport is dominated by a Fickian diffusion mechanism, for $n > 0.5$ the dynamic swelling of the polymers is the main factor. In the case of $n = 0$, the mass transfer is independent of time.

The maximum quantity of water absorbed by THA is 190 g/g for distilled water and 119 g/g for 1.5 M NaCl solution during its 1st hydration cycle (**Figure 1**). For AH hydrogel, these values are as follows: 165 g/g for distilled water and 98 g/g for 1.5 M NaCl solution, for ZH hydrogel they are the lowest: 131 g/g for distilled water and 72 g/g for 1.5 M NaCl solution. The absorption of water by AH, ZH and THA hydrogels was found to be faster in distilled water compared to NaCl. Comparing the obtained results replicated in three steps, it was proved that the first step of hydration was the fastest (data not presented). Swelling of polymer in the case of salt solutions is smaller due to the difference in the osmotic pressure.

Static (batch) tests of absorption of metal complexes with IDS were made by putting 0.1 g of THA, AH and ZH hydrogels and 50 cm^3 of Cu(II) complexes with IDS solution (in the system M(II):IDS = 1:1) into the 100 cm^3 conical flask and shaking mechanically using the laboratory shaker for 1–120 min. The procedure was repeated three times. The samples were shaken using the laboratory shaker type 357 (Elpin Plus). The stirring speed was 180 rpm. The pH was measured using a pH meter CPI-505 (Elmetron, Poland). The concentration of Cu(II) complexes with IDS in the filtrate was determined using the inductively coupled plasma optical emission spectrometer ICP-OES of type 720 ES (Varian, Australia).

The amount of Cu(II) complexes with IDS absorbed on THA, AH and ZH hydrogels was calculated from the difference between the initial concentration and the equilibrium one. The rate of complexes absorption is expressed as percentage of the amount of metal ions absorbed after a certain time related to that required for the state of equilibrium. This can be described as follows (Eq. (3)):

$$S\% = \left(c_0 - c_e\right) / c_0 \times 100\% \tag{3}$$

The absorption capacity (qe_e, mg/g) of THA, AH, ZH hydroirst-order kinetic equation is represented as (Eq. (4):

$$q_e = \left(c_0 - c_e\right) / m_d \times V \tag{4}$$

where c_0 is the initial concentration of M(II) complexes with IDS solution (mg/dm^3), c_t is the concentration of M(II/III) complexes with IDS in the aqueous phase at time t (mg/dm^3), V is the volume of the solution (dm^3) and m_d the mass of the dried hydrogel (g). The examplary results of sorption of the Cu(II)-IDS complexes on THA are presented in **Figures 3** and **4**.

The data show that the amount of Cu(II) ions sorbed by THA hydrogel increases with the increasing of concentration the complexes. The absorption percentage (S%) also increases with increasing concentration. For the Cu(II)-IDS complexes at the concentration 100 mg/dm^3, the equilibrium time was attained in 10 min, whereas equilibrium was reached in about 20 min for the concentration 200 mg/dm^3 and 1 h for 300 mg/dm^3.

http://dx.doi.org/10.5772/63466

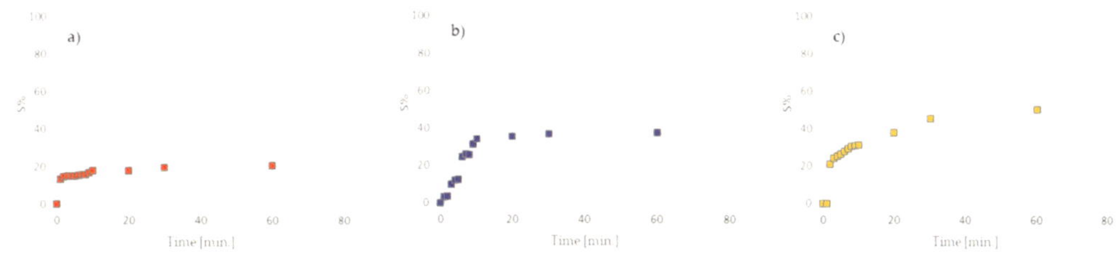

Figure 3. Comparison of $S\%$ of Cu(II) ions in the presence of IDS at different initial concentrations: (a) 100 mg/dm^3, (b) 200 mg/dm^3 and (c) 300 mg/dm^3 on THA hydrogel (0.1 g of THA, 250–500 μm, 100 cm^3 of Cu(II)-IDS solution, pH 6, 298 K).

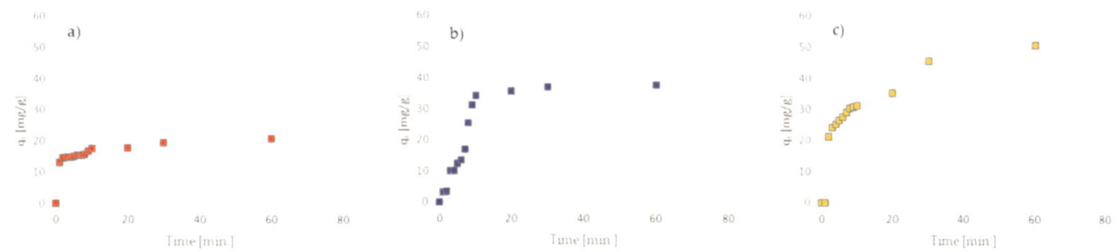

Figure 4. Comparison of absorption kinetics of Cu(II) ions in the presence of IDS at different initial concentrations: (a) 100 mg/dm^3, (b) 200 mg/dm^3 and (c) 300 mg/dm^3 on THA hydrogel (0.1 g of THA, 250–500 μm, 100 cm^3 of Cu(II)-IDS solution, pH 6, 298 K).

For estimation of kinetic parameters the pseudo first order (PFO) and the second order kinetic (PSO) models were used [110, 111]. The pseudo first-order kinetic equation is represented as (Eq. (5))

$$\log\left(q_e - q_t\right) = \log q_e - \left(k_1 t / 2.303\right) \tag{5}$$

where q_e and q_t denote the amount of absorption at equilibrium and at time t (mg/g), respectively; k_1 is the rate constant of the pseudo first order absorption (1/min). Based on the plot of log $(q_e - q_t)$ vs. t the kinetic parameters were calculated. The pseudo second-order model is expressed as (Eq. (6))

$$t / q_t = \left(t / q_e\right) + 1 / k_2 q_e^2 \tag{6}$$

where q_e and q_t denote the amount of absorption at equilibrium and at time t (mg/g), respectively; k_2 is the rate constant of the pseudo second-order absorption (g/mg min). The kinetic parameters were calculated based on the plots of t/q_t vs. t. Comparisons of absorption kinetics of Cu(II) ions in the presence of IDS at different initial concentrations on the THA hydrogel fitted by the pseudo first and pseudo second kinetic models are presented in **Figures 5** and **6**.

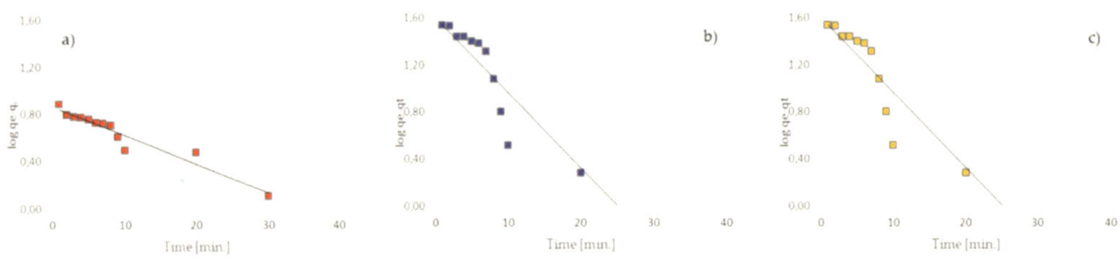

Figure 5. Comparison of absorption kinetics of Cu(II) ions in the presence of IDS at different initial concentrations: (a) 100 mg/dm^3, (b) 200 mg/dm^3 and (c) 300 mg/dm^3 on the THA hydrogel fitted by the pseudo first kinetic model (0.1 g of THA, 250–500 μm, 100 cm^3 of a Cu(II)-IDS solution, pH 6, 298 K).

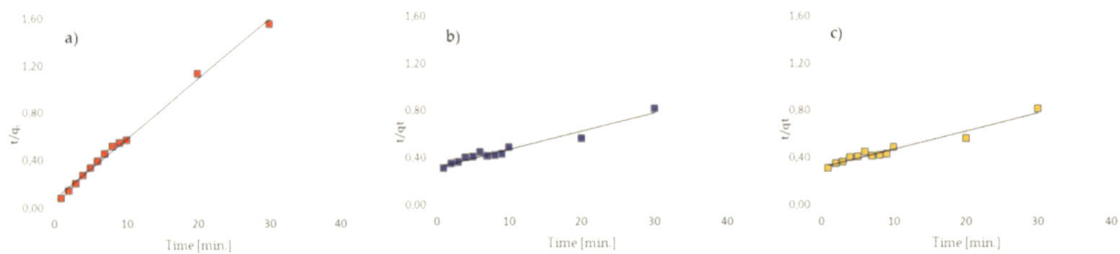

Figure 6. Comparison of absorption kinetics of Cu(II) ions in the presence of IDS at different initial concentrations on THA hydrogel fitted by the pseudo second kinetic model (0.1 g of THA, 250–500 μm, 100 cm^3 of Cu(II)-IDS solution, pH 6, 298 K).

The determined kinetic parameters of the process indicate that its course is consistent with the reaction mechanism typical of the pseudo second-order reaction in the case of Cu(II) complexes sorption with IDS as confirmed by high values of the correlation coefficients and the calculated sorption capacities are in the agreement with the experimental data. The kinetic sorption data were better fitted with the use of the pseudo second kinetic model, which yielded the best correlation coefficients R^2(>0.99) and the rate constant k_2 in the range 0.002–0.035 for the Cu(II)-IDS complexes. The obtained data also show that the k_2 rate constant decreased with the increase of the initial Cu(II)–IDS concentration, which confirmed that the time to reach sorption equilibrium increased with the initial concentration.

Furthermore, the experimental data were analyzed according to the Weber-Morris kinetic (Eq. (7)), i.e., the intraparticle diffusion model (IPD) given below [112]:

$$q_t = k_i t^{1/2} + C \tag{7}$$

where k_i is the intraparticle diffusion rate constant (mg/g min$^{0.5}$), C is the intercept which reflects the boundary layer effect.

The comparison of absorption kinetics of Cu(II) ions in the presence of IDS at different initial concentrations on the THA hydrogel fitted by the pseudo first and pseudo second kinetic as well as Weber-Morris models are presented in **Figures 5** and **7**.

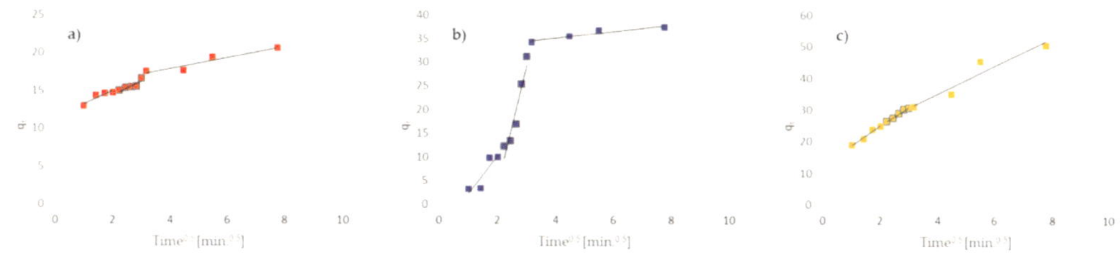

Figure 7. Comparison of absorption kinetics of Cu(II) ions in the presence of IDS at different initial concentrations on THA hydrogel fitted by the intraparticle diffusion model (0.1 g of THA, 250–500 μm, 100 cm^3 of Cu(II)-IDS solution, pH 6, 298 K).

It was also observed that the absorption process can be divided into three stages where the slope corresponds to the absorption rate. In the first stage, the regression line with short intercepts almost passed through the origin, suggesting that the intraparticle diffusion is not the sole rate controlling step. The sharper slope was attributed to the diffusion of studied complexes to the external surface of hydrogels or the boundary diffusion layer. The second stage implied the gradual absorption, where intra-particle diffusion was rate-controlling. As the contact time increases, the effect exerted by the external mass transfer in absorption rate-controlling becomes more and more evident. The intraparticle diffusion slowed down in the third stage due to low concentration of Cu(II)-IDS complexes. It can be concluded that intraparticle diffusion plays a predominant role in the first stage of the sorption process.

THA, AH and ZH hydrogels were also characterized by the pH of the point of zero charge pH_{zpc} determination using the pH drift method. Their pH in 0.01 M NaCl solution was adjusted between 2 and 12 by adding 0.1 M NaOH and 0.1 M HCl. 0.2 g of the THA, AZ and ZH was added to 50 cm^3 of the solution and after 24 h the final pH was measured. As follows from the previous studies pH_{zpc} determined for the THA hydrogel was equal to 6.9 [104]. For AZ and ZH hydrogels those values were equal 7.4 and 6.5. At $pH > pH_{zpc}$ the surface charge will be negative while at $pH < pH_{zpc}$ positive and therefore pH affects the sorption process. In the whole pH range of 2–11, THA was the most efficient for removal of Cu(II)-IDS complexes. Generally, the absorption is more favourable at the pH values 4–10 and decreases at the pH values of 2 and 13. The analogous results were obtained for AH and ZH hydrogels. However, a slight loss of the mass of the used hydrogels was observed. Therefore to examine stability (mass loss) of the used hydrogels, the samples were immersed at room temperature in buffer solutions at pH 1.96, 4.01, 7.0, 9.0 and 11.0 for 3 h. Mass loss was determined by the gravimetric method, as previously. It was found that at pH 11 the mass loss was the largest and equal to 36 and 24% for AZ and ZH hydrogels, respectively. It can be noticed that the swelling ratio of the hydrogels increases at pH greater than 1.96 and decreases sharply for the samples at pH 11.0. At pH 1.96, H-bonding between –OH, –CO and –COOH groups increases the cross-linking density and hence results in small swelling ratio, especially for AH hydrogel. At pH 7.0, the carboxylic groups are dissociated and the hydrogen bonds are disrupted. The electrostatic repulsion between the polymer chains occurs and THA, AH and ZH hydrogels tend to swell. At pH greater than 7.0, the polymeric chains swell rapidly and deprotonation of the basic groups and dissociation of the acidic groups carboxylic groups occurs.

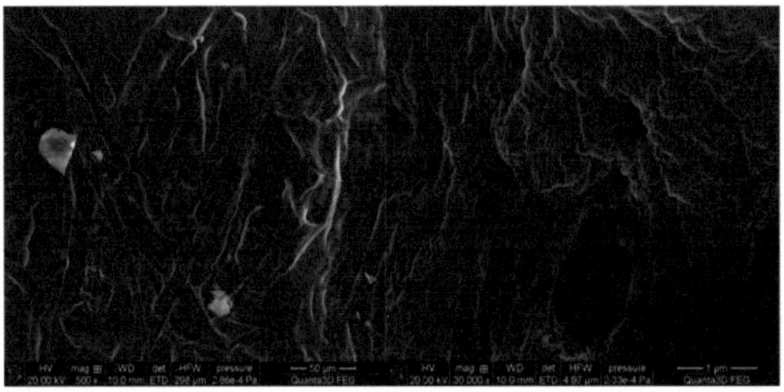

Figure 8. SEM scans of THA before the sorption of Cu(II)–IDS complexes.

For the used hydrogels, scanning electron microscopy images were also recorded using Quanta 3D FEG microscope (FEI). Fourier transform infrared spectra of THA were obtained using the attenuated total reflectance technique (FTIR-ATR) and measured with a FTIR Carry 630 spectrometer (Agilent Technologies). The bands at 3400 cm^{-1} (ν_{as} NH_2), 3190 cm^{-1} (ν_s NH_2) and 1687 cm^{-1} (C=O) are characteristic of the acrylamide unit. The characteristic stretching 2 vibrations at 1453 cm^{-1} connected with the presence of $–CH_2$ group as well as the symmetric and asymmetric stretching vibrations of the carboxylate ion (COO^{-1}) at 1399 and 1553 cm^{-1} are also visible [113, 114].

It should be mentioned that the samples were prepared before the sorption process (**Figure 8**).

7. Conclusions

Cross-linked hydrophilic polymers are capable of absorbing large volumes of waters and salt solutions. Therefore, they can find widespread applications in bioengineering, biomedicine, food industry and water purification as well as in separation processes. They are also applied in the production of slow release of fertilizers. In the present paper, the application of the biodegradable complexing agent, iminodisiccinic acid (IDS) for the sorption of Cu(II), was presented using commercially available hydrogels. It was found that with the increasing phase contact time and concentration the sorption effectiveness increases. The equilibrium state is established at the phase contact time about 10–20 min depending on the complex concentration of the initial solution. The determined kinetic parameters of the process indicate that its course is consistent with the reaction mechanism typical of the pseudo second-order reaction in the case of sorption Cu(II) complexes with IDS as confirmed by high values of the correlation coefficients and the calculated sorption capacities are in the agreement with the experimental data. TerraHydrogel®Aqua, THA hydrogel can find application in the controlled release of fertilizers based on the biodegradable complexing agent.

Author details

Dorota Kołodyńska[1*], Alicja Skiba[2], Bożena Górecka[2] and Zbigniew Hubicki[1]

*Address all correspondence to: kolodyn@poczta.onet.pl

1 Department of Inorganic Chemistry, Maria Curie Skłodowska University, Lublin, Poland

2 Analytical Department, New Chemical Synthesis Institute, Al. Tysiąclecia Państwas Polskiego, Puławy, Poland

References

[1] Ahmed EM, Hydrogel: preparation, characterization, and applications: a review. Journal of Advanced Research. 2015;6:105–121. DOI: 10.1016/j.jare.2013.07.006

[2] Osada Y, Kajiwara K, Fushimi T, Ishida H. Gels Handbook. The Fundamentals. Vol. 1, Academic Press; 2001.

[3] Whitcombe MJ, Chianella I, Larcombe L, Piletsky SA, Noble J, Porter R, Horgan A. The rational development of molecularly imprinted polymer-based sensors for protein detection. Chemical Society Reviews. 2011;40:1547–1571. DOI: 10.1039/C0CS00049C

[4] Li X, Li Q, Su Y, Yue Q ,Gao B, Su Y. A novel wheat straw cellulose-based semi-IPNs superabsorbent with integration of water-retaining and controlled-release fertilizers. Journal of the Taiwan Institute of Chemical Engineers. 2015;55:170–179. DOI: 10.1016/j.jtice.2015.04.022

[5] Lee JW, Kim SY, Kim SS, Lee YM, Lee KH, Kim SJ. Synthesis and characteristics of interpenetrating polymer network hydrogel composed of chitosan and poly(acrylic acid). Journal of Applied Polymer Science. 1999;73:113–120. DOI: 10.1002/(SICI)1097-4628(19990705)73:1

[6] Buwalda SJ, Boere KWM, Dijkstra PJ, Feijen J, Vermonden T, Hennink WE. Hydrogels in a historical perspective: from simple networks to smart materials. Journal of Controlled Release. 2014;190:254–273. DOI: 10.1016/j.jconrel.2014.03.052

[7] Achtenhagen J, Kreuzig R. Laboratory tests on the impact of superabsorbent polymers on transformation and sorption of xenobiotics in soil taking ^{14}C-imazalil as an

example. Science of the Total Environment. 2011;409:5454–5458. DOI: 10.1016/j.scitotenv.2011.09.021

[8] Woodhouse JM, Johnson MS. The effect of gel-forming polymers on seed germination and establishment. Journal of Arid Environmental. 1991;20:375–380. DOI:

[9] Vundavalli R, Vundavalli S, MamathaNakka, SrinivasaRao D. Biodegradable nanohydrogels in agricultural farming—alternative source for water resources. Procedia Materials Science. 2015;10:548–554. DOI: 10.1016/j.mspro.2015.06.005

[10] Kuang J, Yuk KY, Huh KM. Polysaccharide-based superporous hydrogels with fast swelling and superabsorbent properties. Carbohydrate Polymers. 2011;83:284–290. DOI:10.1016/j.carbpol.2010.07.052

[11] Lucas N, Bienaime LN, Belloy C, Queneudec M, Silvestre F, Nava-Saucedo JE. Polymer biodegradation: mechanisms and estimation techniques. Chemosphere. 2008;73:429–442. DOI: 10.1016/j.chemosphere.2008.06.064

[12] Raafat A, Eid M, El-Arnaouty MB. Radiation synthesis of superabsorbent CMC based hydrogels for agriculture applications. Nuclear Instruments and Methods in Physics Research B. 2012;283:71–76. DOI: 10.1016/j.nimb.2012.04.011.

[13] Francis S, Varshne L. Studies on radiation synthesis of PVA/EDTA hydrogels. Radiation Physics and Chemistry. 2005;74:310–316. DOI: 10.1016/j.radphyschem.2005.02.003

[14] Caló E, Khutoryanskiy VV. Biomedical applications of hydrogels: a review of patents and commercial products. European Polymer Journal. 2015;65:252–26. DOI: 10.1016/j.eurpolymj.2014.11.024

[15] Sato R, Noma R, Tokuyama H. Preparation of macroporous poly(N-isopropylacrylamide) hydrogels using a suspension–gelation method. European Polymer Journal. 2015;66:91–97. DOI: 10.1016/j.eurpolymj.2015.01.051

[16] Coutinho DF, Sant S, Shin H, Oliveira JT, Gomez ME, Neves NM, Khademhosseini A, Reis LR. Modified Gellan Gum hydrogels with tunable physical and mechanical properties. Biomaterials. 2010;31:7494–7502. DOI: 10.1010.biomaterials.2010.06.035

[17] Parlak O, Ashaduzzaman M, Kollipara SB, Tiwari A, Turner AP. Switchable bioelectrocatalysis controlled by dual stimuli-responsive polymeric interface, ACS Applied Materials & Interfaces. 2015;7:23837–23847. DOI: 10.1021/acsami.5b06048

[18] Sharma K, Kumar V, Chaudhary B, Kaith BS, Kalia S, Swart HC. Application of biodegradable superabsorbent hydrogel composite based on Gum ghatti-co-poly(acrylic acid-aniline) for controlled drug delivery. Polymer Degradation and Stability. 2016;124:101–111. DOI: 10.1016/j.polymdegradstab.2015.12.021

[19] Kunioka M, Choi HJ. Hydrolytic degradation and mechanical properties of hydrogels prepared from microbial poly(amino acid)s. Polymer Degradation and Stability. 1998;59:33–37. DOI: 10.1016/S0141-3910(97)00181-X

[20] Buchholz FL, Graham AT. Modern Superabsorbent Polymer Technology, New York: Wiley-VCH; 1997. 304 p. ISBN: 978-0-471-19411-8.

[21] Kirwan LJ, Fawell PD, van Bronswijk W. In situ FTIR-ATR examination of poly(acrylic acid) adsorbed onto hematite at low pH. Langmuir. 2003;19:5802–5807. DOI: 10.1021/la027012d

[22] Park K, Chen J, Park H. Hydrogel composites and superporous hydrogel composites having fast swelling, high mechanical strength, and superabsorbent properties, US6271278 B1, 2001.

[23] Othman MBH, Akil HM, Rasib SZMd, Khan A, Ahmad Z. Thermal properties and kinetic investigation of chitosan-PMAA based dual-responsive hydrogels. Industrial Crops and Products. 2015;66:178–187. DOI: 10.1016/j.indcrop.2014.12.057

[24] Panic VV, Madzarevic ZP, Volkov-Husovic T, Velickovic SJ. Poly(methacrylic acid) based hydrogels as sorbents for removal of cationic dye basic yellow 28: Kinetics, equilibrium study and image analysis. Chemical Engineering Journal. 2013;217:192–204. DOI: 10.1016/j.cej.2012.11.081

[25] Witono JR, Noordergraaf IW, Heeres HJ, Janssen LPBM. Water absorption, retention and the swelling characteristics of cassava starch grafted with polyacrylic acid. Carbohydrate Polymers. 2014;103:325–332. DOI: 10.1016/j.carbpol.2013.12.056.

[26] Ullah F, Othman MBH, Javed F, Ahmad Z, Akil HM. Classification, processing and application of hydrogels: a review. Materials Science and Engineering C. 2015;57:414–433. DOI: 10.1016/j.msec.2015.07.053

[27] Kabiri K, Omidian H, Zohuriaan-Mehr MJ, Doroudiani S. Superabsorbent hydrogel composites and nanocomposites: a review. 2012;32:277–289. DOI: 10.1002/pc.21046

[28] Bhattarai N, Gunn J, Zhang M. Chitosan-based hydrogels for controlled, localized drug delivery. Advanced Drug Delivery Reviews. 2010;62:83–99. DOI: 10.1016/j.addr.2009.07.019

[29] Das N. Preparation methods and properties of hydrogel: a review. International Journal of Pharmacy and Pharmaceutical Sciences. 2013;5:112–117. DOI: 10.1016/j.msec.2015.07.053

[30] Gao X, He Ch, Xiao Ch, Zhuang X, Chen X. Biodegradable pH-responsive polyacrylic acid derivative hydrogels with tunable swelling behavior for oral delivery of insulin Polymer. 2013;54:1786–1793. DOI: 10.1016/j.polymer.2013.01.050

[31] Samal SK, Dash M, Dubruel P, Van Vlierberghe S. Smart polymer hydrogels: properties, synthesis and applications. Smart Polymers and Their Applications. 2014:237–270. DOI: 10.1533/9780857097026.1.237

[32] Jyothi AN. Starch graft copolymers: novel applications in industry. Composite Interfaces. 2010;17:165–174. DOI: 10.1163/092764410X490581

[33] Samchenko Y, Ulberg Z, Korotych O. Multipurpose smart hydrogel systems. Advances in Colloid and Interface Science. 2011;168:247–262. DOI: 10.1016/j.cis.2011.06.005

[34] Khan A, Othman MBH, Razak KA, Akil HM. Synthesis and physicochemical investigation of chitosan-PMAA-based dual-responsive hydrogels. Journal of Polymer Research. 2013;20:1–8. DOI: 10.1007/s10965-013-0273-7

[35] Yildiz U, Kemik OF, Hazer B. The removal of heavy metal ions from aqueous solutions by novel pH-sensitive hydrogels. Journal of Hazardous Materials. 2010;183:521–532. DOI: 10.1016/j.jhazmat.2010.07.055

[36] Buenger D, Topuz F, Groll J. Hydrogels in sensing applications. Progress in Polymer Science. 2012;37:1678–1719. DOI: 10.1016/j.progpolymsci.2012.09.001

[37] Koetting MC, Peters JT, Steichen SD, Peppa NA. Stimulus-responsive hydrogels: theory, modern advances, and applications. Materials Science and Engineering R. 2015;93:1–49. DOI:10.1016/j.mser.2015.04.001

[38] Aouada FA, Moura MRd, Mattoso LHP. Biodegradable hydrogel as delivery vehicle for the controlled release of pesticide, In: Stoytcheva M, editor. Pesticides—Formulations, Effects, Fate. InTech; 2011. pp. 81–102. DOI: 10.5772/14139.ch6

[39] Azeem B, KuShaari K, Man ZB, Basit A, Thanh TH. Review on materials & methods to produce controlled release coated urea fertilizer. Journal of Controlled Release. 2014;181:11–21. DOI: 10.1016/j.jconrel.2014.02.020

[40] Guilherme MR, Aouada FA, Fajardo AR, Martins AF, Paulino AT, Davi MFT, Rubira AF, Muniz EC. Superabsorbent hydrogels based on polysaccharides for application in agriculture as soil conditioner and nutrient carrier: a review. European Polymer Journal. 2015;72:365–385. DOI: 10.1016/j.eurpolymj.2015.04.017

[41] Davidson D. Controlled release of agrichemicals using a functionalized carboxymethyl cellulose hydrogel matrix. University of Waterloo: Department of Chemical Engineering; 2012.

[42] Saruchi, Kaith BS, Jindal R, Kumar V. Biodegradation of Gum tragacanth acrylic acid based hydrogel and its impact on soil fertility. Polymer Degradation and Stability. 2015;115:24–31. DOI:10.1016/j.polymdegradstab.2015.02.009

[43] Sempeho SI, Kim HT, Mubofu E, Hilonga A. Meticulous overview on the controlled release fertilizers. Advances in Chemistry. 2014; 2014: 363071. DOI: 10.1155/2014/363071

[44] Senna AM, Carmo JBd, Silva JMSd, Botaro VR. Synthesis, characterization and application of hydrogel derived from cellulose acetate as a substrate for slow-release NPK fertilizer and water retention in soil. Journal of Environmental Chemical Engineering. 2015;3:996–1002. DOI:10.1016/j.jece.2015.03.008

[45] Liang R, Liu M, Wu L. Controlled release NPK compound fertilizer with the function of water retention. Reactive & Functional Polymers. 2007;67:769–779. DOI: 10.1016/j.reactfunctpolym.2006.12.007

[46] Liu MZ, Liang R, Zhan FL, Liu Z, Niu AZ. Preparation of superabsorbent slow release nitrogen fertilizer by inverse suspension polymerization. Polymer International. 2007;56:729–737. DOI: 10.1016/j.reactfunctpolym.2006.12.007

[47] Stahl JD, Cameron MD, Haselbach J, Aust SD. Biodegradation of superabsorbent polymers in soil. Environmental Science and Pollution Research International 2000;7:83–8. DOI: 10.1065/espr199912.014.

[48] Bajpai AK, Giri A. Water sorption behaviour of highly swelling (carboxy methylcellulose-g-polyacrylamide) hydrogels and release of potassium nitrate as agrochemical. Carbohydrate Polymers. 2003;53:271–279. DOI:10.1016/S0144-8617(03)00071-7

[49] Ibrahim S, Nawwar GAM, Sultan M. Development of bio-based polymeric hydrogel: green, sustainable and low cost plant fertilizer packaging material. Journal of Environmental Chemical Engineering. 2016;4:203–210. DOI: 10.1016/j.jece.2015.10.028

[50] Hanafi MM, Eltaib MB, Ahmad MB. Physical and chemical characteristics of controlled release compound fertilizer. European Polymer Journal. 2000;36:2081–2088. PII: S0014-3057(00)00004-5

[51] Han X, Chen S, Hu X. Controlled-release fertilizer encapsulated by starch/polyvinyl alcohol coating. Desalination. 2009;24:21–26. DOI: 10.1016/j.desal.2008.01.047

[52] Sannino A, Demitri C, Madaghiele M. Biodegradable cellulose-based hydrogels: design and applications. Materials. 2009;2:353–373. DOI: 10.3390/ma2020353

[53] Davidson DW, Verma MS, Gu FX. Controlled root targeted delivery of fertilizer using an ionically crosslinked carboxymethyl cellulose hydrogel matrix. SpringerPlus. 2013;2:318. DOI: 10.1186/2193-1801-2-318

[54] Davidson D, Gu FX. Materials for sustained and controlled release of nutrients and molecules to support plant growth. Journal of agricultural and food chemistry. 2012;60:870–876. DOI: 10.1021/jf204092h

[55] Coviello T, Matricardi P, Marianecci C, Alhaique F. Polysaccharide hydrogels for modified release formulations. Journal of Controlled Release. 2007;119:5–24. DOI: 10.1016/j.jconrel.2007.01.004

[56] Xiaoyu N, Yuejin W, Zhengyan W, Lin W, Guannan Q, Lixiang Y. A novel slow-release urea fertiliser: physical and chemical analysis of its structure and study of its release

mechanism. Biosystems engineering. 2013;115:274–282. DOI: 10.1016/j.biosystemseng.2013.04.001

[57] Kashyap PL, Xiang X, Heiden P. Chitosan nanoparticle based delivery systems for sustainable agriculture. International Journal of Biological Macromolecules. 2015;77:36–51. DOI: 10.1016/j.ijbiomac.2015.02.039

[58] Rudzinski WE, Chipuk T, Dave AM, Kumbar SG, Aminabhavi TM. pH-sensitive acrylic-based copolymeric hydrogels for the controlled release of a pesticide and a micronutrient. Journal of Applied Polymer Science. 2003;87:394–403. DOI: 10.1002/app.11382

[59] Li A, Wang A. Synthesis and properties of clay-based superabsorbent composite. European Polymer Journal. 2005;41:1630–1637. DOI: 10.1016/j.eurpolymj.2005.01.028

[60] Wang X, Lu S, Gao C, Xu X, Zhang X, Bai X, Liu M, Wu L. Highly efficient adsorption of ammonium onto palygorskite nanocomposite and evaluation of its recovery as a multifunctional slow-release fertilizer. Chemical Engineering Journal. 2014;252:404–414. DOI: 10.1016/j.cej.2014.04.097

[61] Dragan ES. Design and applications of interpenetrating polymer network hydrogels. A review. Chemical Engineering Journal. 2014;243:572–590. DOI: 10.1016/j.cej.2014.01.065

[62] Santos BRd, Bacalhau FB, Tamires dSP, Souza CF, Faez R. Chitosan-montmorillonite microspheres: a sustainable fertilizer delivery system. Carbohydrate Polymers. 2015;127:340–346. DOI: 10.1016/j.carbpol.2015.03.064

[63] Aider M. Chitosan application for active bio-based films production and potential in the food industry: review. LWT—Food Science and Technology. 2010;43:837–842. DOI: 10.1016/j.lwt.2010.01.021

[64] Baldwin EA, Hagenmaier R, Bai J. Edible Coatings and Films to Improve Food Quality. CRC Press Taylor & Francis Group; 2011.

[65] Gao L, Wang S, Zhao X. Synthesis and characterization of agricultural controllable humic acid superabsorbent. Journal of Environmental Sciences. 2013;25(Suppl.):S69–S76. DOI: 10.1002/humu.22913.

[66] Bouranis DL, Theodoropoulos AG, Drossopoulos JB. Designing synthetic polymers as soil conditioners. Communications in Soil Science and Plant Analysis. 1995;26:1455–1480. DOI: 10.1080/00103629509369384

[67] Tong Z, Yuhai Y, Shihuo Y, Zhongyi H. Superabsorbent hydrogels as carriers for the controlled release of urea: experiments and a mathematical model describing the release rate. Biosystems Engineering. 2009;102:44–50. DOI:10.1016/j.biosystemseng.2008.09.027

[68] Sukhareva LA, Legonkova OA, Yakovlev VS. New concepts in polymer science polymers for packaging and containers in food industry (ed. G.E. Zaikov). Leiden, The Netherlands: Koninklijke Brill NV.; 2008; pp. 71–74. ISBN: 978-90-04-16143-6

[69] Farris S, Schaich KM, Liu L, Piergiovanni L, Yam KL. Development of polyion-complex hydrogels as an alternative approach for the production of bio-based polymers for food packaging applications: a review. Trends in Food Science and Technology. 2009;20:316–332. DOI: 10.1016/j.tifs.2009.04.003

[70] Gómez-Estaca J, de Lacey AL, López-Caballero M, Gómez-Guillén M, Montero P. Biodegradable gelatin-chitosan films incorporated with essential oils as antimicrobial agents for fish preservation. Food Microbiology. 2010;27:889–896. DOI: 10.1016/j.fm. 2010.05.012

[71] Pawar PA, Purwar AH. Bioderadable polymers in food packaging. American Journal of Engineering Research. 2013;02:151–164. e-ISSN: 2320-0847, p-ISSN: 2320-0936

[72] Kunzler JF, Friends GD. US Patent 5,006,622; 1991.

[73] Rinaudo M. Chitin and chitosan: properties and applications. Progress in Polymer Science. 2006;31:603–632. DOI: 10.1016/j.progpolymsci.2006.06.001

[74] Prashanth KVH, Tharanatha RN. Chitin/chitosan: modifications and their unlimited application potential—an overview. Trends in Food Science & Technology. 2007;18:117–131. DOI: 10.1016/j.tifs.2006.10.022

[75] Hoffman AS. Hydrogels for biomedical applications. Advanced Drug Delivery Reviews. 2012;64:18–23. DOI: 10.1016/j.addr.2012.09.010

[76] Saxena AK. Synthetic biodegradable hydrogel (PleuraSeal) sealant for sealing of lung tissue after thoracoscopic resection. The Journal of Thoracic Cardiovascular Surgery. 2010;139:496–497. DOI: 10.1016/j.jtcvs.2008.11.003

[77] Hamidi M, Azadi A, Rafiei P. Hydrogel nanoparticles in drug delivery. Advanced Drug Delivery Reviews. 2008;60:1638–1649. DOI: 10.1016/j.addr.2008.08.002

[78] Klouda L. Thermoresponsive hydrogels in biomedical applications. A seven-year update. European Journal of Pharmaceutics and Biopharmaceutics. 2015;97:338–349. DOI: 10.1016/j.ejpb.2015.05.017

[79] Singh B, Chauhan N. Release dynamics of tyrosine from dietary fiber psyllium based hydrogels for use in Parkinson's disease. Food Research International. 2010;43:1065–1072. DOI:10.1016/j.foodres.2010.01.015

[80] Peppas NA, Bures P, Leobandung W, Ichikawa H. Hydrogels in pharmaceutical formulations. European Journal of Pharmaceutics and Biopharmaceutics. 2000;50:27–46. DOI: 1359-0294'002-00531

[81] Nguyen QV, Huynh DP, Park JH, Lee DS. Injectable polymeric hydrogels for the delivery of therapeutic agents: a review. European Polymer Journal. 2015;72:602–619. DOI: 10.1016/j.eurpolymj.2015.03.016

[82] Kanamala M, Wilson WR, Yang M, Palmer BD, Wu Z. Mechanisms and biomaterials in pH-responsive tumour targeted drug delivery: a review. Biomaterials. 2016;85:152–167. DOI: 10.1016/j.biomaterials.2016.01.061

[83] Casalaro M, Del Bello B, Maellaro E. Hydrogel containing L-valine residues as a platform for cisplatin chemotherapy. Colloids and Surfaces B: Biointerfaces. 2011;88:389–395. DOI: 10.1016/j.colsurfb.2011.07.019

[84] David KI, Jaidev LR, Sethuraman S, Krishnan UM. Dual drug loaded chitosan nanoparticles-sugar-coated arsenal against pancreatic cancer. Colloids and Surfaces B: Biointerfaces. 2015;135:689–698. DOI: 10.1016/j.colsurfb.2015.08.038

[85] Fujioka-Kobayashi M, Ota MS, Shimoda A, Nakahama KI, Akiyoshi K, Miyamoto Y, Iseki S. Cholesteryl group- and acryloyl group-bearing pullulan nanogel to deliver BMP2 and FGF18 for bone tissue engineering. Biomaterials. 2012;33:7613–7620. DOI: 10.1016/j.biomaterials.2012.06.075

[86] Patravale V, Mandawgade S. Novel cosmetic delivery systems: an application update. International Journal of Cosmetics Science. 2008;30:19–33. DOI: 10.1111/j.1468-2494.2008.00416

[87] Chen JJ, Ahmad AL, Ooi BS. Poly(N-isopropylacrylamide-co-acrylic acid) hydrogels for copper ion adsorption: equilibrium isotherms, kinetic and thermodynamic studies. Journal of Environmental Chemical Engineering. 2013;1:339–348. DOI: 10.1016/j.jece.2013.05.012

[88] Chen JJ, Ahmad AL, Ooi BS. Thermo-responsive properties of poly(N-isopropylacrylamide-co-acrylic acid) hydrogel and its effect on copper ion removal and fouling of polymer-enhanced ultrafiltration. Journal of Membrane Science. 2014;469:73–79. DOI: 10.1016/j.memsci.2014.05.062

[89] Wang N, Han Y, Liu Y, Bai T, Gao H, Zhang P, Liu W. High-strength hydrogel as a reusable adsorbent of copper ions. Journal of Hazardous Materials. 2012;213–214:258–264. DOI: 10.1016/j.jhazmat.2012.01.092.

[90] Rahman MA, Rahman MM, Kadohashi K, Maki T, Hasegawa H. Effect of external iron and arsenic species on chelant-enhanced iron bioavailability and arsenic uptake in rice (Oryza sativa L.). Chemosphere. 2011;84:439–445. DOI: 10.1016/j.chemosphere.2011.03.046.

[91] Kaşgöz H, Özgümüs S, Orbay M. Modified polyacrylamide hydrogels and their application in removal of heavy metal ions. Polymer. 2003;44:1785–1793. DOI:10.1016/S0032-3861(03)00033-8

[92] Paulinoa AT, Belfiore LA, Kubota LT, Muniz EC, Tambourgi EB. Efficiency of hydrogels based on natural polysaccharides in the removal of Cd(II) ions from aqueous solutions. Chemical Engineering Journal. 2011;168:68–76. DOI:10.1016/j.cej.2010.12.037

[93] Milosavljević NB, Ristić MD, Perić-Grujić AA, Filipović JM, Štrbac SB, Rakočević ZL, Kalagasidis Krušić MT. Sorption of zinc by novel pH-sensitive hydrogels based on chitosan, itaconic acid and methacrylic acid. Journal of Hazardous Materials. 2011;192:846–854. DOI: 10.1016/j.jhazmat.2011.05.093

[94] Orozco-Guareńo E, Santiago-Gutiérrez F, Morán-Quiroz JL, Hernandez-Olmos SL, Soto V, de la Cruz W, Manríquez R, Gomez-Salazar S. Removal of Cu(II) ions from aqueous streams using poly(acrylic acid-co-acrylamide) hydrogels. Journal of Colloid and Interface Science. 2010;349:583–593. DOI: 10.1016/j.jcis.2010.05.048

[95] Bhattacharyya R, Ray S.K. Adsorption of industrial dyes by semi-IPN hydrogels of acrylic copolymers and sodium alginate. Journal of Industrial and Engineering Chemistry. 2015;22:92–102. DOI: 10.1016/j.jiec.2014.06.029

[96] Fosso-Kankeu E, Mittal H, Mishra SB, Mishra AK. Gum ghatti and acrylic acid based biodegradable hydrogels for the effective adsorption of cationic dyes. Journal of Industrial and Engineering Chemistry. 2015;22:171–178. DOI: 10.1016/j.jiec.2014.07.007.

[97] Sharma R, Kaith BS, Kalia S, Pathania D, Kumar A, Sharma N, Street RM, Schauer C. Biodegradable and conducting hydrogels based on Guar gum polysaccharide for antibacterial and dye removal applications. Journal of Environmental Management. 2015;162:37–45. DOI: 10.1016/j.jenvman.2015.07.044

[98] Lee YK, Lee SY. A colorimetric alginate-catechol hydrogel suitable as a spreadable pH indicator. Dyes and Pigments. 2014;108:1–6. DOI: 10.1016/j.dyepig.2014.04.014

[99] Montesano FF, Parente A, Santamaria P, Sannino A, Serio F. Biodegradable superabsorbent hydrogel increases water retention properties of growing media and plant growth. Agriculture and Agricultural Science Procedia. 2015;4:451–458. DOI: 10.1016/j.aaspro.2015.03.052

[100] Hasegawa H, Rahman MA, Saitou K, Kobayashi M, Okumura C. Influence of chelating ligands on bioavailability and mobility of iron in plant growth media and their effect on radish growth. Environmental and Experimental Botany. 2011;71:345–351. DOI: 10.1016/j.envexpbot.2011.01.004.

[101] Skiba A, Końodyńska D, Tyliszczak B, Pielichowski K, Hubicki Z, Hydrogel superabsorbents as microelements carries. Przemysł Chemiczny. 2015;94:797–801 (in Polish).

[102] Grzmil BU, Kic B, Pyro- and tripolyphosphates and citrates as the complexing agents for micronutrients in liquid fertilizers. Industrial & Engineering Chemistry Research. 2002;41:139–144.

[103] Hyvönen H, Orama M, Saarinen H, Aksela R. Studies on biodegradable chelating ligands: complexation of iminodisuccinic acid (ISA) with Cu(II), Zn(II), Mn(II) and

Fe(III) ions in aqueous solution. Green Chemistry. 2003;5:410–414. DOI: 10.1039/B303372B

[104] Pinto ISS, Neto IFF, Soares HMVM. Biodegradable chelating agents for industrial, domestic and agricultural applications—a review. Environmental Science and Pollution Research. 2014;21:11893–11906. DOI: 10.1007/s11356-014-2592-6

[105] Kołodyńska D, Hubicka H, Hubicki Z. Studies of application of monodisperse anion exchangers in sorption of heavy metal complexes with IDS. Desalination. 2009;239:216–228. DOI:10.1016/j.desal.2008.02.024

[106] Kołodyńska D. Iminodisuccinic acid as a new complexing agent for removal of heavy metal ions from industrial effluents. Chemical Engineering Journal. 2009;152:277–288. DOI: 10.1016/j.cej.2009.05.002

[107] Knepper TP. Synthetic chelating agents and compounds exhibiting complexing properties in the aquatic environment. Trends in Analytical Chemistry. 2003;22:708–724. DOI: 10.1016/S0165-9936(03)01008-2

[108] Kołodyńska, Skiba, Superabsorbents as effective materials for adsorption of metal complexes with biodegradable complexing agent. International Conference Industrial Waste & Wastewater Treatment & Valorisation, 21–23 May 2015.

[109] Wallace WE, Fischer DA, Efimenko K, Wu WL, Genzer J, Polymer chain relaxation: surface outpaces bulk. Macromolecules. 2001;34:5081–5082. DOI: 10.1021/ma002075t

[110] Kaşgoz H, Durmus, A. Dye removal by a novel hydrogel-clay nanocomposite with enhanced swelling properties. Polymers for Advanced Technologies. 2008;19:838–845. DOI: 10.1002/pat.1045

[111] Liu J, Li Q, Su Y, Yue Q, Gao B. Characterization and swelling-deswelling properties of wheat straw cellulose based semi-IPNs hydrogel. Carbohydrate Polymers. 2014;107:232–40. DOI: 10.1016/j.carbpol.2014.02.073

[112] Weber WJ, Morri JC. Kinetics of adsorption on carbon from solution. Journal of the Sanitary Engineering Division. 1963;89:31–60.

[113] Magalhães ASG, Neto MPA, Bezerra MN, Ricardo NMPS, Feitosa JPA, Application of FTIR in the determination of acrylate content in poly(sodium acrylate-co-acrylamide) superabsorbent hydrogels. Química Nova. 2012;35:1464–1467. DOI:10.1590/S0100-40422012000700030

[114] Özeroglu C, Birdal A, Swelling properties of acrylamide-N,N′-methylene bis(acrylamide) hydrogels synthesized by using meso-2,3-dimercaptosuccinic acid-cerium(IV) redox couple, eXPRESS Polymer Letters. 2009;3:168–176. DOI: 10.3144/expresspolymlett.2009.22

Chapter 5

Design, Characterization, and Environmental Applications of Hydrogels with Metal Ion Coordination Properties

Viviana Campo Dall' Orto and
Juan Manuel Lázaro-Martínez

Additional information is available at the end of the chapter

http://dx.doi.org/10.5772/62902

Abstract

In this chapter, we discuss the design and synthesis of hydrogels and related polymeric materials with metal ion coordination properties, with the aim to review the main synthetic strategies used in the area. Then, we focus on the solid-state nuclear magnetic resonance (ss-NMR) spectroscopic technique due to its importance as a structural elucidation tool in both powdered and hydrated state, with emphasis on cross-polarization magic angle spinning (CP-MAS) and high-resolution magic angle spinning (HRMAS). Also, we explain different adsorption models, with the aim to present the methods most commonly used to analyze the uptake properties of hydrogel materials toward metal ions or organic compounds. Finally, we will discuss the applications of these materials for the removal of heavy metal ions and organic compounds, in terms of efficiency in the uptake of these ions and the different techniques commonly used to study the coordination process and the generation of reactive oxygen species (ROS) from hydrogen peroxide (H_2O_2). The main aim is to provide scientists with a review of the spectroscopic techniques most commonly used for bulk and surface characterization of non-soluble materials.

Keywords: Polymers, coordination, ss-NMR, metal complexes, H_2O_2 activation

1. Introduction

Various human activities lead to increase in the concentrations of heavy metal ions in the environment. For example, the effluents from electrical and plastic industries contain copper and cadmium ions, which are toxic and harmful, even at low concentrations, not only to humans but also to plants and animals because they are not biodegradable [1]. An effective and versatile method to remove these heavy metal ions is adsorption [2]. Thus, in the last years, many research groups have worked on the design of functionalized polymers for the development of effective and economic adsorbents for the removal of these toxics from effluents. In this context, polymeric materials with polyampholyte and polyelectrolyte characteristics have interesting properties such as acid–base behavior and coordination of inorganic and organic compounds [3]. Polymeric materials with polyampholyte characteristics consist of monomers which can have positive and negative charges, whereas those with polyelectrolyte characteristics may possess electric charge of one sign. Both can be synthesized by conventional techniques of radical polymerization [4].

Regarding the textile industry, most of the pigments used today are poorly biodegraded or resistant to environmental conditions. Thus, there is a growing need to remove these pigments from its effluents and a growing demand for new physical, chemical, and/or biological methods to reduce their concentrations [5].

With the aim to fight this pollution panorama, scientists have developed different smart systems using copper complexes of inorganic or organic material for H_2O_2 activation, thereby generating reactive oxygen species (ROS) for oxidative degradation of the pollutant. There are numerous examples of Cu(II)/organic ligand complexes as homogeneous systems in which amino acids [6], carboxylic acid [7], and Schiff bases are used as chelating agents [8], since the coordination of the metal ion produces an increase in the catalytic activity of Cu(II). In this area, heterogeneous catalysts, obtained by modifying the surface of particulate materials (such as silica, clays, and diverse polymeric materials) with Cu(II), where its efficiency is strongly dependent on the synthetic method used in their preparations, are actively being used.

In turn, the macromolecular complexes generated between ligands and transition metal ions have been widely studied because they can model the complexation of metal ions with biological ligands present in the active site of enzymes that can be emulated with carboxylic, hydroxyl, and imidazole groups present in the polymer matrix [9, 10]. Particularly, copper proteins are a group of enzymes which have copper ion as a cofactor. According to the coordination mode in the copper centers, a subclassification is carried out taking into account the geometry of the complex or the number of Cu(II) ions [11].

Nowadays, the use of biohydrogels is preferred to reduce the amount of vinyl or acryl monomers used in the preparation of synthetic hydrogels, and thus to decrease the impact on the environment. However, in general, natural materials are modified with chemical crosslinking molecules to enhance their mechanical strength and applications. Furthermore, it is interesting to obtain biomimetic Cu(II) systems that are resistant to the attack by free radicals and adverse conditions of pH and temperature, which make their recovery and reutilization possible [12–15].

2. Synthesis of polyelectrolyte and polyampholyte hydrogels

Ionic polymers contain covalent and ionic bonds, the latter being responsible for their acid–base and coordination properties. This class of materials is divided into two groups: polyelectrolytes, which have anionic or cationic groups, and polyampholytes, which contain both groups. For polyampholytes, the charged or ionizable groups can be located in the same or different monomer units. From the chemical point of view, polyampholytes are copolymers consisting of weak acidic and basic monomers or strong acidic and basic monomers as well as combinations of them, where the net charge and the charge distribution along the polymer chain is mainly controlled by pH changes occurring in the solution wherein the polymeric matrix is dissolved or swelled depending on the soluble behavior of the material. The net charge and the effect of the different function groups in terms of acid–base properties can be studied through potentiometric titration and the determination of the Z-potential in different conditions [16]. The first polyampholytes with weak basic and acidic groups were synthesized from acrylic acid (AA) or methacrylic acid (MAA) and 2-vinylpyridine (2VP) in the 1950s by the research group of Morawetz and Katchalsky as indicated below (AIBN: azo-bis-isobutyronitrile) (**Scheme 1**) [17]:

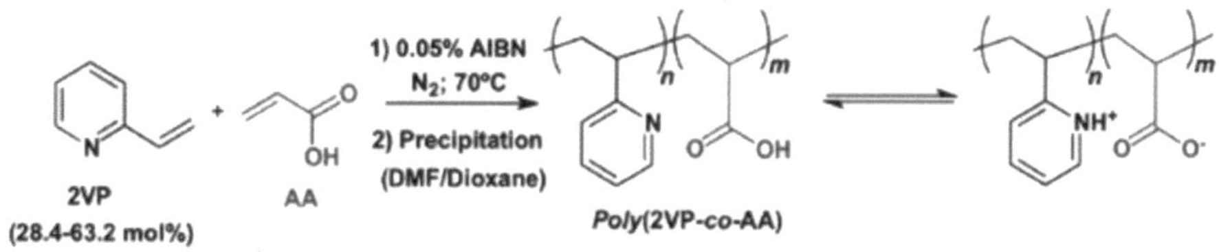

Scheme 1. Synthesis of polyampholytes materials from acrylic acid (AA) and 2-vinylpyridine (2VP).

Also, the vinyl pyridine compound can be replaced by any other vinyl basic monomers such as diethylamino-ethyl methacrylate. Using the same strategy, polymers containing sulfonic acids with vinyl or styrene residues with *N*-substituted allylamines were obtained to generate polyampholytes with strong acidic and basic groups [3]. Usually, the radical copolymerization of the acidic or basic monomers results in polymers with a statistical distribution of molecular weight due to the different reactivity of the monomers used. A classic example is the copolymerization of 2VP and MAA. The ability to ionize or to form hydrogen bonds between the monomers can greatly affect the copolymerization reaction [3].

More recently, Annenkov et al. [18] noted that the polymerization of AA or MAA with vinyl imidazole leads to a polymer contaminated with free imidazole monomers due to acid–base interactions between them [18]. This reaction leads to a deviation from the classic mechanisms of polymerization because some of the monomers are coordinated with polymer chains, decreasing the concentration of the monomers and stimulating competing reactions, such as template polymerization. Regarding this point, our research group obtained polyampholyte hydrogels from MAA, ethylene glycol diglycidyl ether (EGDE, a diepoxy compound), and imidazole (IM) or 2-methylimidazole (2MI) monomers (*poly*(EGDE-MAA-IM or 2MI)).

However, the synthetic yield was lower than that of those obtained in the synthesis of related polyelectrolyte *poly*(EGDE-MAA) and *poly*(EGDE-2MI) materials (**Scheme 2**) [16, 19].

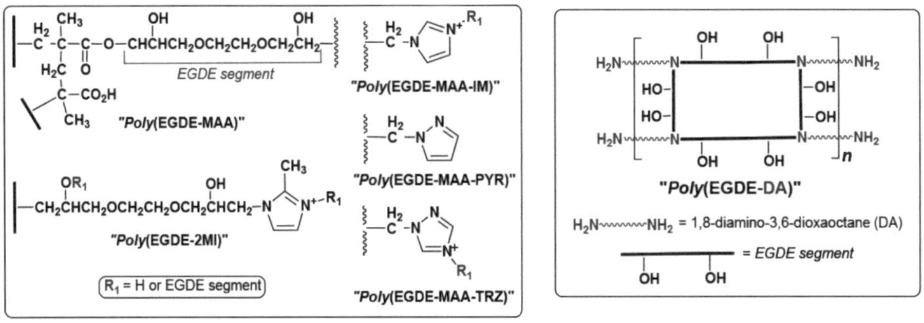

Scheme 2. Chemical structures of *poly*(EGDE-MAA), *poly*(EGDE-2MI), *poly*(EGDE-MAA-IM), *poly*(EGDE-MAA-PYR), *poly*(EGDE-MAA-TRZ), and *poly*(EGDE-DA).

Chromatographic methods led to the observation that during the evolution of the synthetic procedure of *poly*(EGDE-MAA-2MI), a new chemical compound (X) arose and remained through the synthesis with the polymer material (TLC Silica gel 60, Cl_2CH_2:CH_3OH 9:1, $R_{f\,X}$: 0.2, $R_{f\,EGDE}$: 0.8, $R_{f\,2MI}$: 0 and $R_{f\,MAA}$: 0.5). The liquid ^{1}H-NMR spectrum in Cl_3CD shows that the isolated compound, through chromatographic methods, was an ionic pair formed between 2MI (p*K*a = 7.18) and MAA (p*K*a = 4.66) monomers during the reaction. However, even when the yield of the reaction was low (45–50%) for *poly*(EGDE-MAA-IM or 2MI), it was possible to obtain functionalized polymer materials in only one simple synthetic step where IM and MAA residues were attached to a same polymer backbone with no remaining free IM molecules since all the monomers used in its synthesis were removed with acetonitrile solvent and alkaline solution during the washing step. Three parallel reactions take place in these polymer materials: radical polymerization of MAA, which gives linear segments, such as *poly*(MAA), condensation between EGDE and MAA monomers, and reaction of some epoxy units with IM, which results in N_1-substituted IM units. These N_1-substituted azole units may react with other oxirane rings to give rise to N_1,N_3-disubstituted IM units. Furthermore, linear *poly*(MAA) segments interact with azole moieties of the EGDE-IM fragments, giving some kind of interpolymeric complex, which results in an interesting system with interpenetration of linear and crosslinked polymers with application in different areas [19]. Based on this observation, our research group prepared new hydrogel materials from triazole (p*K*a = 2.39) and pyrazole (p*K*a = 2.5) molecules with a synthetic yield of around 80–90%, since these azole compounds have the advantage of not forming an ionic pair with MAA (**Scheme 2**) [16].

This evidence indicates that the best strategy to carry out the synthesis of copolymers by radical polymerization of vinyl monomers with acid–base properties is the use of the sodium salts of the corresponding acids [18]. In this way, the reproducibility of the chemical reaction is achieved, and the sodium salts of the carboxylic acids are readily obtained after precipitation from ethanolic solutions, by the addition of equimolar amounts of sodium hydroxide. Another alternative is to carry out the synthesis of soluble polyampholytes in aqueous media using the copolymer prepared from 2-(1-imidazolyl)ethyl methacrylate (ImEMA) and tetrahydropyranyl methacrylate (THPMA), which is subsequently unprotected in the final step to generate the polyampholyte matrix as indicated below (MTS: 1-methoxy-1-trimethyl-siloxy-2-methyl-1-propene, TBABB: tetrabutylammonium bibenzoate) (**Scheme 3**) [20]:

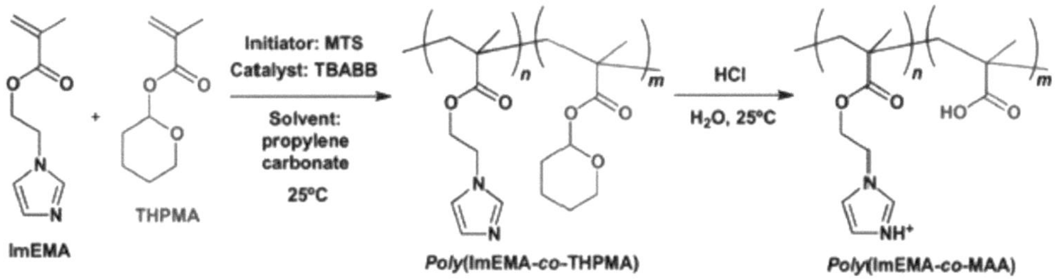

Scheme 3. Modern strategy to synthesize polyampholytes materials carrying both imidazole and carboxylic acid groups.

Significantly, these procedures require the use of monomers where the acid function is protected. The combined use of imidazole derivatives and carboxylic acids allows obtaining polymers with targeting application, such as catalysts, resins, and ion exchange matrices for solid phase extraction, which are widely applied in analytical and organic synthetic chemistry. These are materials that respond to changes in pH and ionic strength of the medium, being suitable as matrices for the uptake of inorganic and organic compounds and for controlled release of drugs and proteins [21]. Furthermore, the coordination processes may be studied during the uptake of Cu(II) ions in polyampholyte systems bearing 2-methyl-5-vinylpyridine and AA due to the catalase-like activity of the Cu(II) hydrogel toward H_2O_2 decomposition [9].

In turn, azole heterocyclic systems are interesting systems given their wide distribution in synthetic and natural compounds. In particular, our research group synthesized macromolecules containing imidazole in the structure due to the catalytic activity of this heterocyclic compound in a wide range of hydrolytic enzymes. The imidazole ring is present in most of the enzymes as part of the histidine amino acid residue, being partially responsible for catalytic activity with synergistic effect of other groups in the active site such as carboxylic acid, hydroxyl, or sulfhydryl residues. In addition, these imidazole-containing polymers have been used as anticorrosion agents [22], for protein separation [23], and as models to understand the biological activity of proteins involved in Alzheimer's disease or prion infections [24]. Furthermore, the absence of toxicity in some of these materials makes them good candidates for their use in the engineering of artificial tissues. With this purpose, Casolaro et al. synthesized polyampholytes containing methacrylate-modified L-histidine (MA-His) as outlined below (TEA: triethylamine, EBA: *N,N′*-ethylenebis-acrylamide, and APS: ammonium peroxodisulfate) (**Scheme 4**) [25]:

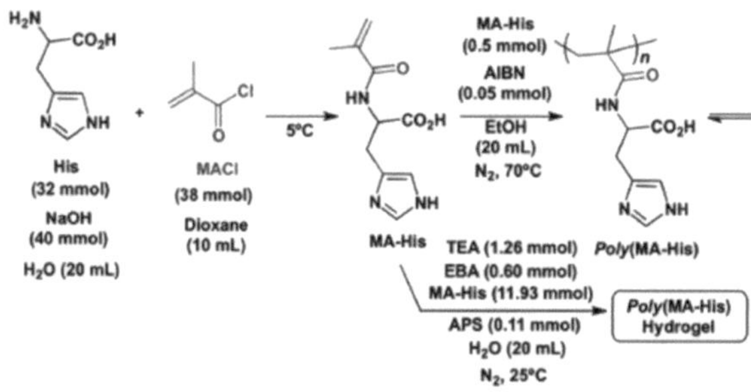

Scheme 4. Synthetic strategy for the synthesis of Poly(histidine) hydrogel materials.

The *Poly*(histidine) hydrogel is also currently being studied for its use in the controlled release of bisphosphonates to improve the bioavailability of these active drugs, together with the benefit that the polymer material is not toxic to osteoblasts [26].

Continuing the development of new polymeric materials, monomers containing epoxy groups in its structure are widely used to cause thermosetting epoxy resins, given the high degree of crosslinking that can be obtained as the curing agent used. Furthermore, the amino or carboxyl residues present in the different monomers have the chemical ability to produce the opening of epoxy groups. The reason why the imidazole molecules are highly effective for use as curing agents when added to epoxy systems is due to the fact that they can catalyze the homopolymerization of epoxide groups via a *poly*-O-etherification mechanism that leads to thermostable materials [27]. In particular, the hydrogel *poly*(EGDE-IM or 2MI) can be synthesized upon the opening of the epoxy group in the EGDE molecule by the imidazole molecules ("A") followed by a proton migration ("B"); then, an N_1-substituted imidazole unit produces the opening of another oxirane ring, allowing the *poly*-O-etherification mechanism between the alkoxide intermediate species and EGDE molecules ("C"), as indicated below for any diepoxy (DE) compound and imidazole (**Scheme 5**) [16, 27]:

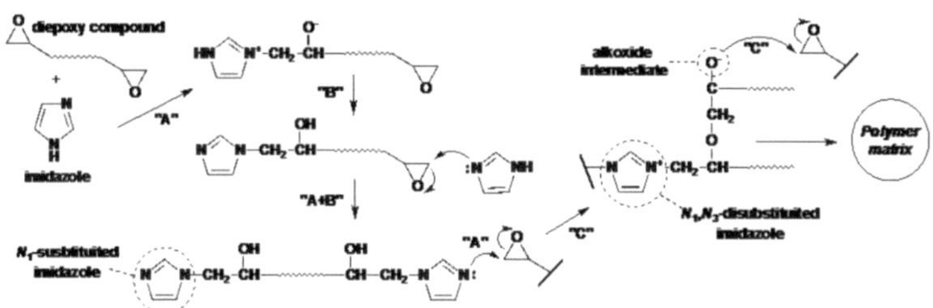

Scheme 5. Synthetic strategy for the synthesis of hydrogel materials containing imidazole groups.

http://dx.doi.org/10.5772/62902

In this way, our research group prepared an interesting polyelectrolyte hydrogel *poly*(EGDE-IM) (**Scheme 2**) bearing 55% of N_1-monosubstituted imidazole units able to be protonated or neutral, depending on the pH of the contact solution, and 45% of N_1,N_3-disubstituted imidazole units with a permanent positive charge in all the pH range, according to the measurement of the zeta potential (**Figure 1**) and potentiometric titration [16]. In contrast, the hydrogel *poly*(EGDE-MAA-IM) has a positive charge at pH values lower than 8.0 and a negative charge at pH values higher than 8.0. Thus, the polyampholyte hydrogels *poly*(EGDE-MAA-IM or 2MI) with an isoelectric point of 8.0 present a high loading capacity for bovine serum albumin (730 mg g^{-1}), together with a good desorption profile, which makes these materials acceptable for eventual applications in formulations for controlled release of proteins [19]. Even when the chemical nature of both hydrogels is quite different, the maximum loading capacity (q_m) values for copper ion uptake are around 60–70 mg g^{-1} due to the presence of the imidazole ring as the most active ligand in the coordination of the metal ion, since the *poly*(EGDE-MAA) material bearing hydroxyl and carboxyl groups takes up only 1 mg of copper per gram of polymer [10]. In the case of *poly*(EGDE-MAA), the zeta potential was zero at pH values below 4, becoming negative at pH higher than 4, as expected for a weak polyelectrolyte (**Figure 1**) [16].

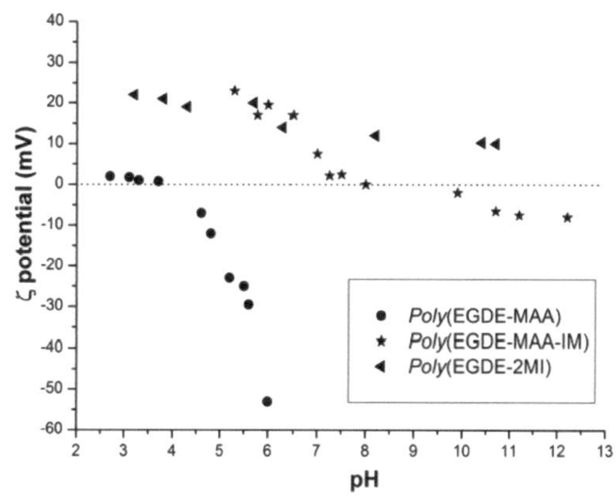

Figure 1. Effect of pH on the zeta (ζ) potential for the indicated materials [14, 16]. ζ potential measurements were performed with a zeta potential analyzer ZetaPlus from Brokhaven Instruments Corporation at 25°C and constant ionic strength of 10^{-3} M KCl. Each polymer suspension (0.25 g L^{-1}) was dispersed and shaken on a magnetic stirrer. The pH was adjusted using KOH or HCl, and the pH of the final supernatant was measured before and after the ζ potential measurements. The optical unit contains 35 mW solid-state laser red (660 nm wavelength). ζ potential was measured using a 16 V cm^{-1} electric field, 15 mA current, and 21 count times.

More traditional strategies for the design of macroporous ion exchange resins consist in the radical polymerization of glycidylmethacrylate (GMA) and ethylene glycol dimethacrylate (EGDMA) since the resulting material, *poly*(GMA-*co*-EGDMA), is versatile to be functionalized with amine groups [28]. In this sense, the group of Driessen immobilized azole ligands in *poly*(GMA-O-*co*-EGDMA) and in the sulfur derivative *poly*(GMA-S-*co*-EGDMA) after the opening of the oxirane and thiirane ring, respectively, give rise to different hydrogels with varied chemical structures (**Scheme 6**) [29]. The uptake of Cu(II) ions indicated that the maximum loading capacities of these materials were 25 and 39 mg per gram of material for IM-SH and TRZ-SH hydrogels, respectively (**Scheme 2**). Then, the authors also synthesized similar resins by using bis-imidazole derivative compounds as a new class of azole compounds to increase the density of coordination sites. These materials showed that the uptake of Cu(II) ions was around 42 mg per gram of material, without a significant increase in the q_m compared to the above-mentioned resins containing imidazole, pyrazole, triazole, and tetrazole. However, highly Cu(II)-selective resins were obtained (**Scheme 6**) [30].

Finally, another strategy to synthesize hydrogels is the use of diepoxy (DE) with diamine (DA) compounds, which gives rise to hydrogels with high capacities for metal ion uptake. In this way, the linear chain containing --DE-DA-DE-DA-- is crosslinked by free DE molecules since the nitrogen site in the linear polymeric chain can still produce the opening of new oxirane rings [28]. In particular, *poly*(EGDE-DA) (DA = 1,8-diamino-3,6-dioxaoctane) presents a q_m for copper of 151 mg g^{-1} (**Scheme 2**) [12].

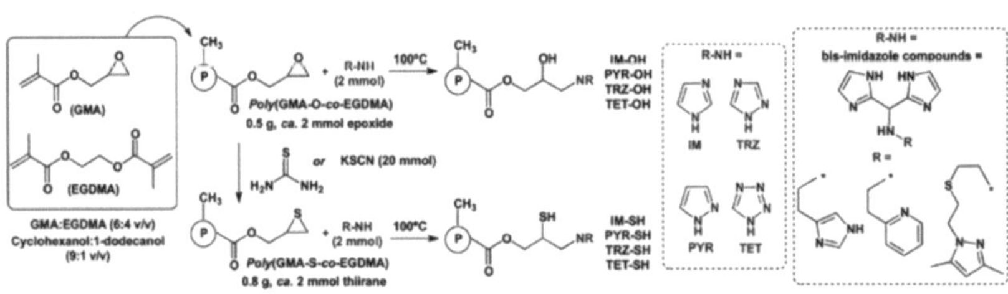

Scheme 6. Azole-modified oxirane and thiirane resins.

2.1. Natural materials

Regarding biopolymers, chitosan is one of the most widely used for the treatment of wastewater containing heavy metal ions as well as starch and cellulose. Chitosan is a *poly*(D-glucosamine) with a variable content of acetylation of the amine group since it is obtained from the deacetylation of chitin (*poly*(*N*-acetylglucosamine)). The problem that arises with the use of chitosan is that it is partially soluble in acidic solution. For that reason, to increase its chemical stability and resistance to acidic and alkaline medium as well as to increase its pore size, mechanical strength, and biocompatibility, it is chemically crosslinked with different molecules such as EGDE, glutaraldehyde, and epichlorohydrin [31]. It is important to point out that although the chemical stability is increased with the crosslinking, the q_m is reduced due to the chemical modification of the reactive sites (amino and hydroxyl groups) involved in the uptake of different metal ions. Particularly, the q_m capacity for copper ion in chitosan is 80 mg g^{-1}, whereas that for the modified chitosan with glutaraldehyde, epichlorohydrin, and EGDE is 60, 62, and 45 mg g^{-1}, respectively. Also, once these hydrogels are saturated with copper ions, they can be easily removed with EDTA solution or with acidic solution as in other Cu(II) polymer complexes and be reused many times.

Another natural polymer source to create hydrogels for the removal of pollutants together with the immobilization of enzymes is cellulose. Cellulose is a linear polymer of (1→4)-β-O-

http://dx.doi.org/10.5772/62902

glucopyranose linkage between the glucose units and with the same chemical modifications as in chitosan. Cellulose can be grafted with acrylamide or AA, increasing the partition coefficient and retention capacity of the hydrogel if the grafted cellulose is hydrolyzed. The graft polymerization of acrylamide onto cellulose presents a metal ion uptake of around 80–90% for chrome, manganese, nickel, and lead [32].

3. Characterization techniques

3.1. Nuclear magnetic resonance (NMR)

Concerning the characterization of the chemical structure of the synthesized hydrogels or any other polymeric compound, only a few spectroscopic methods, such as FT-IR, Raman, and NMR, may bring substantial chemical information. Particularly, the non-soluble behavior of these compounds prevents studying them through the common spectroscopic techniques used in the structural elucidation of soluble organic compounds. Thus, solid-state NMR (ss-NMR) experiments are used to analyze in detail the chemical structure of hydrogels and non-soluble materials in general. This technique will be briefly explained below.

Conventional liquid or solution state ^{1}H- and ^{13}C-NMR spectra are formed by narrow and well-resolved signals containing molecular information that can be interpreted by chemists. However, similar experiments performed in solid samples produce very broad signals, which can be up to several kHz or MHz, which prevents obtaining accurate information by direct observation of the spectra. This broadening also implies loss of sensitivity, especially when low-abundance nuclei such as ^{13}C (1.1%) are studied. The difference in the form of solid and liquid lines comes from the different mobility of the molecules. In the liquid state or in solution, molecules are reoriented very quickly, averaging anisotropic interactions, whereas in solid samples, this does not occur. Thus, special techniques should be applied to obtain high-resolution spectra of solids.

To resolve the structures of complex molecules, mainly chemical shifts (δ) together with the scalar coupling (*J*) are used. This information is obtained from NMR experiments in solution. For solid-like powder samples, these parameters are masked due to the presence of heavy anisotropic interactions such as dipole coupling and chemical shift anisotropy, which cause the widening of the signals between 0 and 50 kHz, unlike 5 kHz and 0–200 Hz for the δ and *J* values obtained in the liquid state, respectively [33, 34].

In fact, for spins $I = ½$, the Hamiltonian (H) can be expressed as follows: $H = H_z + H_j + H_d + H_{cs}$, where H_z represents the Zeeman nuclear spin interaction with the applied magnetic field, H_{cs} is the chemical shift interaction arising from the magnetic fields induced by electrons, H_j is the scalar coupling interaction (*J*) between linked nuclei or through related chemical bonds, and H_d is the dipolar coupling interaction for each nuclear spin. In homonuclear and heteronuclear coupling, the Hamiltonians depend on the orientation of the molecule with respect to the direction of the external magnetic field. Generally, solid powder-like samples contain various crystals with random orientations where the anisotropic interactions produce a characteristic pattern for solids since all the different molecular orientations in the sample will result in different absorption values [34]. Therefore, the information is inaccessible due to the lack of resolution of the spectrum, and it is necessary to use special techniques to increase the resolution. This is the difference with NMR in the liquid state, where the rapid movement of the molecules causes the average of the anisotropic interactions to zero to yield the isotropic chemical shift. In the solid state, the frequency of absorption for a particular crystal has spatial dependence, but all interactions have the same spatial dependence for the second term (ω) described below: $\omega \propto \frac{1}{2}(3\cos^2\theta - 1)$, where θ represents the angle between the main axis of the interaction of the tensioner and the static magnetic field B_o. This spatial dependence can be used for our convenience to obtain a high-resolution ^{13}C NMR spectrum in the solid state for example. In the 1950s, the pioneers Lowe [35] and Eades [36] showed that the widening caused by the anisotropic interactions could be averaged to zero when the sample is physically rotated about an angle θ = 54.74°. This angle was called "magic angle" (θ_m) because the spectrum obtained had narrow signals that resembled those obtained in solution. From its discovery, this technique was known as rotation at the magic angle spinning (MAS), and it is widely used in experiments in the solid state. The routine rotation speeds are between 10 and 30 kHz and up to 80–100 kHz with technological advances that have been made in this area in the last years. This is why when a sample is rotated at a spinning rate greater than the anisotropic interaction, at the magic angle, all crystals seem to have the same orientation, and the θ_m dipolar interactions are averaged to zero, reducing the width of the signals. Unfortunately, this technique is not perfect, since such widths for each signal ^{13}C resonance frequency are around 50 Hz in the solid state, but this also depends on the characteristics of the samples, being narrower when the sample is crystalline than when it is amorphous, like hydrogels.

Additionally, the sensitivity of X nuclei that exhibit low isotopic abundance, such as ^{13}C, ^{15}N, or ^{29}Si, which are typical nuclei studied through ss-NMR, can be improved by increasing the signal through the transfer of the magnetization of the abundant nuclei (^{1}H in general) toward the X nuclei. In solids, this technique is known as cross-polarization (CP) [37], which is optimal under certain radiofrequency field, known as the Hartmann–Hahn condition. This condition is described as follows: $\gamma_H B_1$ (^{1}H) = $\gamma_X B_1$(X), where γ corresponds to the gyromagnetic ratio and B_1 to the field of spin lock for each nucleus, respectively. The result of combining CP with MAS (CP-MAS) is the right strategy to obtain a high-resolution spectrum in the solid state, with adequate sensitivity for different nuclei. Therefore, it is a very valuable tool used to characterize and study insoluble materials, heterogeneous catalysts, polymorphic compounds, and pharmaceutical formulations in the solid state.

The presence of compounds containing naturally abundant nuclei, such as protons, usually generates strong interactions (either homonuclear or heteronuclear between the ^{1}H and ^{13}C, ^{15}N, ^{29}Si, and ^{31}P). This leads to a broadening of the signals that cannot be completely averaged, making it necessary to carry out additional techniques, including homo- and heteronuclear decoupling sequences of radiofrequency pulses to average residual dipolar interactions under MAS conditions. The ^{1}H-^{13}C decoupling sequences most commonly used are SPINAL-64, Two-Pulse Phase-Modulation, and continuous-wave among others, used in the area of polymeric materials or pharmaceutical compounds [38]. In summary, the combination of MAS, CP, and heteronuclear decoupling is used for the acquisition of the ^{13}C-NMR experiment which is referred to as solid-state ^{13}C CP-MAS (**Figure 2D**).

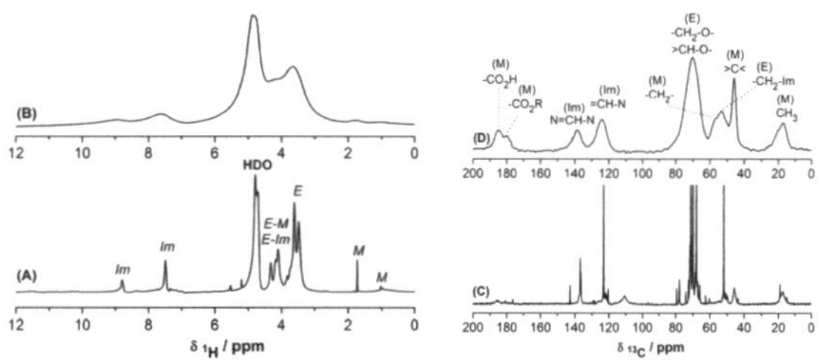

Figure 2. ^{1}H HRMAS spectra of 600 MHz for *poly*(EGDE-MAA-IM) acquired in a 4-mm HRMAS probe swelled in D_2O at a spinning rate of 4 (A) and 0 kHz or "static conditions" (B). 150 MHz ^{13}C-NMR spectrum for *poly*(EGDE-MAA-IM) acquired in a 4-mm HRMAS probe swelled in D_2O at a spinning rate of 4 kHz (C) and ^{13}C CP-MAS in a 3.2 mm ss-NMR probe at a spinning rate of 10 kHz (D). The assignment of the NMR signals corresponds to imidazole (Im), polymerized MAA (M), and EGDE (E) segments [16, 39]. All the ss-NMR experiments were performed at room temperature in a Bruker Avance III HD Ascend 600 MHz spectrometer.

In addition, the ^{13}C CP-MAS spectra can be edited to assist in the assignment of the NMR signals observed. Some of the edition techniques frequently used are the cross-polarization with polarization inversion (CPPI) [40] and non-quaternary suppression (NQS) experiments [34]. In the former, the quaternary (>C<) and methyl ($-CH_3$) carbons remain as positive signals, the methylene carbons are negative or have inverted signals, and methyne (>CH-) carbons remain in the baseline without being observed. In NQS experiments, only the quaternary and methyl carbons are visualized.

Chemists are also interested in obtaining quantitative information related to the monomer composition in copolymer materials or in any modification made in the polymer structure. However, this information is difficult to obtain because it is necessary to apply ^{13}C direct polarization techniques (^{13}C DP), which require the use of the NMR spectrometer for a long time, and because not all the polymeric powders give rise to an adequate signal-to-noise ratio. In contrast, ^{13}C DP spectra can be very useful to provide quantitative information related to the crystalline and amorphous amounts in crystalline or semicrystalline polymers. For instance, our research group studied semicrystalline *poly*(ethylenimine) hydrochloride systems ($[-CH_2-CH_2-NH(HCl)-]_n$) due to the different environments of the crystalline and amorphous domains where the polymeric segments in the crystal lattice are confined. We observed two well-resolved ^{13}C resonance signals at 44.5 and 42.9 ppm, being the signals associated with the amorphous and crystalline regions, respectively [41].

Regarding ^{1}H-NMR spectra, it is necessary to point out that these kinds of simple experiments are not determined as routine in the solid state because the dipolar coupling among protons is higher than the commonly spinning speed obtained by the commercial NMR probes (10–35 kHz). However, partially or well-resolved ^{1}H spectra can be obtained at high spinning rate (>60 kHz) or with particularly high-power decoupling techniques at moderate spinning rate. In general, proton spectra of solid polymer samples consist in wide lines and in some cases an overlapping of wide and sharp lines at static conditions [34].

Fortunately, for hydrogels, there is an experiment called high-resolution magic angle spinning (HRMAS), where the ^{1}H-^{1}H dipolar interactions can be partially averaged, rendering liquid-like ^{1}H-NMR spectra similar to those observed for liquids. In this technique, the material is swelled with deuterated solvents, making it possible to expand the analysis of hydrogel compounds and tissue samples [42, 43]. Another advantage of this technique is that the sample is not spun at high rates because at 2–4 kHz the residual ^{1}H–^{1}H dipolar couplings, which were partially averaged by the swelling produced by the solvent, are eliminated (**Figure 2**). The deuterated solvents that can be used are the same as in the liquid state since the HRMAS probe is designed with a deuterium channel (^{2}H) to lock the NMR signal. The sample is placed in special zirconia rotors with different volumes as in ss-NMR, with the difference that they are designed with cylindrical spacers to contain the sample. In this way, HRMAS experiments allow chemists to study hydrogels or swelled samples as in liquid-state NMR (**Figure 2A** and **B**).

Although all these examples are very interesting, the ss-NMR technique has to achieve some sensitivity limits that make this spectroscopic tool not able to analyze small chemical modifications of materials or encapsulation of molecules among others. However, these sensitivity problems can be partially resolved with the use of dynamic nuclear polarization (DNP) experiments for solid samples [44–46]. In this technique, the powder sample is impregnated with a biradical solution and introduced in a zirconia rotor which will be spun at the magic angle at ~90–100 K. At this temperature, the polarization transfer is increased from the electron of the radical molecules to the proton network of the material under study, allowing the polarization of the spins in the material under study through spin diffusion. To polarize the electron of the radical molecules, microwave irradiation is on throughout the experiment. For protons, the maximum theoretical enhancement achievable is given by the gyromagnetic ratios (γ_{e^-}/γ_{1H}), being around 660.

3.2. X-ray photoelectron spectroscopy (XPS)

Several spectroscopic analyses exist for surface characterization, but the most commonly used to conduct this experiment is based on the irradiation of the surface of the sample with monochromatic X radiation, called X-ray photoelectron spectroscopy (XPS), also known as electron spectroscopy for chemical analysis (ESCA) [47]. This spectroscopic technique allows identifying all the elements of the periodic table, except hydrogen and helium, but more importantly it allows determining the oxidation state of an element and species to which it is attached, thus providing valuable information on the electronic structure of the molecules.

http://dx.doi.org/10.5772/62902

Due to the short penetration power of the electrons, this technique only provides information on a surface layer of a thickness of 20–50 Å. The most important and valuable application of XPS is for the qualitative analysis of the surfaces of solids, such as metals, alloys, polymers, semiconductors, and heterogeneous catalysts. It also allows quantifying each element on the surface, having into account their sensitivity factors.

XPS records the kinetic energy of the emitted electrons after a monochromatic X-ray beam of known energy (hv) affects the surface of the testing sample. This causes the release of an electron from a K orbital with specific energy E_b. The reaction can be represented as follows: $A + hv \rightarrow A^{+*} + e^-$, where A can be an atom, a molecule or an ion, and A^{+*} is an electronically excited positively charged ion. The kinetic energy (E_k) of the emitted electron is measured in the electron spectrometer. In this way, the electron binding energy (E_b) can be calculated by the following equation: $E_b = hv - E_k - w$. In this equation, w is called the work function of the spectrometer, a correction factor of the electrostatic environment in which the electron is formed and recorded. The value of w can be determined by several methods. The E_b is characteristic of an electron orbital of the atom, wherein the electron was released.

As a result, an XPS spectrum is a graph of the number of emitted electrons (or electron beam power) based on E_b. The high background level observed occurs because with each characteristic peak there is a queue due to the ejected electrons that have lost some of their energy in inelastic collisions inside the solid sample. These electrons have less kinetic energy than equivalent non-dispersed electrons and therefore appear at higher binding energies.

Simultaneously, Auger electron emissions are originated throughout the interaction with X-ray. The Auger process emissions are generated during the relaxation of the excited ion A^{+*} after interacting with a beam of monochromatic X-ray photons, where an Auger electron is emitted at the same time as an ion A^{++} is generated on the surface of the material under study. These emissions are described according to the type of orbital transitions involved in the production of the electron (KLL, LMM, or MNN Auger transitions). The applications of XPS in hydrogels and copper–hydrogel complexes will be discussed in the last part of this chapter.

3.3. Other characterization techniques

Another technique for the study of heterogeneous catalysts or hydrogel-containing paramagnetic centers is electron paramagnetic resonance (EPR), also called electronic spin resonance (ESR). This technique is based on the absorption of electromagnetic radiation in the microwave region of a sample with paramagnetic electronic properties which is subjected to a magnetic field. However, this magnetic field is not as intense as in NMR. EPR allows obtaining information about the different geometries adopted by paramagnetic ions when they are part of either biological complexes or biomimetic systems, as well as of heterogeneous catalysts [10, 11]. An important difference with any other spectroscopic technique is that in the EPR spectra, the first derivative absorption line is plotted. The number and intensity of the EPR lines depend on the interaction between the unpaired electron spin (S = 1/2) and the nuclear spin (I, i.e. I_{Cu} = 3/2). The application and examples of EPR experiments will be discussed later in this chapter.

Other complementary techniques for the study of surfaces include the following:

- Scanning electron microscopy (SEM) and atomic force microscopy (AFM), which can provide information on the physical microstructure by images of morphology and topography.
- Energy dispersive X-ray diffraction (EXD) experiments, which allow obtaining a semiquantitative elemental composition of the surface of the material observed by SEM microscopy and analyzing the distribution of a particular element such as spreading of a metal ion in a particular section of interest.
- N_2 adsorption isotherms, which allow measuring the specific surface area by using the nitrogen adsorption isotherm BET [48]. The advantage of this approach is that it allows estimating the surface area and pore dimensions of different materials together with the interior texture of particles.

4. Metal ion uptake equilibria and characterization techniques

To analyze the sorption processes involved in the uptake of different inorganic (metal ions) or organic compounds (dyes, proteins, pollutants, among others), different models can be used, and these will be discussed in this section.

When a gas or solute in a solution affects a solid surface, it can either bounce or remain attached (*adsorbed*). The sorbate can diffuse on the surface, remain attached, undergo a chemical reaction, or be dissolved in the solid (this last process is known as *absorption*). Two different behaviors can be distinguished: physisorption and chemisorption, although intermediate situations are often found.

In physisorption, the molecules remain attached to the surface of the sorbent by means of van der Waals forces (dipolar interactions, dispersion, and/or induction). In chemisorption, the molecules remain attached to the surface, forming a strong covalent binding, and the chemisorption enthalpies are higher than the physisorption enthalpies and are generally exothermic processes that stop after monolayer formation on the surface. Chemisorption involves the breakdown and formation of bonds. This is why the chemisorbed molecule does not preserve the same electronic structure as in the former phase.

The equilibrium of metal ion uptake can be explored by analyzing the results of the adsorption isotherm for the polymeric adsorbent at a given temperature. Several isotherm models are generally used to fit the experimental data of the adsorption of ions or any other molecule on particles by nonlinear regression. These models include the Langmuir adsorption isotherm, the Freundlich equation, the Temkin model, and the Dubinin–Radushkevich isotherm, each of which is described below.

The Langmuir adsorption isotherm is derived from theoretical models, provides information on uptake capabilities, and reflects the usual equilibrium process behavior but does not provide information about the mechanistic aspects of adsorption. The equation is $q_e = \frac{q_m C_e}{K_L + C_e}$,

http://dx.doi.org/10.5772/62902

where q_e is the adsorption capacity in equilibrium with the corresponding C_e (M), which is the concentration of metal ion, q_m is the maximum adsorption capacity, and K_L is the equilibrium constant of dissociation. In some cases, the linear form is preferred to simplify the analysis of the experimental data [31]. The Langmuir equation assumes a homogeneous surface of the adsorbent (a flat surface), a single site per molecule (monolayer), and no interaction between the adsorbed species and adjacent active sites [49]. The change in free energy related to the ion uptake can be calculated from: $\Delta G = -R\ T \ln K_a$, where R (8.314 x 10^{-3} kJ mol^{-1} K^{-1}) is the gas constant, T is the thermodynamic temperature, and K_a (M^{-1}) is the equilibrium constant of association ($1/K_L$). The negative sign of this thermodynamic parameter indicates the spontaneous nature of the reaction.

The Freundlich equation is described as $q_e = K_F\left(C_e^{\frac{1}{n}}\right)$. This model assumes that the surface is heterogeneous in the sense that the adsorption energy is distributed, and the surface topography is patchwise [50]. The sites with the same adsorption energy are grouped together into one patch (the adsorption energy here is the energy of interaction between the adsorbate and the adsorbent). Each patch is independent of each other (i.e., there is no interaction between patches), and the Langmuir equation is applicable for the description of equilibrium of each patch. In fact, this isotherm can be theoretically derived supposing that the surface has different types of adsorption sites. K_F can be defined as the sorption or distribution coefficient and represents the amount of sorbed molecules/ions onto the polymer surface normalized by the equilibrium concentration and $1/n$ is a measure of surface heterogeneity. When the value of n becomes larger than about 10, the adsorption isotherm approaches a so-called irreversible isotherm, because the concentration needs to go down to an extremely low value before the adsorbate molecules desorb from the surface [50].

The Temkin model is empirically derived and assumes that the heat of adsorption (which is a function of temperature) of all molecules in the layer will decrease linearly rather than logarithmically with coverage [51]. It allows estimating the equilibrium constant and the adsorption heat: $q_e = \frac{R\ T}{b_T} \ln(K_T C_e)$.

The Dubinin–Radushkevich isotherm is a semi-empirical equation which was originally developed for subcritical vapors in microporous solids, where the adsorption process follows a pore-filling mechanism: $q_e = q_m e^{-B_D \epsilon^2}$, where q_m is the maximum amount of adsorbate that can be adsorbed in micropores, and B_D is a constant related to the energy.

Liquid-phase adsorption data can also be analyzed by the equation below, where the amount adsorbed corresponding to any adsorbate concentration is assumed to be a Gaussian function of the Polanyi potential (ε): $\epsilon = R\ T \ln\left(1 + \frac{1}{C_e}\right)$, where C_e represents the solute concentration at equilibrium (g solute per gram of solution) [52]. This equation is generally applied to express the adsorption mechanism with a Gaussian energy distribution onto a heterogeneous surface. The approach is usually applied to distinguish the physical and chemical adsorption of metal ions by means of the equation $E = \frac{1}{\sqrt{2B_D}}$. The parameter E is the mean free energy of sorption [53], whose magnitude is a way to estimate the type of sorption process. In the case that E is lower than 8 kJ mol^{-1}, weak physical forces, such as van der Waals and hydrogen bonding, may affect the sorption mechanism. If this value is between 8 and 16 kJ mol^{-1}, the sorption process can be explained by ion exchange. If this value is higher than 16 kJ mol^{-1}, the sorption process can be explained by other chemical reactions such as coordination [54].

The isotherm parameter sets with statistical support can be determined by nonlinear regression, using the algorithm based on the Gauss–Newton method. An error function can be used to evaluate the fit, the second-order corrected Akaike information criterion (AIC):

$$AIC = N\ln\left(\frac{RSS}{N}\right) + 2P$$

$$AIC_c = AIC + \frac{2P(P+1)}{N-P-1}$$

where P is the number of parameters in the model, N the number of data points, and RSS the residual sum of squares [55]. Δ AIC represents the difference in AIC values between two competing models. The associated Akaike weights for the better and worse models are:

$$W_{better} = \frac{1}{1+e^{-\frac{1}{2}\Delta AIC}}$$

$$W_{worse} = \frac{e^{-\frac{1}{2}\Delta AIC}}{1+e^{-\frac{1}{2}\Delta AIC}}$$

The Akaike weights provide information about the strengths of evidence supporting the two competing models. The ratio of the two Akaike weights, W_{better}/W_{worse}, is termed as the evidence ratio and represents the relative likelihood favoring the better of two competing models. As reported by other authors, an evidence ratio greater than 20 would indicate extremely strong evidence favoring the better model [56]. Another fundamental aspect of the uptake of metal ions or organic molecules for industrial application is the knowledge of the kinetic parameters ruling this process and the rate-controlling step of the sorption process. For this reason, the adsorption capacity of the adsorbent is usually studied as a function of time. The sorption mechanism could be controlled either by a chemical reaction or by diffusion processes, such as pore and film diffusion. If the experiments are performed in a well-stirred batch system, the thickness of the boundary layer surrounding the particle should be minimal, and boundary layer resistance or film diffusion should not be major rate-controlling factors [57]. Besides, the particles can have mesopores and macropores, expected to be accessible for different metal ions and small organic molecules in general. If the contact time necessary to achieve equilibrium conditions is short, it might initially indicate that the adsorption of the studied cations is a chemical-reaction-controlled process [58–60]. For example, the *poly*(EGDE-MAA-2MI) hydrogel with polyampholyte properties (**Scheme 1**) was tested as adsorbent for the removal of Pb(II) and Cd(II) from aqueous solutions. The metal ion uptake equilibration end point could be estimated in 10 hours for Cd(II), even if the 80% of the load was reached in less than 3 hours. For Pb(II), the equilibrium was reached in a shorter time. These results evidenced a chemical-reaction-controlled process. In this case, a network expansion was predicted as a result of repulsive interaction between the fixed positive charges. The same kinetics profile was found for Cu(II) and Co(II) ions, with this type of adsorbents [61].

Different kinetic models can be used to fit the experimental adsorption data by nonlinear regression. The Elovich equation is based on a general second-order reaction mechanism for heterogeneous chemisorption processes and is formulated as: $\frac{dq_t}{dt}=\alpha e^{-\beta q_t}$, where q_t is the amount of adsorbed ion at the contact time t [57, 58]. After integration and application of the boundary conditions, for $q_t = 0$ at $t = 0$ and $q_t = q_t$ at $t = t$, the equation becomes [57] $q_t=\frac{1}{\beta}\ln(1+(\alpha\beta t))$. Teng and Hsieh proposed that constant α is the initial adsorption rate, and β is related to the extent of surface coverage and the activation energy involved in chemisorption [62]. This equation assumes that the active sites of the sorbent are heterogeneous in nature and therefore exhibit different activation energies for chemisorption [57]. Another explanation for this form of kinetic law involves a variation of the energetics of chemisorption with the extent of surface coverage [62].

The modified Freundlich model was originally developed by Kuo and Lotse: $q_t=k_F C_o(t^{1/m})$, where k_F is the apparent adsorption rate constant, C_o is the initial ion concentration, and m is the Kuo–Lotse constant [63]. Bache and Williams indicated that the energy of adsorption decreases exponentially with increasing surface saturation when the adsorption fits the Freundlich equation. Everett suggested an increase in the perturbation potential as a consequence of the interactions between the species at close distance [64]. In concordance with this theory, k_F usually decreases as C_o increases. This modified Freundlich model can describe a surface-diffusion-controlled process different from intraparticle diffusion. Instead, when the estimated m parameter is close to 2, the kinetics is controlled by the intraparticle diffusion in the pores, involving the movement of the adsorbing ion along the walls of the less accessible spaces of the adsorbent and the diffusion into the solid itself.

In addition, most metal ion sorption system models in the literature are based on reaction kinetics using pseudo first-order kinetics or pseudo second-order kinetics. The pseudo first-order model suggests that one ion sorbs to one active site. The integrated equation by applying the boundary conditions, for $q_t = 0$ at $t = 0$ and $q_t = q_t$ at $t = t$, is $q_t=q_e(1-e^{-kt})$, where q_e is the amount of cadmium sorbed at equilibrium as well as the equilibrium adsorption capacity of the sorbent, and k is the rate constant. A series of pseudo first-order terms involves different types of binding sites, and in each case, the stoichiometric ratio between the binding site and the adsorbate molecule or ion is 1:1. The pseudo second-order model is described by

$$q_t = \frac{q_e^2 k\ t}{1 + (q_e k\ t)}$$

$$h = q_e^2 k,$$

where *k* is the rate constant of the pseudo second-order equation, and *h* is the initial adsorption rate. The pseudo second-order kinetic model presupposes that each metal ion binds to the active sites on the surface in a 1:2 stoichiometric ratio [57, 65]. The experimental kinetic results of Cd(II) and Pb(II) uptake on *poly*(EGDE-MAA-2MI) hydrogel fitted different models, depending on the initial metal ion concentration. For lower concentration levels, the best model was the modified Freundlich. At higher concentration values, Cd(II) uptake followed the Elovich model, and Pb(II) the pseudo first-order model with two parameters. It could be concluded that the kinetic models that better fitted these results were those that described variations of the energetics of chemisorption with the extent of coverage due to interactions between the involved species [61].

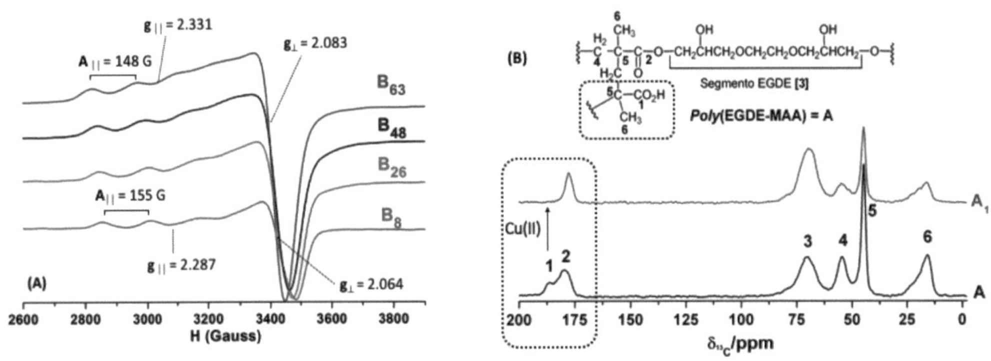

Figure 3. X-band EPR spectra for the solid Cu(II) *poly*(EGDE-MAA-IM) complex (**Scheme 1**) with different copper ion content (*B*x = x mg copper per gram of complex as determined from the direct measurement of the solids through X-ray fluorescence (XRF) in an advant XP+ thermo electron spectrometer) (A) [10]. ^{13}C CP-MAS for *poly*(EGDE-MAA) and its Cu(II)-complex (**Scheme 1**) with 1 mg Cu(II) $g^{-1} = A_1$ (B) [66]. EPR measurements of the Cu(II)-complexes were performed at X-band on a Bruker EMX plus spectrometer at 20°C. The ss-NMR experiments were performed at room temperature in a Bruker Avance-II 300 MHz spectrometer. The different Cu(II) complexes were obtained from the adsorption of Cu^{2+} ions on the different polymers by batch adsorption experiments. Each material (0.1000 g) was placed in contact with 2 mL of cupric sulfate ($CuSO_4$) solution in a 4–100 mM concentration range. The resulting suspensions were shaken in a thermostatic bath at 25°C for 24 hours. Then, the samples were centrifuged, filtered, and dried at 60°C for 24 hours.

Although all these analyses allow the complete characterization in terms of physicochemical aspects, they do not allow obtaining information related to example with the ligands involved in the uptake of metal ions. For that reason, complementing them with spectroscopic techniques can bring important information related to chemical aspects associated with the

http://dx.doi.org/10.5772/62902

coordination process in different conditions. Our research group, for example, studied the coordination sphere of copper ions in the synthetic Cu(II) hydrogel complexes by using a combination of ss-NMR and EPR techniques [10, 66]. The EPR spectra of isolated Cu(II) centers with a minimum distance of 10–15 Å between each paramagnetic nucleus provide more information because the values of parallel hyperfine coupling constant ($A_{||}$) and parallel *g*-factor ($g_{||}$) obtained from the hyperfine structure of the Cu(II) complexes differ depending on the geometry and the ligands of the paramagnetic entity (**Figure 3A**). In general, the EPR parameters show that when oxygen becomes more active in the uptake of copper ions, the $A_{||}$ values decrease, and the $g_{||}$ values increase as the amount of ions in the polymeric structure increases. However, EPR spectroscopy must be used together with ss-NMR or any other spectroscopic technique due to the broad range of materials and chemical compositions, because the elucidation of the chemical sphere of Cu(II) centers based on EPR does not result in a clear-cut result [10]. For Cu(II) proteins and any other Cu(II) materials, the Peisach–Blumberg plots can be very useful to access to the coordination sphere of the paramagnetic center [10, 67]. On the other hand, the ss-NMR technique retrieves information of the ligands involved in the coordination of paramagnetic ions due to the enhancement in the relaxation behavior of the different nuclei present in the polymer matrix (^{1}H, ^{13}C, ^{15}N, etc.), avoiding the detection of resonance signals. In particular, ^{13}C CP-MAS experiments are effective to study the preferences in the uptake of Cu(II) ions at different concentrations of the metal ions in polyelectrolyte and polyampholyte materials. However, some experimental conditions, such as the contact time used in the cross-polarization step in ^{13}C CP-MAS experiments, must be taken into account before setting the acquisition parameters. For example, for a Cu(II) hydrogel (*poly*(EGDE-MAA-IM)) bearing carboxylic acid as ligand for copper ion, it is possible to observe how the signal corresponding to the carbon of these groups is affected after the uptake of the paramagnetic ion and is thus not detected (**Figure 3B**) [10, 66]. This strategy to study the coordination of copper ions can also be used to understand the uptake of zinc [68], mercury [69], cobalt [13], and samarium [70] ions in polymer matrixes.

The uptake of Cu(II) ions allows that different ligands distributed in the same or different polymeric chains of the adsorbent material participate in the coordination process, modifying the polymer properties. Particularly, the glass transition temperatures (T_g) are increased as a consequence of the crosslinking induced by the metal ion [12, 68, 71]. However, the presence of metal ions produces a lower thermal stability of the complex, as a result of the weakening in the chemical bonds after the coordination of the metal ion that produced changes in the electron density [10, 66]. For example, the T_g value obtained for the polyelectrolyte hydrogel *poly*(EGDE-DA) (**Scheme 2**) was 120.0°C, while that for the Cu(II) hydrogel complex containing 131 mg of Cu(II) per gram of polymer was 169.5 °C. This shift to higher temperature values was due to the stabilization of the *d*-electron of the copper ion on coordination, as expected, and it was also another fact of the crosslink between the polymer chains and the copper ion. For the Co(II) hydrogel complex, the T_g value was 134.1 °C, and it was lower than in the case of copper, in concordance with the low amount of cobalt ions adsorbed to the polymer material (23 mg of Co(II) per gram of polymer), reducing the crosslinking between the polymer chains and cobalt ions [12].

Finally, the polymeric particles loaded with cations were cut into slices to see if these compounds had access to the entire network or were only attached to the surface. In all the cases, they reached the interior binding sites of these particles, and it was impossible to distinguish a core from the exposed surface. These results, together with the swelling properties of these materials, confirm the fact that they are hydrogels. The water molecules and solutes have access to the bulk of the swelled particles. The materials absorb water and dissolved ions. This was demonstrated by the fact that "water adsorption surface" determined for *poly*(EGDE-MAA-2MI) and its copper complex was 287 and 346 $m^2 g^{-1}$, respectively, clearly indicating the absorption of water molecules [12, 16].

5. Environmental application of hydrogels and metal ion hydrogel complexes

Both natural and synthetic polymeric materials can be used to remove organic and inorganic pollutants given their high load capacities associated with the functional groups present in their chemical structures. Adsorption is the easiest method to remove substances from effluents, but a second residue is generated because pollutants are adsorbed to the material used. Specifically, hydrogels can be used to concentrate different kinds of industrial dyes. Regarding inorganic contaminants adsorbed in different materials, heavy metal ions can be desorbed under acidic conditions from the matrix where they are retained. In these conditions, the metal ions are concentrated, which allows their recovery. Some of the materials that can be used for metal ion uptake have been described throughout this chapter.

On the contrary, organic pollutants can be adsorbed in the first instance, and then oxidative methods where ROS are generated from H_2O_2 can achieve the partial or complete mineralization to CO_2 and H_2O. These catalytic systems are usually called advanced oxidative technologies for wastewater treatment, and some of them will be covered in this section.

Hydrogen peroxide (H_2O_2) is a powerful oxidant used in the degradation of pollutants combined with catalysts and/or UV light to give rise to reactive species such as hydroxyl radical. With respect to the oxidation of aromatic hydrocarbons, Fenton-like reactions (Fe(III)/H_2O_2) are widely used but are only effective in acidic conditions [72]. In contrast, Cu(II)/H_2O_2 systems can be used for similar purposes but in a broader pH range. In turn, the catalytic

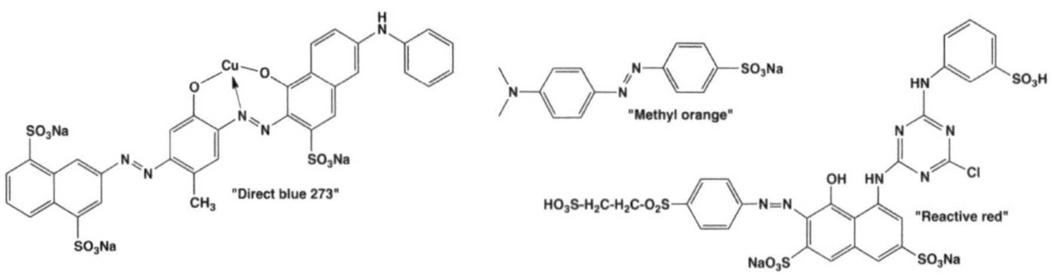

Scheme 7. Chemical structures of some dyes used in the textile industry.

activity of the copper ion in the activation of H_2O_2 with concomitant generation of hydroxyl radicals is enhanced after coordination with pyridine, organic acids, and other chelating agents, but the recovery of these complexes is expensive because they are soluble in water. For these reasons, heterogeneous catalysts are an attractive alternative for decolorization, combining effectiveness, ease of recovery, and reuse potential. Also, transition metals supported on alumina and silica have proved to be more efficient for the activation of H_2O_2 than homogenous catalysts. In addition, complexes of Cu(II) with alumina or chitosan have been used to remove color from industrial waste [73, 74], Cu(II) complexes immobilized on silica particles have been used as catalysts in the hydroxylation of phenols from H_2O_2 [75] and non-soluble Cu(II) chitosan/H_2O_2 systems have been used to degrade anthraquinone and azo dye compounds commonly found in the textile industry [76]. Textile dyes are considered the most common industrial pollutants in waters. In addition, modern dyes are stable to the ineffective conventional treatment methods performed on wastewater (**Scheme 7**). This results in an intensely colored discharge that is released from the factory with a direct negative impact on the environment. Both Cu(II) chitosan and any other modified chitosan coordinated with copper ions can act as efficient heterogeneous catalysts for the activation of H_2O_2, in which radical species (mainly hydroxyl radicals, OH•) are generated, and dyes and other organic pollutants are concomitantly degraded to CO_2 and H_2O. This process is highly dependent on the efficiency in the oxidative capacity of the catalyst, susceptibility of the pollutant, and the amount of H_2O_2 added [76].

However, even when these systems are very effective, some precautions must be taken into account to ensure the catalytic activity. To not affect the structure of the catalyst, the initial concentration of H_2O_2 ($[H_2O_2]_o$) needs to be controlled. As an example, synthetic Cu(II) *poly*(EGDE-MAA-IM or 2MI) hydrogels are stable up to a $[H_2O_2]_o$ = 50 mM since higher concentrations affect the structure of the catalyst (**Scheme 2**) [14, 15]. In addition, at higher $[H_2O_2]_o$, the H_2O_2 molecules combine with OH• radical species to generate superoxide radicals ($O_2^{\bullet -}$) with lower oxidation potential. Moreover, the use of high concentrations of H_2O_2 produces its self-decomposition to O_2 and H_2O [15].

There are some experimental techniques that can be done to do an exhaustive characterization of each Cu(II)-supported material or any other metal complex. Once the coordination behavior and the amount and distribution of the metal ion have been studied, the activation of H_2O_2 can be explored. However, it is also useful to explore the stability of the metal complexes in the experimental conditions in which the catalyst will be used (pH, ionic strength, temperature, etc.).

The reaction with 4-aminoantipyrine is usually used for the detection of free radical species and can be tested in different samples as an easy screening [15, 77]. Nevertheless, it is not able to discern which free radical species are generated through the activation process of H_2O_2 on the catalytic surface. For the correct identification of the radical species generated in diverse Cu(II) complex/H_2O_2 systems, EPR measurements should be made with a spin trap molecule. In this way, a stable radical is formed from the incubation of 5,5-dimethyl-1-pyrroline *N*-oxide (DMPO) solution with a sample obtained from the reaction mixture (Cu-hydrogel/H_2O_2) where the radical (R•) species are generated, thus allowing its detection and identification through EPR spectroscopy. The EPR spectra obtained depend on the R• nature in the medium since the spectral lines characterize each DMPO-R adduct. As an example, **Figure 4** shows the EPR spectrum obtained by the reaction of DMPO (1 M) with an aliquot from the supernatant of a mixture reaction containing 10 mg of Cu(II) *poly*(EGDE-MAA-IM) hydrogel (q_m Cu(II): 60 mg g^{-1}), $[H_2O_2]_o$ = 25 mM and a final volume of phosphate buffer of 10 mL (pH = 7.0).

The Co(II)-*poly*(EGDE-DA)/H_2O_2 heterogeneous system also produced free radicals that diffused to the solution and were detected by spin trapping experiments with DMPO ($[H_2O_2]_o$ = 63 mM) (**Figure 4E**). The simulation and fits of the experimental spectrum allowed establishing the presence of two species: DMPO-OOH (93%) and DMPO-OH (7%) adducts from anion superoxide ($O_2^{\bullet -}$) and hydroxyl radical (OH•), respectively [12]. In addition, for the coexistence of both hydroxyl and superoxide radical species in cobalt-complexes, it is possible to add the enzyme superoxide dismutase to only visualize the DMPO-OH adduct and to indirectly evidence the presence of the DMPO-OOH (DMPO + $O_2^{\bullet -}$) adduct in the mixture [12].

It is important to note, that, in some cases, ROS can oxidize DMPO and/or the organic matrix, giving rise to nitroxide-like radical and/or carbon-centered radical, respectively, with a characteristic DMPO adduct depending on the reactivity of the system [15].

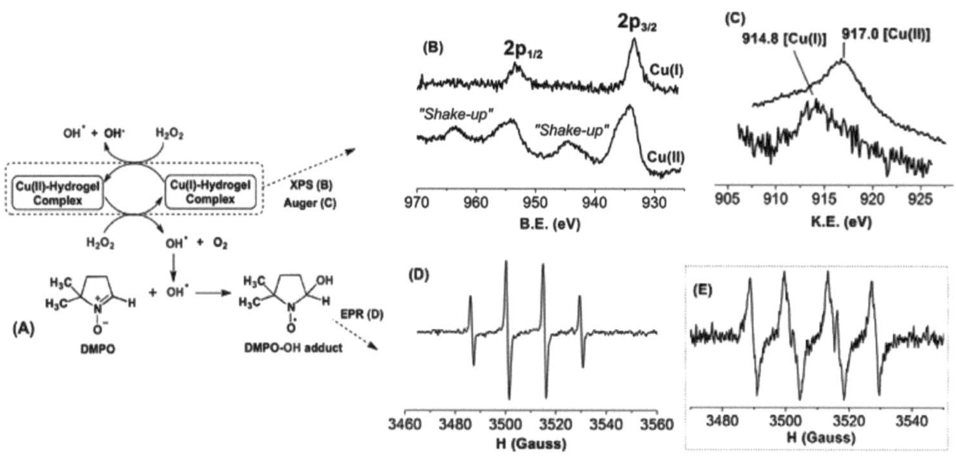

Figure 4. Schematic representation of the inner sphere mechanism where the OH• radicals are generated due to the interconversion between Cu(II)⇔Cu(I) in a Cu(II)- *poly*(EGDE-MAA-IM) hydrogel as a heterogeneous catalyst (A), Cu *2p* region of the XPS spectra (B), Auger spectra for Cu(I) or Cu(II)-hydrogel complexes (C), and the corresponding X-band EPR spectrum of a DMPO-OH adduct obtained from a mixture reaction as indicated (D) [14, 15]. X-band EPR spectra of DMPO-OOH and DMPO-OH adducts obtained from a Co(II)-*poly*(EGDE-DA)/H_2O_2 heterogeneous system as indicated (E) [12]. EPR measurements were performed at X-band on a Bruker EMX plus spectrometer. XPS spectra were collected using a physical electronics PHI 5700 spectrometer.

Then, when the generation of free radical species is identified, it is possible to access to the catalytic system which can describe the mechanism involved in the generation of these species. Particularly, the Cu(II) complexes have been extensively studied, and the most acceptable catalytic cycle involves the reduction of Cu(II) to Cu(I) induced by the action of the H_2O_2, which is converted to OH• radical (inner sphere mechanism, **Figure 4**). In this aspect, the XPS technique is a valuable tool to analyze changes in the oxidation state of metal centers. For example, the synthetic Cu(II) *poly*(EGDE-MAA-IM) hydrogel system, containing 63 mg of Cu(II) per gram of polymer (**Scheme 2**), after being in contact with a $[H_2O_2]_o$ = 25 mM for 1 hour, produces the reduction of Cu(II) to Cu(I), which can be visualized because the conversion of Cu(I) to Cu(II) is low with the decrease in the consumption of H_2O_2. The Cu *2p* region in the XPS spectrum is different after the treatment with H_2O_2 because characteristic shake-up satellite structures of Cu(II) are not observed. However, the Auger line for copper has to be measured with a short irradiation time due to a photoreduction of Cu(II) to Cu(I). With the correct acquisition of the Auger line for copper, it is possible to prove the presence of Cu(I) species generated by the action of H_2O_2 in a Cu(II) hydrogel system (**Figure 4**) [14]. Furthermore, it has been observed that the ultrahigh vacuum in which the sample is exposed prior to the XPS measurement can produce the reduction of Cu(II) to Cu(I) in Cu(II) *poly*(ethylenimine) complexes [41].

In addition, the catalytic performance of the catalyst can be studied with the azo dye methyl orange as a model compound because its absorbance can be easily monitored as a function of time through UV-visible spectroscopy at 465 nm (**Scheme 7**). In general, the concentration of methyl orange in the solution where the catalyst and H_2O_2 are also present decreases as a consequence of two parallel processes: surface adsorption on the catalyst and oxidative degradation, both following pseudo first-order kinetics [5, 7, 12, 14, 15].

Regarding other metal ion complexes, cobalt complexes can be very useful for soft oxidative procedures in organic chemistry. There are only a few reports related to Co(II) complexes supported in non-soluble polymeric structures, although a Co(II)-crosslinked polyacrylamide can be mentioned as a selective catalyst for the oxidation of olefins and alkyl halides with H_2O_2 in aqueous media [78]. The difference with copper complexes is that the metal ion is converted to Co(III) from the corresponding Co(II) in the presence of H_2O_2 [12, 13, 78].

Regarding environmental applications, our research group observed that the Co(II) *poly*(EGDE-DA) hydrogel/H_2O_2 system is not as reactive for the degradation of methyl orange as the corresponding Cu(II) system supported in the same polymeric structure (**Scheme 2**). It is important to remark that the maximum loading capacities for Co(II) and Cu(II) ions are estimated in 15 and 151 mg g^{-1}. However, although a Co(II) hydrogel/H_2O_2 system can achieve 87% of oxidation of methyl orange in 110 minutes, the Cu(II) hydrogel/H_2O_2 system can do so in 10 minutes. Even when the amount of Co(II) ions is low, the catalytic importance makes this kind of system of interest since the Co(II) can produce superoxide radicals with a lower oxidative potential [12, 13].

6. Remarks

The coordination of a polymeric ligand by a transition metal ion is an efficient way to obtain processable materials with unique and valuable properties. Polymer networks offer new possibilities to scientists for the creation of artificial materials. In recent years, hydrogels with chelating ligands have attracted the attention of industrial applications. In particular, polyelectrolyte and polyampholyte hydrogels have become of great interest in the macromolecular chemistry area due to their versatility as excellent adsorbents of chemical compounds. Stimulus-sensitive hydrogels are used in a variety of novel applications, including controlled drug delivery, immobilized enzyme systems, separation processes, fuel cells, and sensor development.

The development of polymers containing nitrogen remains a growing area because of their applications in the destabilization of negative colloids in effluents and water clarification, electrophoretic depositions, recovery of heavy metal ions or exchange resins for ions, and the mimicking of active sites of enzymes.

Currently, the main goals for the material science community are the design and synthesis of new hydrogels containing ligand for the uptake of heavy metal ions to reduce the direct impact of this industrial waste on the environment. The characterization of the different non-soluble polymeric structures is limited to the spectroscopic techniques in the solid state, being ss-NMR the principal characterization tool for bulk analysis. However, some sensitivity problems can be resolved with new polarization techniques such as DNP-NMR. In addition, some other surface characterization techniques such as X-ray photoelectron spectroscopy can be used. However, the results provide only information limited to the interface area and not from the bulk content. Particularly, the activation of H_2O_2 from the corresponding Cu(II) and Co(II) hydrogels, obtained from the uptake of Cu(II) or Co(II) ions, can be successfully used with H_2O_2 for the degradation of azo dyes, to reduce the impact of both inorganic or organic pollutants.

Author details

Viviana Campo Dall' Orto and Juan Manuel Lázaro-Martínez*

*Address all correspondence to: lazarojm@ffyb.uba.ar

Universidad de Buenos Aires, Facultad de Farmacia y Bioquímica & IQUIFIB-CONICET, Junín 956 (1113), Ciudad Autónoma de Buenos Aires, Argentina

References

[1] Vijayaraghavan K, Jegan J, Palanivelu K, Velan M. Batch and column removal of copper from aqueous solution using a brown marine alga Turbinaria ornata. Chem Eng J. 2005;106(2):177–84.

[2] Bailey SE, Olin TJ, Bricka RM, Adrian DD. A review of potentially low-cost sorbents for heavy metals. Water Res. 1999;33(11):2469–79.

[3] Kudaibergenov SE. Polyampholytes: Synthesis, Characterization and Application. Springer US; 2002.

[4] Lowe AB, McCormick CL. Synthesis and solution properties of zwitterionic polymers. Chem Rev. 2002;102(11):4177–89.

[5] Janus M, Morawski A. New method of improving photocatalytic activity of commercial Degussa P25 for azo dyes decomposition. Appl Catal B Environ. 2007;75(1–2):118–23.

[6] Lin T, Wu C. Activation of hydrogen peroxide in copper(II)/amino acid/H2O2 systems: Effects of pH and copper speciation. J Catal. 2005;232:117–26.

[7] Shah V, Verma P, Stopka P, Gabriel J, Baldrian P, Nerud F. Decolorization of dyes with copper(II)/organic acid/hydrogen peroxide systems. Appl Catal B Environ. 2003;46(2): 287–92.

[8] Meng X, Zhu J, Yan J, Xie J, Kou X, Kuang X, et al. Studies on the oxidation of phenols catalyzed by a copper(II)-Schiff base complex in aqueous solution under mild conditions. J Chem Technol Biotechnol. 2006;81:2–7.

[9] Bekturov EA, Kudaibergenov SE, Sigitov VB. Complexation of amphoteric copolymer of 2-methyl-5-vinylpyridine-acrylic acid with copper (ll) ions and catalase-like activity of polyampholyte-metal complexes. Polymer (Guildf). 1986;27:1269–72.

[10] Lázaro-Martínez JM, Monti GA, Chattah AK. Insights into the coordination sphere of copper ion in polymers containing carboxylic acid and azole groups. Polymer (Guildf). 2013;54(19):5214–21.

[11] Solomon EI, Sundaram UM, Machonkin TE. Multicopper oxidases and oxygenases. Chem Rev. 1996;96(7):2563–606.

[12] Lombardo Lupano LV, Lázaro Martínez JM, Piehl LL, Rubín de Celis E, Torres Sánchez RM, Campo Dall' Orto V. Synthesis, characterization, and catalytic properties of cationic hydrogels containing copper(II) and cobalt(II) ions. Langmuir. 2014;30(10): 2903–13.

[13] Lombardo Lupano LV, Lázaro Martínez JM, Piehl LL, Rubin de Celis E, Campo Dall' Orto V. Activation of H2O2 and superoxide production using a novel cobalt complex based on a polyampholyte. Appl Catal A Gen. 2013;467:342–54.

[14] Lázaro Martínez JM, Rodríguez-Castellón E, Sánchez RMT, Denaday LR, Buldain GY, Campo Dall' Orto V. XPS studies on the Cu(I,II)–polyampholyte heterogeneous catalyst: An insight into its structure and mechanism. J Mol Catal A Chem. 2011;339(1–2):43–51.

[15] Lázaro Martínez JM, Leal Denis MF, Piehl LL, de Celis ER, Buldain GY, Campo Dall' Orto V. Studies on the activation of hydrogen peroxide for color removal in the presence

of a new Cu(II)-polyampholyte heterogeneous catalyst. Appl Catal B Environ. 2008;82(3–4):273–83.

[16] Lázaro Martínez JM, Chattah AK, Torres Sánchez RM, Buldain GY, Campo Dall' Orto V. Synthesis and characterization of novel polyampholyte and polyelectrolyte polymers containing imidazole, triazole or pyrazole. Polymer (Guildf). 2012;53(6): 1288–97.

[17] Alfrey T, Morawetz H. Amphoteric polyelectrolytes. I. 2-Vinylpyridine—methacrylic acid copolymers. J Am Chem Soc. 1952;74(2):436–8.

[18] Annenkov VV, Danilovtseva EN, Tenhu H, Aseyev V, Hirvonen SP, Mikhaleva AI. Copolymers of 1-vinylimidazole and (meth)acrylic acid: Synthesis and polyelectrolyte properties. Eur Polym J. 2004;40(6):1027–32.

[19] Leal Denis MF, Carballo RR, Spiaggi AJ, Dabas PC, Campo Dall' Orto V, Martínez JML, et al. Synthesis and sorption properties of a polyampholyte. React Funct Polym. 2008;68(1):169–81.

[20] Krasia TC, Patrickios CS. Synthesis and aqueous solution characterization of amphiphilic diblock copolymers containing carbazole. Polymer (Guildf). 2002;43(10):2917–20.

[21] Kumar A, Lahiri SS, Singh H. Development of PEGDMA: MAA based hydrogel microparticles for oral insulin delivery. Int J Pharm. 2006;323(1–2):117–24.

[22] Jang J, Ishida H. A study of corrosion protection on copper by new polymeric agents: Silane-modified imidazoles. Corros Sci. 1992;33(7):1053–66.

[23] Jilge G, Sébille B, Vidal-Madjar C, Lemque R, Unger KK. Optimisation of fast protein separations on non-porous silica-based strong anion exchangers. Chromatographia. 1993;37(11):603–7.

[24] Kowalik-Jankowska T, Ruta-Dolejsz M, Wiśniewska K, Łankiewicz L, Kozłowski H. Copper(II) complexation by human and mouse fragments (11–16) of β-amyloid peptide. J Chem Soc, Dalton Trans. 2000:4511–9.

[25] Casolaro M, Ito Y, Ishii T, Bottari S, Samperi F, Mendichi R. Stimuli-responsive poly(ampholyte)s containing L-histidine residues: Synthesis and protonation thermodynamics of methacrylic polymers in the free and in the cross-linked gel forms. Express Polym Lett. 2008;2(3):165–83.

[26] Casolaro M, Bottari S, Cappelli A, Mendichi R, Ito Y. Vinyl polymers based on l-histidine residues. Part 1. The thermodynamics of poly(ampholyte)s in the free and in the cross-linked gel form. Biomacromolecules. 2004;5(4):1325–32.

[27] Barton JM, Hamerton I, Howlin BJ, Jones JR, Liu S. Studies of cure schedule and final property relationships of a commercial epoxy resin using modified imidazole curing agents. Polymer (Guildf). 1998;39(10):1929–37.

[28] Kwok AY, Qiao GG, Solomon DH. Interpenetrating amphiphilic polymer networks of poly(2-hydroxyethyl methacrylate) and poly(ethylene oxide). Chem Mater. 2004;16(26):5650–8.

[29] van Berkel PM, Driessen WL, Reedijk J, Sherrington DC, Zitsmanis A. Metal-ion binding affinity of azole-modified oxirane and thiirane resins. React Funct Polym. 1995;27(1):15–28.

[30] van Berkel PM, Punt M, Koolhaas GJAA, Driessen WL, Reedijk J, Sherrington DC. Highly copper(II)-selective chelating ion-exchange resins based ion bis(imidazole)-modified glycidyl methacrylate copolymers. React Funct Polym. 1997;32(2):139–51.

[31] Wan Ngah WS, Endud CS, Mayanar R. Removal of copper(II) ions from aqueous solution onto chitosan and cross-linked chitosan beads. React Funct Polym. 2002;50(2): 181–90.

[32] Nada AAM, Alkady MY, Fekry HM. Synthesis and characterization of grafted cellulose for use in water and metal ions sorption. BioResources. 2008;3:46–59.

[33] Blanc F, Copéret C, Lesage A, Emsley L. High resolution solid state NMR spectroscopy in surface organometallic chemistry: Access to molecular understanding of active sites of well-defined heterogeneous catalysts. Chem Soc Rev. 2008;37:518–26.

[34] Harris RK. Nuclear Magnetic Resonance Spectroscopy: A Physiocochemical View. London: Logman Scientific and Technical; 1986.

[35] Lowe IJ. Free induction decays of rotating solids. Phys Rev Lett. 1959;2(7):285–7.

[36] Andrew ER, Bradbury A, Eades RG. Removal of dipolar broadening of nuclear magnetic resonance spectra of solids by specimen rotation. Nature. 1959;183(4678): 1802–3.

[37] Hartmann SR, Hahn EL. Nuclear double resonance in the rotating frame. Phys Rev. 1962;128(5):2042–53.

[38] Fung BM, Khitrina K, Ermolaev K. An improved broadband decoupling sequence for liquid crystals and solids. J Magn Reson. 2000;142(1):97–101.

[39] Lombardo Lupano LV, Lázaro-Martínez JM, Vizioli NM, Torres DI, Campo V, Orto D, et al. Synthesis of water-soluble oligomers from imidazole, ethyleneglycol diglycidyl ether, and methacrylic acid. An insight into the chemical structure, aggregation behavior and formation of hollow spheres. Macromol Mater Eng. 2016;301(2):167–81.

[40] Wu XL, Zilm KW. Complete spectral editing in CPMAS NMR. J Magn Reson Ser A. 1993;102(2):205–13.

[41] Lázaro-Martínez JM, Rodríguez-Castellón E, Vega D, Monti GA, Chattah AK. Solid-state studies of the crystalline/amorphous character in linear poly(ethylenimine hydrochloride) (PEI·HCl) polymers and their copper complexes. Macromolecules. 2015;48(4):1115–25.

[42] Power WP. High resolution magic angle spinning – applications to solid phase synthetic systems and other semi-solids. Annu Reports NMR Spectrosc. 2003;51(03): 261–95.

[43] Higashi K, Yamamoto K, Pandey MK, Mroue KH, Moribe K, Yamamoto K, et al. Insights into atomic-level interaction between mefenamic acid and Eudragit EPO in a supersaturated solution by high-resolution magic-angle spinning NMR spectroscopy. Mol Pharm. 2014;11(1):351–7.

[44] Maly T, Debelouchina GT, Bajaj VS, Hu K, Joo C, Mak-Jurkauskas ML, et al. Dynamic nuclear polarization at high magnetic fields. J Chem Phys. 2008;128(5):052211.

[45] Lesage A, Lelli M, Gajan D, Caporini MA, Vitzthum V, Miéville P, et al. Surface enhanced NMR spectroscopy by dynamic nuclear polarization. J Am Chem Soc. 2010;132(44):15459–61.

[46] Zagdoun A, Casano G, Ouari O, Lapadula G, Rossini AJ, Lelli M, et al. A slowly relaxing rigid biradical for efficient dynamic nuclear polarization surface-enhanced NMR spectroscopy: Expeditious characterization of functional group manipulation in hybrid materials. J Am Chem Soc. 2012;134(4):2284–91.

[47] Wagner CD, Riggs WM, Davis LE, Moulder JF, Muilenberg GE. Handbook of X-ray photoelectron spectroscopy. In Muilenberg GE, editor. Minnesota: Perkin-Elmer Corporation; 1979.

[48] Brunauer S, Emmett PH, Teller E. Adsorption of gases in multimolecular layers. J Am Chem Soc. 1938;60(1):309–19.

[49] Langmuir I. The adsorption of gases on plane surfaces of glass, mica and platinum. J Am Chem Soc. 1918;40(9):1361–403.

[50] Do DD. Adsorption analysis: Equilibria and kinetics, Vol. 2. London: Imperial College Press; 1998.

[51] Aharoni C, Sparks DL. Rates of Soil Chemical Processes. In Sparks DL and Suarez DL, editors. Madison, WI; Soil Science Society of America 1991. 1–18 p.

[52] Hsieh C-T, Teng H. Langmuir and Dubinin–Radushkevich analyses on equilibrium adsorption of activated carbon fabrics in aqueous solutions. J Chem Technol Biotechnol. 2000;75(11):1066–72.

[53] Hobson JP. Physical adsorption isotherms extending from ultrahigh vacuum to vapor pressure. J Phys Chem. 1969;73(8):2720–7.

[54] Gübbük IH, Güp R, Ersöz M. Synthesis, characterization, and sorption properties of silica gel-immobilized Schiff base derivative. J Colloid Interface Sci. 2008;320:376–82.

[55] Hadi M, Samarghandi MR, McKay G. Equilibrium two-parameter isotherms of acid dyes sorption by activated carbons: Study of residual errors. Chem Eng J. 2010;160(2): 408–16.

[56] Burnham KP, Anderson DR, Huyvaert KP. AIC model selection and multimodel inference in behavioral ecology: Some background, observations, and comparisons. Behav Ecol Sociobiol. 2011;65(1):23–35.

[57] Cheung C, Porter J, Mckay G. Sorption kinetic analysis for the removal of cadmium ions from effluents using bone char. Water Res. 2001;35(3):605–12.

[58] Pérez-Marín AB, Zapata VM, Ortuño JF, Aguilar M, Sáez J, Lloréns M. Removal of cadmium from aqueous solutions by adsorption onto orange waste. J Hazard Mater. 2007;139(1):122–31.

[59] Ho Y. The kinetics of sorption of divalent metal ions onto sphagnum moss peat. Water Res. 2000;34(3):735–42.

[60] Loukidou MX, Zouboulis AI, Karapantsios TD, Matis KA. Equilibrium and kinetic modeling of chromium(VI) biosorption by Aeromonas caviae. Colloids Surfaces A Physicochem Eng Asp. 2004;242(1–3):93–104.

[61] Copello GJ, Diaz LE, Campo Dall' Orto V. Adsorption of Cd(II) and Pb(II) onto a one step-synthesized polyampholyte: Kinetics and equilibrium studies. J Hazard Mater. 2012;217–218:374–81.

[62] Teng H, Hsieh C-T. Activation energy for oxygen chemisorption on carbon at low temperatures. Ind Eng Chem Res. 1999;38(1):292–7.

[63] Kuo S, Lotse EG. Kinetics of phosphate adsorption and desorption by hematite and gibbsite. Soil Sci. 1973;116(6):400–406.

[64] Bache BW, Williams EG. A phosphate sorption index for soils. J Soil Sci. 1971;22(3):289–301.

[65] Ho Y., McKay G. Pseudo-second order model for sorption processes. Process Biochem. 1999;34(5):451–65.

[66] Lázaro Martínez JM, Chattah AK, Monti GA, Leal Denis MF, Buldain GY, Campo Dall' Orto V. New copper(II) complexes of polyampholyte and polyelectrolyte polymers: Solid-state NMR, FTIR, XRPD and thermal analyses. Polymer (Guildf). 2008;49(25):5482–9.

[67] Peisach J, Blumberg WE. Structural implications derived from the analysis of electron paramagnetic resonance spectra of natural and artificial copper proteins. Arch Biochem Biophys. 1974;165(2):691–708.

[68] Andersson M, Hansson Ö, Öhrström L, Idström A, Nydén M. Vinylimidazole copolymers: Coordination chemistry, solubility, and cross-linking as function of Cu2+ and Zn2+ complexation. Colloid Polym Sci. 2011;289(12):1361–72.

[69] Sun Z, Jin L, Zhang S, Shi W, Pu M, Wei M, et al. An optical sensor based on H-acid/layered double hydroxide composite film for the selective detection of mercury ion. Anal Chim Acta. 2011;702(1):95–101.

[70] Fang X, Chen R, Xiao L, Chen Q. Synthesis and characterization of Sm(III)–hyperbranched poly(ester-amide) complex. Polym Int. 2011;60(1):136–40.

[71] Belfiore LA, Graham H, Uedat E. Ligand field stabilization in nickel complexes that exhibit extraordinary glass transition temperature enhancement. Macromolecules. 1992;25:2935–9.

[72] Kang N, Lee DS, Yoon J. Kinetic modeling of Fenton oxidation of phenol and monochlorophenols. Chemosphere. 2002;47(9):915–24.

[73] Gemeay AH, Salem MA, Salem IA. Activity of silica-alumina surface modified with some transition metal ions. Colloids Surfaces A Physicochem Eng Asp. 1996;117:245–52.

[74] Salem IA. Kinetics of the oxidative color removal and degradation of bromophenol blue with hydrogen peroxide catalyzed by copper(II)-supported alumina and zirconia. Appl Catal B Environ. 2000;28(3–4):153–62.

[75] Ray S, Mapolie SF, Darkwa J. Catalytic hydroxylation of phenol using immobilized late transition metal salicylaldimine complexes. J Mol Catal A Chem. 2007;267(1–2):143–8.

[76] Šuláková R, Hrdina R, Soares GMB. Oxidation of azo textile soluble dyes with hydrogen peroxide in the presence of Cu(II)–chitosan heterogeneous catalysts. Dye Pigment. 2007;73(1):19–24.

[77] Ettinger M, Ruchhoft C, Lishka R. Sensitive 4-aminoantipyrine method for phenolic compounds. Anal Chem. 1951;23(12):1783–8.

[78] Tamami B, Ghasemi S. Modified crosslinked polyacrylamide anchored Schiff base–cobalt complex: A novel nano-sized heterogeneous catalyst for selective oxidation of olefins and alkyl halides with hydrogen peroxide in aqueous media. Appl Catal A Gen. 2011;393(1–2):242–50.

In Situ-Forming Cross-linking Hydrogel Systems: Chemistry and Biomedical Applications

Xiangdong Bi and Aiye Liang

Additional information is available at the end of the chapter

http://dx.doi.org/10.5772/63954

Abstract

With the development of chemical synthetic strategies and available building blocks, *in situ*-forming hydrogels have attracted significant attention in the biomedical fields over the past decade. Due to their distinct properties of easy management and minimal invasiveness via simple aqueous injections at target sites, *in situ*-forming hydrogels have found a broad spectrum of biomedical applications including tissue engineering, drug delivery, gene delivery, 3D bioprinting, wound healing, antimicrobial research, and cancer research. The objective of this chapter is to provide a comprehensive review of updated research methods in chemical synthesis of *in situ*-forming cross-linking hydrogel systems and their diverse applications in the biomedical fields. This chapter concludes with perspectives on the future development of *in situ*-forming hydrogels to facilitate this multidisciplinary field.

Keywords: chemical cross-linking, free radical polymerization, *in situ*-forming hydrogel, biomedical applications, hydrophilic polymers

1. Introduction

Hydrogels are a class of three-dimensional (3D) cross-linked polymeric structures capable of holding large amounts of water or biological fluids in their swollen state [1]. The first report of water-swollen cross-linking polymer network of 2-hydroxylethyl methacrylate and ethylene glycol dimethacrylate for applications of contact lenses was published by Wichterle and Lim in 1960 [2]. Due to their high water content and structure similarity to natural extracellular matrix (ECM) as well as their biodegradability and low immunogenicity, hydrogels have gained considerable interest in biomedical and pharmaceutical fields over the past few decades, especially with the development of a wide variety of chemical building blocks and various synthetic strategies in polymer chemistry, organic chemistry, and bioconjugation chemistry. Up to date, hydrogels have been applied in a broad spectrum of biomedical applications including tissue engineering, 3D bioprinting, drug delivery, gene delivery, would healing, antimicrobial research, and cancer research [3]. Among these applications, *in situ*-forming hydrogels have been intensively investigated because of their easy formulation to encapsulate bioactive ingredients and/or cells into hydrogel by mixing followed by a cross-linking process. They can be delivered into the desired sites through minimal invasive injection, which improves patient compliance and curative efficacy [4, 5]. They can also be used in surgery as tissue adhesive to seal tissue defects [6].

With regard to biomedical applications, *in situ*-forming hydrogels must first meet the basic requirement of biocompatibility to provide the appropriate macro- and microenvironment for cell proliferation and tissue growth. Therefore, it is of great importance to develop *in situ*-forming hydrogels with minimized immunological rejection through thoughtful selection of the materials and incorporation of physicochemical and biological cues to replicate the natural ECM. Porosity of the hydrogel should be considered when selecting the hydrogel materials as it affects the viability and proliferation of cells in 3D culture as well as the drug release profile for delivery. In addition, rigidity of the *in situ*-forming hydrogel is crucial due to its influence on cell differentiation [7] and its mechanical strength to support 3D constructs in bioprinting and tissue engineering.

There are two strategies to synthesize *in situ*-forming hydrogels. One strategy is the polymerization of small molecules in the presence of initiators and cross-linkers. The other strategy is to directly cross-link either natural or synthetic hydrophilic polymers [3]. In general, synthetic polymers are hydrophobic and mechanically stronger compared to natural polymers, which results in slow degradation but high durability in hydrogels. Another property of synthetic polymers is their inert cellular environment that prohibits active cell binding, which results in low cell viability. To compensate, bioactive compounds such as peptides or growth factors need to be incorporated into the hydrogel network [8]. Natural polymers, on the other hand, have the advantages of low toxicity and biodegradability but their mechanical properties are weaker. The opposite properties between synthetic and natural polymers need to be balanced through optimal design for specific hydrogel applications [9].

The most used natural polymers for *in situ* hydrogels are hyaluronic acid (HA), collagen, gelatin, alginate, chitosan, fibrin, etc. Some commonly used synthetic polymers are poly(ethylene glycol) (PEG), poly(acrylic acid) (PAA), poly(acrylamide) (PAM), poly(vinyl pyrrolidone) (PVP), poly(vinyl alcohol) (PVA), poly(lactic acid) (PLA), and poly(lactic-co-glycolic acid) (PLGA). Detailed information about each polymer will not be covered in this chapter. Interested readers may refer to review articles for more information [10–12].

2. Chemistry of *in situ*-forming cross-linking hydrogels

Since hydrogels are simply hydrophilic polymer networks cross-linked in some fashion to produce an elastic structure, any strategy that can produce a cross-linked network could be used for *in situ* hydrogel synthesis [9]. Hydrogels can be classified into different categories based on various parameters such as preparation method, ionic charge, and mechanical and structural characteristics [13]. Based on cross-linking mechanisms, hydrogels can be divided into physically cross-linked hydrogels and chemically cross-linked hydrogels. Physically cross-linked networks are often called "reversible" or "physical" gels, as they can be dissolved by changing environmental conditions such as pH, ionic strength, or temperature [3]. They possess temporary connections either through polymeric chain entanglement or physically induced gelation through stimuli such as ion-ion interaction, hydrogen bonding, thermo-induced gelation, complementary binding, inclusion complex formation, and hydrophobic interactions [9]. Physically cross-linked hydrogels are of great interest for encapsulation of bioactive substance and cells [14], although they are not covered in this chapter. Chemically cross-linked hydrogels are also called "permanent" or "chemical" gels which are networks cross-linked by covalent bonds through chemical reactions to achieve cross-linking of macromolecular chains in solution. In order for in-depth discussion, this chapter will focus on chemically cross-linked *in situ* hydrogels by reviewing various chemical cross-linking strategies to synthesize *in situ*-forming hydrogels and their updated biomedical applications.

To be suitable for biomedical applications, it is preferred that the cross-linking occurs in aqueous and mild conditions. The reactions should not damage the cells or biofunctional molecules in the hydrogel matrix. Reactions employed should not generate toxic side products and should not require high temperatures or heavy metal catalysts that are toxic to the cells [15].

Chemical cross-linking is the most common and highly efficient method for the formation of *in situ*-forming hydrogels that have excellent mechanical strength. This section will review the updated chemistry and various cross-linking mechanisms in hydrogel synthesis including free radical polymerization, reactions of complementary groups, and enzyme-catalyzed reactions.

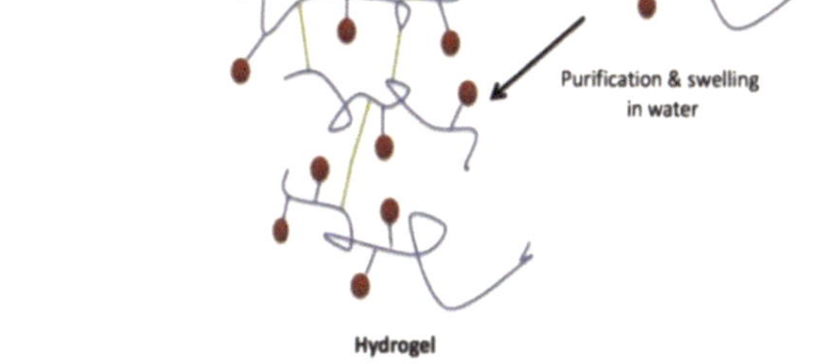

Figure 1. Hydrogels by 3D polymerization.

There are two commonly used strategies to prepare chemically cross-linked hydrogels which are illustrated in **Figures 1** and **2** [3]. The first strategy is called "3D polymerization," which is achieved by polymerization of hydrophilic molecules, such as acrylates and vinylic monomers, in the presence of multifunctional cross-linkers. The drawback of 3D polymerization is the significant amount of unreacted monomers and other small molecules, which could be toxic and have to be removed by extensive purification processes. The second strategy is to directly crosslink hydrophilic polymers so that extensive purification can be avoided as there are not as many toxic small molecules remained in the system. Water-soluble polymers such as PAA, PVA, PEG, PAM, and polysaccharides are commonly used systems for biomedical and pharmaceutical applications due to their nontoxicity and biocompatibility [3].

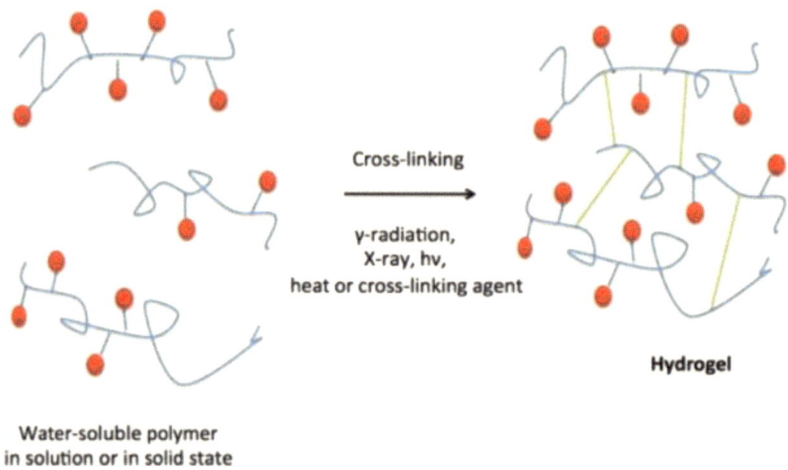

Figure 2. Hydrogels by cross-linking of biopolymers.

2.1. Cross-linking by free radical polymerization

Free radical polymerization is the most commonly used cross-linking strategy for hydrogel synthesis due to its advantages over other polymerization methods. First, it is highly reactive, which results in polymers of high molecular weights and cross-linking density. Second, free radical polymerization tolerates a variety of functional groups and it occurs in mild conditions, even in aqueous conditions. This makes it a facile approach for cross-linking hydrogel synthesis [16]. Free radical polymerization can be further classified as homopolymerization, copolymerization, and multipolymer interpenetrating polymeric hydrogels by hydrogel composition [9].

2.1.1. Hydrogels by homopolymerization

Homopolymeric hydrogels are hydrogel systems originated from polymerization of a single monomer species in the presence of an initiator and a cross-linker. For example, homopolymerization of *N*-acryloylglycinamide through radical polymerization with 2,2′-azo-bisisobutyronitrile (AIBN) as an initiator and *N*,*N*'-methylene-bis-acrylamide (MBAm) as a chemical cross-linker yields a hydrogel [17]. Similarly, *N*-vinyl-2-pyrrolidone is homopolymerized in the presence of a free radical initiator AIBN and a cross-linker MBAm to produce a pH-responsive hydrogel for *in vitro* delivery of propranolol hydrochloride [18].

2.1.2. *Hydrogels by copolymerization*

Copolymerization hydrogels are synthesized by polymerization of two or more different monomers with at least one hydrophilic component. Depending on the structure of the polymer chain, copolymers are further classified into random, block, or alternating copolymers. A recent report described the synthesis of pH temperature dual stimuli-responsive smart hydrogel for drug release. In this research, PEG was reacted with methyl ether methacrylate to afford methacrylate terminated PEG, which copolymerized with *N*,*N*'-dimethylaminoethyl methacrylate. The copolymer solution is mixed with α-cyclodextrin (α-CD) to form a pH-thermo dual sensitive hydrogel allowing the release of a model drug 5-fluorouracil to be effectively controlled by temperature and pH [19]. Copolymerization allows for the combination of properties from both ingredients with varied ratios.

2.1.3. *Hydrogels of interpenetrating polymer networks*

Interpenetrating polymer networks (IPNs) and semi-interpenetrating polymer networks (semi-IPNs) have emerged as innovative biomaterials for biomedical applications. IPNs are a family of hydrogels that contain two independent hydrogel components with each component being a cross-linked hydrogel from synthetic and/or natural polymers. Semi-IPNs contain two independent hydrogels with one being cross-linked hydrogel and the other noncross-linked hydrogel [9]. The purpose of two hydrogel components is to provide tuned physical properties or stimuli responsiveness. For example, Reddy et al. fabricated cyclotriphosphazene-based IPN hydrogels through free radical polymerization of mono (methacryloyl-2-ethoxy)-pentakis(N^1,N^1-dimethylpropane-1,3-diamino)-cyclotriphosphazene and acrylamide, and cross-linked by MBAm in the presence of pectin. The synthesized semi-IPN hydrogels exhibited dual responsiveness for pH and temperature which triggered the delivery of 5-fluorouracil [20].

2.2. Cross-linking by chemical reaction of complementary groups

In situ-forming hydrogels can be prepared by direct cross-linking of natural or synthetic polymers through chemical reactions of complementary functional groups. In consideration of biomedical applications, chemical reactions employed in cross-linking hydrogels should be achievable in aqueous solutions without generating toxic by-products. The reaction should be very efficient with few reactants and active functional groups remaining. For that purpose, click chemistry, Michael additions, thiol-ene/yne coupling, Diels-Alder reaction, disulfide formation, Schiff-base formation, and epoxide reactions are examples of suitable reactions which will be discussed in detail. A summary of cross-linking reactions via complementary groups is depicted in **Table 1**.

Entry	Reaction	Complementary groups	Conditions	Ref.
1	Click reaction	Alkyne + azide	Cu(I), aqueous	[23–25]
		Alkyne + azide	aqueous, 37 °C	[26]
		Oxanorbornadiene + azide	aqueous, 37 °C	[22, 27]
		Cyclooctyne + azide	aqueous, 37 °C	[28–31]
2	Michael addition	Maleimide + thiol	aqueous, 37 °C	[15, 36]
		Vinyl sulfone + thiol	aqueous, 37 °C	[37]
		Acrylate + thiol	aqueous, 37 °C	
		Methacrylate + thiol	aqueous, 37 °C	
3	Thiol-ene/yne coupling	Alkene + thiol	aqueous, 37 °C	[38–43]
		Norbornene + thiol	Photo, radical catalyzed	
		Alkyne + thiol		
4	Diels-Alder reaction	Furan + maleimide	aqueous, 37 °C	[44, 45]
		Tetrazine + norbornene	aqueous, 37 °C	[46]
		Tetrazine + *trans*-cyclooctene	aqueous, 37 °C	[47, 48]
5	Disulfide formation/ exchange	Thiol-thiol	aqueous, H_2O_2	[49, 50]
		Pyridyl disulfide + thiol	aqueous, 37 °C	[52, 53]
6	Epoxide coupling	Epoxide + amine	aqueous, 37 °C	[54, 55]
		Epoxide + hydroxyl	acidic or basic	[56]
		Diepoxide + amine	aqueous, carbon black	[57]
7	Schiff-base formation	Aldehyde + amine	aqueous, 37 °C	[59]
		Aldehyde + hydrazide	aqueous, 37 °C	[61]
		Aldehyde + hydroxylamine	aqueous, 37 °C	[60, 61]
8	Condensation	Amine or hydroxyl + acid derv.	aqueous, 37 °C	[33]
		Amine or hydroxyl + isocyanate	DMSO, 35 °C	[62]
		Boronic acid + amine/hydroxyl	aqueous, pH 4.8	[65]
	Genipin coupling	Amines + genipin	0.1 M acetic acid, 4 °C	[63]
	Photo-induced crosslink	Alkenes	UV, photoinitiator	[64]
9	Staudinger-ligation	Azide + ester derivative of triphenylphosphine	aqueous, $BaCl_2$, 37 °C	[70]
	Tetrazole-photoclick	Alkene + tetrazole	aqueous, UV light	[71]
	Quadricyclane-ligation	Quadricyclane + Ni bis(dithiolene)	aqueous, pH 4.5	[72]

Table 1. Summary of cross-linking reactions of complementary groups.

2.2.1. Click reactions

Click reactions refer to the cycloaddition of azide and alkyne to form a linkage through a triazole ring. It was proposed by Sharpless and co-workers in 2001 as copper-catalyzed azide-alkyne cycloaddition (CuAAC), aiming for efficient chemical synthesis with minimized byproducts and purification effort [21]. Over the past decade, click chemistry has gained significant application in a broad range of chemical synthesis of small molecules, polymers, dendrimers, biomacromolecules, and bioconjugation. Click chemistry has been widely used in cross-linking hydrogels due to its high selectivity and efficiency without generating by-products in aqueous conditions, plus the bioorthogonality of the components without interaction with the environment of biological or biomedical systems [22]. To date, click reactions of different versions have been developed not only to build materials that are biologically compatible, highly functional and organized in structure, but also to produce highly complex patterns of biofunctionalities within a single cellular scaffold [23]. For example, HA-based hydrogels were cross-linked by CuAAC reaction to produce a thermo-responsive hydrogel with tailorable mechanical properties [23, 24]. Kaga et al. fabricated "clickable" hydrogels using dendron polymer-based triblock polymers [25].

To avoid the toxicity of copper, metal-free click chemistry has been developed to eliminate heavy metal residue in hydrogel matrix. Truong et al. synthesized chitosan-PEG hydrogels by copper-free azide-alkyne click reaction [26]. Chitosan and HA-based hydrogel were cross-linked after functionalization by oxanorbornadiene and azide [22, 27] (**Table 1**, entry 1). Another good example of metal-free click reaction is strain-promoted azide-alkyne cycloaddition (SPAAC) that involves a difluorinated cyclooctyne moiety. Due to the ring strain and the electron-withdrawing difluoride, the alkyne functionality is greatly activated for a cycloaddition without a catalyst [28]. This reaction has been shown to be very efficient with high chemoselectivity even for *in vivo* applications, making it suitable for *in situ* cross-linking hydrogel [29, 30]. DeFrost et al. synthesized PEG-based hydrogel by cross-linking PEG-tetraazide and a difluorinated cyclooctyne functionalized cleavable peptide. The reaction occurred in aqueous conditions at 37 °C in the presence of cells, which enabled independent and *in situ* tuning of biochemical properties of biomaterials [31]. It is worth noticing that the cross-linking of azide and alkyne functionalized polymers with metal-free click reactions is normally too slow to be used in most *in situ* applications [29]. In order for this cross-linking strategy to be useful, the gelation kinetics needs to be improved by increasing the electron deficiency of alkynes and electron density of azide structures. Interested readers may refer to review articles for more information [32, 33].

2.2.2. Michael addition

Michael addition is a 1,4-addition of nucleophiles to α,β-unsaturated ketones or esters. It occurs in high efficiency under aqueous conditions without any side products, making it a suitable approach for cross-linking in hydrogel synthesis. The common nucleophiles are macromolecules that are functionalized with multiple terminal amine or thiol groups, which cross-link with electrophilic macromolecules functionalized with alkene groups with adjacent electron withdrawing groups, such as vinyl sulfone, acrylate, or methacrylate [15]. Surfactants can be used to promote the kinetics of the Michael addition when the nucleophilic and the electrophilic macromolecules exhibit significant differences in hydrophilicity [34].

Like the cross-linking of alkyne and azide in SPAAC, the cross-linking kinetics of Michael addition depends greatly on the electron deficiency of the alkenes. Our research group has systematically studied the gelation kinetics and mechanical property of poly(amidoamine) (PAMAM) dendrimer-HA cross-linking hydrogels. PAMAM dendrimers are a family of synthetic polymers with well-defined structures and ample surface groups for conjugation of bioactive functionalities. They are widely used as a platform to deliver bioactive molecules into biological systems due to their water solubility, nontoxicity and nonimmunogenicity [35]. The combination of PAMAM and HA allows for easy chemical modification of dendrimer structures to modulate the physical and mechanical properties of the cross-linking hydrogel. In this research, PAMAM dendrimers are functionalized with maleimide, vinyl sulfone, acrylic, methacrylic, and a normal alkene group. When alkene functionalized PAMAM dendrimers are cross-linked with thiolated HA at experimental concentration, the gelation time displayed a large range from 8 seconds to 18 hours, and modulus from 36 to 183 Pa depending on the alkene group attached to the dendrimer. ^{1}H NMR study revealed that the gelation time is governed by the electron deficiency of alkenes [36]. Introduction of a RGD peptide in hydrogel greatly enhanced the cell attachment, viability, and proliferation of both bone marrow stem cells and human umbilical vein endothelial cells [37].

2.2.3. Thiol-ene/yne coupling

Thiol-ene reaction is a radical-mediated mechanism at room temperature and in aqueous conditions even in the presence of biological cargos such as proteins or cells, thus making it a good technique for hydrogel cross-linking. Unlike the traditional free radical polymerization, the radical thiol-ene reactions are relatively not oxygen sensitive [38]. The radicals that initiate the thiol-ene reaction can be generated using thermal, oxidation-reduction, or photochemical process based on initiator selection [39]. Fairbanks et al. devised photoinitiated thiol-ene reaction between four-armed PEG tetra-norbornene and dicysteine-terminated peptide to form *in situ*-forming hydrogels [40]. In another study, tetra-acetylene functionalized PEG and pentaerythritol are cross-linked in the presence of trimethylamine under moderate temperatures to form a robust hydrogel network [41]. Interested readers are referred to other recent articles for more information [42, 43].

2.2.4. Diels-Alder reaction

The Diels-Alder reaction is a robust cross-linking strategy for biopolymer-based hydrogels as it is rapid, efficient, versatile, and selective. It proceeds with high efficiency in aqueous conditions for hydrogel cross-linking or covalent immobilization of functional biomolecules. The most commonly employed functional groups for Diels-Alder cross-linking are furan and maleimide groups. As an example, furan and maleimide functionalized HA were cross-linked in 2-(*N*-morpholino)ethanesulfonic acid buffer at various volume ratios for delivery of dexamethasone which is an adipogenic factor [44]. Using the Diels-Alder reaction, Fisher et al. cross-linked furan-modified HA with bismaleimide enzyme-cleavable peptide cross-linkers to study MDA-MB-231 breast cancer invasion [45]. Recently, inverse electron demand Diels-Alder reaction has been employed in hydrogel cross-linking with tetrazine as the diene for cycloaddition with an alkene or alkyne (**Table 1**, entry 4). Desai et al. synthesized tetrazine and norbornene functionalized alginate to form cross-linking hydrogel using inverse electron demand Diels-Alder reaction without the external input of energy, cross-linkers, or catalyst [46]. Tetrazine and *trans*-cyclooctene are also a good pair of functional groups for inverse demand Diels-Alder reactions. Jung and others have employed tetrazine and *trans*-cyclooctene for protein conjugation using a tobacco mosaic virus template assembled with hydrogel microparticles for protein sensing applications [47]. Zhang et al. reported the interfacial bioorthogonal cross-linking of tetrazine modified HA with bis-*trans*-cyclooctene cross-linker to produce pattern biomaterials through diffusion-controlled gelation at the liquid-gel interface [48]. Similar to SPAAC, the cycloaddition between tetrazine and cyclooctene is also promoted by the strain in the *trans*-cyclic structure and the rate constant was determined to be $k_2 > 10^5\ M^{-1}\ S^{-1}$ [48].

2.2.5. *Disulfide formation/exchange*

Disulfide bonds are usually formed from oxidation of thiol groups. For example, thiolated HA can be chemically synthesized with varied degrees of thiolation and cross-linked through oxidation in the air or using hydrogen peroxide [49]. In another report, HA and gelatin were chemically modified using 3,3'-dithiobis(propionic hydrazide) followed by treatment of dithiothreitol. The thiol derivatives of HA and gelatin bearing thiol groups were mixed to form a disulfide cross-linking hydrogel in the presence of hydrogen peroxide. The hydrogel can be degraded by hyaluronidase [50]. Zhang et al. synthesized elastin-like polypeptide hydrogels for wound repair by disulfide bond cross-linking in the presence of ultraviolet (UV) light [51].

Macromolecules with pyridyl disulfide can react with thiol functionalized polymer through disulfide exchange, eliminating pyridine-2-thione as a by-product. Kannan et al. developed PAMAM dendrimer-PEG hydrogels for the sustained release of amoxicillin through disulfide exchange of a pyridyl disulfide functionalized PAMAM dendrimer generation 4 and an eight-armed thiolated PEG [52]. Similarly, an HA-based cleavable hydrogel was synthesized by cross-linking pyridyl disulfide functionalized HA with PEG dithiol [53]. The disulfide exchange reaction has been found to exhibit fast cross-linking kinetics and cytocompatibility as hydrogels can be synthesized in minutes under physiological pH in the presence of many cell types, with tunable rheological and physical properties. The limitation of the disulfide cross-linking approach is that the hydrogels may exhibit low stability by degradation especially in the presence of hyaluronidase or reducing agents such as glutathione [53].

2.2.6. *Epoxide coupling*

Water-soluble epoxides are highly reactive electrophiles that readily react with nucleophiles such as amines, alcohols, and even carboxylic acids, and the reactions are not oxygen sensitive. Due to the difference in nucleophilicity, the reaction rate is fast with amines and slow with alcohols. PEG diepoxide, 1,2,3,4-diexpoxbutane, and 1,4-butandiol diepoxide are commonly used epoxide sources [15]. HA was reported to react with diepoxide to form cross-linking hydrogels under either basic or acidic conditions [54, 55]. However, epoxides may suffer from some degree of hydrolysis under basic conditions. Binetti and others cross-linked PVA with PEG diglycidylether (PEGDGE) through epoxide coupling to form PVA/PEG hydrogels for injectable nucleus replacement [56]. Calvert et al. made epoxy hydrogels as hydrogel sensors for glucose by epoxide coupling of PEGDGE and Jeffamine in aqueous conditions [57].

2.2.7. *Schiff-base reaction*

A Schiff base is usually achieved by reaction of amines, hydrazides, or hydroxylamines with aldehydes or ketones to form an imine, hydrazone, or oxime linkage. Schiff-base formation can occur in aqueous conditions without using extra chemicals or catalysts. It also displays controllable reaction rates depending on the pH. Therefore, it becomes a facile approach to produce *in situ*-forming hydrogels. The disadvantage of Schiff-base reaction is that aldehyde-containing compounds could affect bioactive factors or extracellular matrix molecules through reaction with amine groups [58]. In one study, aldehyde functionalized alginate was prepared by reaction of oxidized alginate and borax, followed by cross-linking with amine groups in gelatin. The hydrogel formation time ranged from a couple of seconds to less than one minute by varying the concentration of the components [59]. Similarly, aldehyde and oxyamine functionalized PEG was cross-linked through Schiff-base chemistry [60]. The hydrazone linkage is known to be labile and reversible due to hydrolysis, which may cause instability to hydrogels. A recent report demonstrated a hydrazone bond that was 15-fold more stabilized than regular hydrazone by fine tuning of the charge distribution over the hydrazone moiety, thus producing a stable hydrogel for tissue engineering [61]. The oxime linkage exhibits better hydrolytic stability than the hydrazone or imine linkage [58].

2.2.8. *Cross-linking by other reactions*

Other chemical reactions suitable for hydrogel cross-linking include condensation reactions that involve hydroxyl or amino groups reacting with carboxylic acid derivatives such as activated esters or isocyanates [33, 62], and genipin coupling through which nucleophiles such as amine or alcohol-containing polymers react with a natural cross-linker genipin [63]. Reactants that contain a photo-activated functional group can be cross-linked by photoirradiation [64]. The reaction between boronic acid and amine or hydroxyl group has been employed as a cross-linking approach to make pH responsive hydrogels [65].

2.2.9. *Cleavable cross-linking*

It is worth noticing that the stability of cross-linking is not always preferable in some applications when the encapsulated payload in hydrogel needs to be released at the target site. In that case, reversible cross-linking with labile linkages can be introduced so that the hydrogel is broken down in response to stimuli and the cargo can be released for function. The most common cleavable cross-links are cleavable to pH, photo, redox, enzyme, etc. Interested readers should refer to relevant articles for details [33, 66].

2.2.10. Biorthogonal chemistry

Despite the various cross-linking strategies for synthesis of different hydrogel networks and immobilization of functional cues within a cellular scaffold, techniques are needed to introduce different functionalities in specific locations and at different times to produce spatiotemporally complex and yet well-defined biochemical cues in hydrogels for specific applications or studies [67]. For that purpose, a strategy has been established to use two or more orthogonal reactions in a sequential fashion with the first reaction to form the cross-linking hydrogel and the second or later reactions to introduce biochemical functionalities. **Figure 3** illustrates how bioorthogonal strategy increases spatiotemporal diversity in hydrogel networks.

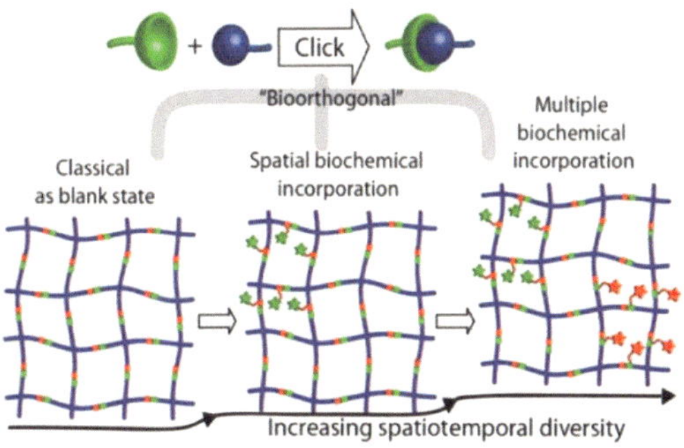

Figure 3. Bioorthogonal chemistry to increase spatiotemporal diversity in hydrogels [67].

The reactions employed in bioorthogonal chemistry must be nontoxic to cells. In addition, the reaction should be selective and the fidelity of the reaction should not be affected by other functionalities present. Among the cross-linking reactions for *in situ*-forming hydrogels, the azide-alkyne click chemistry, the SPAAC, thiol-ene reaction and thiol-based Michael addition are widely used to achieve orthogonality due to their virtues of simplicity, mild reaction conditions, availability of functional precursors, and versatility for spatiotemporal manipulation. For example, DeForest et al. demonstrated the bioorthogonal chemistry of a four-armed PEG-tetraazide and a degradable peptide which was double functionalized with dicyclooctyne and a pendant allyl moiety. The cross-linking occurred between the dicyclooctyne in peptide and the azide in PEG through SPAAC. The allyl moiety in the hydrogel matrix allowed for subsequent postgelation, photopatterning, or attachment of bioactive molecules such as RGD sequence at a different time and location through orthogonal photoinitiated thiol-ene reaction [68]. This hydrogel system created niches for 3T3 fibroblast cell culture and displayed a remarkable impact on dynamic cellular behaviors such as cell attachment, proliferation, and morphology in defined regions controlled by photoexposure. This system demonstrated the possibility to manipulate cell functions by spatially tuning the material properties through chemistry. In another study, a tough hydrogel was synthesized by two orthogonal cross-linking reactions that reinforce each other. First, a loose network was formed through an electron-demand Diels-Alder reaction of norbornene-functionalized chitosan and ditetrazine-functionalized PEG. Subsequent postfunctionalization through thiol-ene cross-linking of a four-armed PEG-tetraalkyne and a linear PEG-dithiol resulted in a dense and biocompatible hydrogel network that exhibited high viability in 3D culture of human mesenchymal stem cells [69].

In spite of all the recent advancements, bioorthogonal strategy is still open for expansion in chemical synthesis of biomaterials for specific applications. Staudinger ligation, tetrazole photoclick reaction, and quadricyclane ligation are new orthogonal reactions that are potentially useful for synthesizing *in situ*-forming hydrogels with bioorthogonality [70–72] (**Table 1**, entry 9). In addition, reactions facilitated by thermal- or photochemical stimuli also hold potential as bioorthogonal reactions for hydrogels. Thiol-yne photoclick reactions are yet to be investigated for bioorthogonal chemistry in hydrogel [73]. Reactions of ultrasound-mediated reversible cycloadditions are also candidates for bioorthogonal chemistry [74].

2.3. Cross-linking by enzyme catalysis

Enzymes can be used to cleave or establish a chemical bond with greater efficiency than other methods. Due to their short reaction times and specificity, enzymes have been used for catalytic cross-linking to form hydrogels without interference with other chemical functional groups in macromolecules. Enzymatic cross-linking occurs under mild reaction conditions such as an aqueous environment, a neutral pH, or mild temperatures [58]. The most studied enzymes to prepare hydrogels are horseradish peroxidase, transglutaminase, tyrosinase, phosphopantetheinyl transferase, and lysyl oxidase [75]. Themolysin, galactosidase, and esterase have also been used to prepare *in situ* hydrogels through the enzymatic cross-linking reactions or decreased aqueous solubility of the polymer or the compound [15].

Overall, enzyme catalyzed reactions are a new approach in hydrogel formation. The reactions provide exceptional control over hydrogel formation, promoting higher complexity, noncytotoxicity and noninvasiveness. Despite the major advantages of enzymatic reactions, the challenges of this approach include instability of some of the enzymes and the insufficient mechanical properties of the gels formed. Interested readers may refer to more recent reports for a comprehensive review [76, 77].

3. Biomedical applications of *in situ*-forming cross-linking hydrogels

Hydrogels are suitable for biomedical applications due to their favorable properties such as biocompatibility, structural similarity to ECM, and biodegradability. Among the hydrogel family, *in situ*-forming hydrogels are specifically attractive as they can be injected by minimally invasive techniques and exhibit sol-to-gel phase transition under external physical or chemical cross-linking [78]. Among any possible applications, we reviewed the most promising applications of hydrogels, such as tissue engineering, 3D bioprinting, drug delivery, gene delivery, antimicrobial hydrogels, wound healing, and cancer research.

3.1. Tissue engineering

The objective of tissue engineering is the fabrication of living parts for the body due to the tremendous need for organs and tissues [79]. Hydrogels have been widely used in tissue engineering because their properties are similar to those of natural ECM, which is essential for these purposes [48]. Many strategies employ material scaffolds to engineer tissues. These scaffolds serve as a synthetic ECM to provide a 3D architecture for cells and direct the growth and formation of desired tissues. Those scaffolds can be used for space filling agents, bioactive molecule delivery, and cell/tissue delivery.

Hydrogels made from natural polysaccharides are ideal scaffolds, as they resemble the native ECM of tissues which are comprised of various glycosaminoglycans. An injectable electroactive and antioxidant hydrogel based on a tetra-aniline functional copolymer and α-CD has exhibited excellent biodegradation, cell proliferation, and regeneration properties in tissue engineering [80]. *In situ*-forming metal-free chitosan/hyaluronan hydrogels also showed biocompatible and biodegradable properties and are good for soft tissue engineering. The hydrogel could support cell survival and proliferation of human adipose-derived stem cells (hADSCs) [27].

One major drawback of hydrogels is the lack of mechanical strength; hence, maintaining and improving the mechanical integrity of the processed scaffolds has become a key issue regarding 3D hydrogel structures [81]. Many researchers focus on the development of hydrogels using synthetic biomaterials that may have enhanced mechanical properties. Schuurman et al. prepared gelatin-methacrylamide (GelMA) hydrogels with tunable mechanical properties by manipulating the cross-linking parameters [82]. Hu et al. designed and synthesized a library of supramolecular hydrogels inspired by collagen. Those hydrogels exhibited promising potential for tissue engineering as they mimic properties of collagen [83]. Readers may refer to reviews of hydrogel scaffolds for tissue engineering [13, 76].

Lack of vascular networks in engineered tissues thicker than 200 μm puts a limitation on the nutrition of encapsulated cells. Therefore, advances in tissue engineering *in vitro* have generated a necessity for a parallel development and design of 3D vascular networks [84]. Current approaches in vascular network bioengineering mainly use natural hydrogels as embedding scaffolds that have drawbacks of poor mechanical stability and suboptimal durability. The search for improved hydrogels has become a priority in tissue engineering. Chen et al. developed a vascular network generated in photo-cross-linkable GelMA hydrogels, in which human blood-derived endothelial colony-forming cells and bone marrow mesenchymal stem cells generate extensive capillary-like networks *in vitro* [85].

3.2. 3D Bioprinting

3D bioprinting technologies enable the automated biofabrication of cell-laden constructs through the layer-by-layer deposition of biochemicals (termed as "bioinks") both *in vitro* and *in vivo*. Among the wide range of biofabrication techniques available to generate cellular constructs for tissue engineering, 3D bioprinting is currently one of the most fascinating because of its ability to print multiple biomaterials, cells, and bioinks in precise spatial locations with high resolution and accuracy, which is important to the precise shape of the 3D structures as well as the self-development of cells in the structures [86, 87].

The full implementation of bioprinting strongly depends on the development of novel biomaterials exhibiting fast-cross-linking kinetics, appropriate printability, cell-compatibility, and biomechanical properties. Different biomaterials have been used in 3D bioprinting, such as alginate [88], HA, PEG [89], gelatin, collagen [90], and thermo-responsive biodegradable polyurethane [91]. Various cells have been printed on those materials, such as tumor cells, neural cells, and stem cells [88–91]. Hsieh et al. investigated two thermo-responsive water-based biodegradable polyurethane dispersions (PU1 and PU2) [91]. The stiffness of the hydrogel could be easily fine-tuned by the solid content of the dispersion. Neural stem cells (NSCs) were embedded into the polyurethane dispersions before gelation and showed excellent proliferation and differentiation in 25–30% PU2 hydrogels. Therefore, the newly developed 3D bioprinting technique involving NSCs embedded in the thermo-responsive biodegradable polyurethane ink offers new possibilities for future applications of 3D bioprinting in neural tissue engineering.

Photo-cross-linkable hydrogels are attractive materials for bioprinting as they provide fast polymerization under cell-compatible conditions and exceptional spatiotemporal control over the gelation process. Photo-cross-linkable GelMA and PEG dimethacrylate (PEGDA) were used as example hydrogels to demonstrate the feasibility and effectiveness of the approach. An array of GelMA/PEG hydrogels encapsulating human periodontal ligament stem cells (PDLSCs) with a gradient of material composition was printed, and the responses of human PDLSCs in the hydrogel array were investigated. The approach may be helpful for human PDLSCs-ECM screening and other cell-ECM systems. Please refer to the review for most recent developments on 3D bioprinting of photo-cross-linkable biodegradable hydrogels for tissue engineering [86].

3.3. Drug delivery

The use of hydrogels in drug delivery applications has been a significant subject in recent research due to the unique physical properties of hydrogels, the biodegradability, and the changes in gel structure according to environmental stimuli such as temperature, pH, and/or ionic strength. Their highly porous structure can easily be tuned by controlling the density of cross-links in the gel matrix and the affinity of the hydrogels for the aqueous environment in which they are formed. Their porosity also permits the loading of drugs into the gel matrix and subsequent drug release through the gel network [92]. The hydrogel systems have been developed for delivery of biomolecules ranging from small molecular drugs to large biomacromolecules such as nucleic acids, peptides, and proteins [93, 94]. Readers are directed to some recent reviews on drug delivery [95, 96].

There are several important areas in the field of hydrogels for drug delivery. First, development of *in situ*-forming hydrogel systems for drug delivery is an area of great interest. Fan et al. developed *in situ* injectable biocompatible poly(glutamic acid)-based hydrogels that are potential candidates in cell encapsulation and drug delivery [97].

Second, the development of strategies to increase the loading rate and capacity and to control the release of drugs from the hydrogel is an important area of interest. There are generally two ways to load drugs into hydrogels, e.g., soaking formed hydrogels in drug solution and forming hydrogels in the presence of drugs. Both ways have their limitations and often result in a modest amount of drugs loaded into hydrogels. Ray et al. prepared an IPN hydrogel based on PVA networking with PAA (PVA-co-PAA)/NaCl microspheres. The hydrogels were loaded with diltiazem hydrochloride (DL) and showed comparatively higher DL entrapment (79%), better control over DL release up to 24 h, and were more effective in reducing blood pressure to 40.1% [98]. Josef et al. investigated a composite gel system formulated from microemulsions (ME) embedded in alginate hydrogels. These hydrogels appeared to be a promising drug delivery system since they were capable of loading several hydrophobic compounds with a wide range of aqueous solubility and exhibited a release of 6–8 hours in water [99].

Third, there is a great interest in the development of stimuli-responsive hydrogels that are sensitive to temperature, pH, sugar, ionic strength, etc. These hydrogels are good candidates for controlled delivery systems that would release drugs to match a patient's physiological needs at the proper time and/or site [100]. Zhou et al. synthesized a series of pH-temperature dual stimuli-responsive copolymers [19]. Altering the temperature and the pH values of the environment could effectively control the release of the model drug. Li et al. synthesized pH and glucose dually responsive injectable hydrogels through the dynamic covalent imine bond and phenylboronate ester based on phenylboronic modified chitosan and oxidized dextran [101]. The rapid gelation and biocompatible cross-linking chemistry were appropriate for the incorporation of drug molecules and cells by *in situ* gel formation. Either decreasing the pH from physiological to mildly acidic or increasing the concentration of glucose around the gels could accelerate the rate of drug release from the gels. This hydrogel system potentially represented a versatile material platform for anticancer drug delivery.

3.4. Gene delivery

Gene delivery via hydrogels provides a fundamental tool for a variety of clinical applications including regenerative medicine, gene therapy for inherited disorders, and drug delivery [102]. Hydrogels serve the purpose of gene delivery by preserving activity of viral or nonviral vectors and shielding vectors from any host immune response. Hydrogels can also be injectable and environmentally responsive. Therefore, hydrogels hold great promise for gene delivery. There are two major areas that attracted great attention, vectors and biomaterial delivery system optimization.

Hydrogels used in gene delivery often need higher strength for extended use in order for internalization and transgene expression to occur. A hydrogen bonding strengthened hydrogel was prepared by radical copolymerization of PEG methacrylated β-CD and 2-vinyl-4,6-diamino-1,3,5-triazine monomer [103]. Kidd et al. investigated the delivery of lentiviral gene therapy vectors from fibrin hydrogels containing hydroxyapatite nanoparticles that can interact with both fibrin and the lentivirus [104]. The interaction of the hydroxyapatite with the fibrin may stabilize the hydrogels that will influence the rate of cell infiltration and vector release. The interaction of hydroxyapatite with lentiviral particles can enhance and localize gene transfer within the hydrogel. These studies demonstrate the potential of fibrin hydrogels to serve as a material support for regenerative medicine and as a vehicle for the localized delivery of lentiviral vectors *in vivo*.

By choosing the proper vector and biomaterial system, gene delivery can be controlled for improved transgene expression. The interactions between scaffolds and vectors should be optimized so that vectors are adequately retained, but undergo dissociation, which enable vectors to interact with nearby cells and internalize [102]. The diffusion rate must be balanced to maintain spatial localization of gene transfer. The addition of interconnected macropores into the hydrogels can further increase the probability that infiltrating cells will internalize vectors and thereby improve transgene expression.

3.5. Wound healing

Wound healing is a complex process that implies equilibrium between inflammatory and vascular activity in the connective tissue and epithelial cells. The regenerative process needs the assistance of important elements to activate the natural processes of angiogenesis, activation of growth factors, and regeneration in a well-structured and biomimetic sequential process [105]. Hydrogel wound dressing has been widely researched because hydrogels promote wound healing by moisture retention to maintain a homeostatic environment. Ajji et al. prepared hydrogel wound dressings composed of PVP, PEG, and agar [106]. The gel fraction increased with increasing PVP, and decreased with increasing PEG. The hydrogel dressings could also be considered a good barrier against microbes. Reyes-Ortega et al. reported a new system based on the sequential release of two complementary bioactive components for application in the healing of compromised wounds [105]. The internal layer was a highly hydrophilic and biodegradable film loaded with the proangiogenic, anti-inflammatory, and antibacterial peptide, proadrenomedullin N-terminal 20 peptide for release. The more stable and less hydrophilic external layer was loaded with resorbable nanoparticles of bemiparin to promote the activation of growth factors and to provide a good biomechanical stability and controlled permeability of the bilayer dressing. This system demonstrated high efficacy of the early steps of the regenerative process in the wound site.

Traditional hydrogel dressings are inconvenient in applications as they need some degree of expertise and cause pain during changes. The thermo-sensitive hydrogels avoid the necessity of repeated and complicated application. Lee et al. investigated the ability of a thermo-sensitive hydrogel made of a triblock copolymer, PEG-PLGA-PEG, with TGF-β1 to treat the wound surface [107]. Results showed that the thermo-sensitive hydrogel provided excellent wound dressing activity and delivered plasmid TGF-β1 to promote wound healing in a diabetic mouse model. Hassan et al. developed a stem cell hydrogel system, in which the hADSCs were encapsulated *in situ* in the water-soluble, thermo-responsive hyperbranched PEG-based copolymer with multiple acrylate functional groups in combination with thiolated HA [108]. The hADSCs were successfully encapsulated *in situ* with high cell viability for up to 7 days in hydrogels and secrete proangiogenic growth factors with low cytotoxicity. This stem cell hydrogel system could be an ideal living dressing system for wound healing applications.

Supramolecular hydrogels are formed by noncovalent cross-linking of polymeric chains in water and can be developed specifically for biomedical applications [95]. Supramolecular hydrogels prepared by incorporating uranium chelating agents to eliminate uranium ions from the radionuclides contaminated wound sites of mice [109]. D-Glucosamine-based supramolecular hydrogels assist wound healing and prevent the formation of scars [110].

3.6. Antimicrobial hydrogels

The infectious diseases caused by pathogenic microorganisms such as bacteria, viruses, and parasites are still a public health problem despite the major development in health care and medical technology. Treatment with conventional antibiotics of infectious diseases often leads to the development of antibiotic resistance [111]. Recently, a new strategy to treat infectious diseases has been developed using antimicrobial hydrogels. The hydrogels act on the entire cellular membrane, which leads to cell membrane rupture, followed by a leakage of cytoplasmic contents and cell death. Different types of antimicrobial hydrogels have been developed in recent research.

Some hydrogels possess antimicrobial properties that include natural and synthetic polymeric hydrogels, and peptide-based hydrogels. Mohamed et al. prepared hydrogels by chitosan cross-linked with different amounts of pyromellitimide benzoyl thiourea moieties [112]. The hydrogels were extremely porous and exhibited a higher antibacterial activity and antifungal activity. The swelling ability of hydrogels and their antimicrobial activity increased with cross-linking density. Peng et al. developed novel cellulose-based hydrogels that showed superabsorbent property, high mechanical strength, good biocompatibility, and excellent antimicrobial efficacy against *Saccharomyces cerevisiae* [113]. The results showed possible use of these hydrogels for hygienic application. Synthetic polymeric hydrogels often have good mechanical properties and have been widely studied. Liu et al. developed a series of *in situ*-forming antimicrobial and antifouling hydrogels generated from cationic polycarbonate and four-armed PEG [114]. Peptide-based antimicrobial hydrogels with excellent inherent antibacterial activity have also been reported in recent years. Salick et al. designed a β-hairpin hydrogel scaffold based on the self-assembling of 20-residue peptide MAX1 that possessed intrinsic broad-spectrum antibacterial activity [115].

Despite the tremendous ability of antimicrobial hydrogels in breaking down multidrug resistant microbes, the interactions between antimicrobial polymers and microbial cell membranes are nonspecific which, in most cases, cause mammalian cell death above certain concentrations [111]. One solution is to combine antibiotics and antimicrobial hydrogels so that less antimicrobial hydrogel is used and the associated toxicity is minimized.

Another type of interesting hydrogels contains antimicrobial metal nanoparticles. The use of silver ions and silver nanoparticles in hydrogels has obtained substantial advances in wound treatment [111]. The silver nanoparticles supported within PVA/cellulose acetate/gelatin was successfully synthesized. The hydrogels have antimicrobial activity against various fungi and bacteria [116]. The toxicity of silver and other metal salts is a disadvantage for this type of hydrogels. Efforts have been made to reduce the toxicity.

3.7. Cancer research

Hydrogels have been used in cancer research among many other applications. However, many drugs are hydrophobic and cannot be efficiently loaded and released from hydrogels. There are two ways to improve the loading and releasing, incorporating hydrophobic domains into hydrogels and introducing nanoparticles which encapsulate hydrophobic compounds [99].

Recent trends have indicated significant and growing interest in developing nanocomposite hydrogels (NCH) for various biomedical applications. NCH are hydrated polymeric networks, cross-linked with each other and/or with nanostructures [117]. Some of the nonocarriers have been successfully incorporated in gel networks, such as carbon nanotubes, ME, dendrimers, metal, ceramic, and polymeric nanoparticles [99, 117]. The NCH has nanocarrier stabilization, shape regulation, improved composite viscoelasticity, and mechanical properties with optimized drug release kinetics on top of the conventional hydrogel characteristics [118]. Abdel-Bar et al. developed a cisplatin ME hydrogel for controlled cisplatin release and improved cytotoxicity with decreased side effects [119]. The NCH containing nano-sized carriers allowed a zero order drug release for 14 days and enhanced cytotoxicity. The higher animal survival rate and lower tissue toxicities proved the decreased toxicity of cisplatin nanocomposite compared to its solution. This system could help in achieving better outcomes and quality of life during use of chemotherapy for cancer treatment by intraperitoneal administration.

Another great interest of hydrogels in cancer research focuses on the cancer cell invasion in hydrogels. Fisher et al. studied the HA-based cross-linked hydrogels in breast cancer cell invasion [45]. The results showed that increased crosslink density correlates with decreased breast cancer cell invasion whereas incorporation of enzyme-cleavable sequences within the peptide cross-linker enhances invasion. This study provides a platform that recapitulates variable tissue properties and elucidates the role of the microenvironment in cancer cell invasion by independently tuning the mechanical and chemical environment of ECM mimetic hydrogels. Zhang et al. created covalently cross-linked hydrogel materials through a rapid reaction at the gel-liquid interface [48]. The interfacial cross-linking was then used to encapsulate prostate cancer cells. The cells obtained 99% viability, proliferated readily, and formed aggregated clusters. Such *in vitro* models are critically needed for drug testing and discovery.

4. Conclusions and perspectives

In this chapter, we have discussed recent progress in polymerization, various strategies for cross-linking of natural and synthetic biopolymers for preparation of *in situ*-forming hydrogels, and updated applications in biomedical fields. There are a number of points that should be emphasized for future direction of design and synthesis of *in situ*-forming hydrogels for biomedical applications.

From a chemistry point of view, the reactions selected should occur in an aqueous environment under mild reaction conditions without damaging the encapsulated biofunctional molecules and cells. The chemicals used for cross-linking such as monomers, initiators, cross-linkers, and catalysts should be carefully selected to minimize toxicity to cells. Attention should also be paid to biocompatibility between polymers and incorporated bioactive species such as cells and proteins. Introduction of reactive functional groups to hydrogel materials will surely promote the cross-linking of hydrogels, but the active groups may also show off-target reactivity to incorporated bioactive species or cells. For example, amino groups and thiol groups on proteins can react with vinyl sulfone and acrylate groups used in Michael addition, aldehyde groups used in Schiff-base formation, or diepoxide groups in coupling reaction for hydrogel cross-linking. These undesired reactions may damage proteins, reduce drug efficacy, or induce immunogenicity. Therefore, reactions are preferred to be highly efficient and selective for cross-linking so that all groups intended for cross-linking are reacted, but no undesired reactions occur between the hydrogel and incorporated biofunctionalities or cells.

Because of the widespread biomedical applications, research on *in situ*-forming hydrogels is increasingly intensive. However, the potential of hydrogels has not been fully explored yet. There are future challenges in each area of the applications. For example, one of the most important challenges in tissue engineering is how hydrogels can be used to stimulate the blood vessel network formation through angiogenic factors and endothelial cells in the desired tissue [120]. In addition, many tissues such as bone and muscle require high mechanical properties that most current hydrogels lack. The mechanical properties of hydrogels originate from the intrinsic rigidity of the polymers and the cross-linking density. Therefore, higher mechanical properties may be achieved by increasing the component of synthetic polymers in hydrogel. Orthogonal cross-linking reactions could be employed to enhance mechanical properties by increasing cross-linking density. Other strategies to develop strong hydrogels include double networks for which two polymeric networks with contrasting properties are combined to achieve much higher mechanical properties than the two independent networks, topological gels in which the long polymer chains are topologically interlocked by cross-linkers to achieve high tensility, and nanocomposite hydrogels in which the main hydrogel networks are incorporated with high-strength inorganic nanostructures [121]. In drug delivery, hydrogels will become a large portion of drug delivery systems in the future. Hydrogel systems that administer drugs with a controlled release rate at the desired sites are to be investigated. Specifically, more research on hydrogels for delivery of therapeutic proteins and peptides is expected [3].

In general, a deeper understanding of material properties for the development of *in situ*-forming hydrogels that replicate the complex nature of tissue will facilitate these efforts. To meet these goals, both physicochemical and biological cues should be applied with spatio-temporal control in hydrogel. Novel hydrogel materials such as stimuli-responsive smart hydrogels, hydrogels prepared with bioorthogonality, enzyme-mediated cross-linking, or a combination of diverse chemistries should result in hydrogels that are precisely controllable, diverse, and biomimetic in order to perform specific requirements of biomedical applications in the future.

Acknowledgements

This research is financially supported by the NSF EPSCoR RII Track 1 cooperative agreement awarded to the University of South Carolina (NSF EPSCoR Cooperative Agreement number EPS-0903795).

Author details

Xiangdong Bi* and Aiye Liang

*Address all correspondence to: xbi@csuniv.edu

Department of Physical Sciences, Charleston Southern University, Charleston, South Carolina, USA

References

[1] Tran NQ, Joung YK, et al. Supramolecular hydrogels exhibiting fast in situ gel forming and adjustable degradation properties. Biomacromolecules 2010;11:617–625.

[2] Wichterle O, Lim D. Hydrophilic gels for biological use. Nature 1960;185:117–118.

[3] Calo E, Khutoryanskiy VV. Biomedical applications of hydrogels: a review of patents and commercial products. Eur. Polym. J. 2015;65:252–267.

[4] Martinez-Sanz E, Ossipov DA, et al. Bone reservoir: Injectable hyaluronic acid hydrogel for minimal invasive bone augmentation. J. Controlled Release 2011;152:232–240.

[5] Thornton AJ, Alsberg E, et al. Shape retaining injectable hydrogels for minimally invasive bulking. J. Urol. 2004;172:763–768.

[6] Duarte AP, Coelho JF, et al. Surgical adhesives: Systematic review of the main types and development forecast. Prog. Polym. Sci. 2012;37:1031–1050.

[7] Engler AJ, Sen S, et al. Matrix elasticity directs stem cell lineage specification. Cell 2006;126:677–689.

[8] Spiller KL, Maher SA, et al. Hydrogels for the repair of articular cartilage defects. Tissue Eng. B-Rev. 2011;17:281–299.

[9] Ahmed EM. Hydrogel: preparation, characterization, and applications: a review. J. Adv. Res. 2015;6:105–121.

[10] Rice JJ, Martino MM, et al. Engineering the regenerative microenvironment with biomaterials. Adv. Healthcare Mater. 2013;2:57–71.

[11] Thiele J, Ma YJ, et al. 25th Anniversary article: designer hydrogels for cell cultures: a materials selection guide. Adv. Mater. 2014;26:125–148.

[12] Slaughter BV, Khurshid SS, et al. Hydrogels in regenerative medicine. Adv. Mater. 2009;21:3307–3329.

[13] Van Vlierberghe S, Dubruel P, et al. Biopolymer-based hydrogels as scaffolds for tissue engineering applications: a review. Biomacromolecules 2011;12:1387–1408.

[14] Hennink WE, van Nostrum CF. Novel crosslinking methods to design hydrogels. Adv. Drug Deliv. Rev. 2012;64:223–236.

[15] Ko DY, Shinde UP, et al. Recent progress of in situ formed gels for biomedical applications. Prog. Polym. Sci. 2013;38:672–701.

[16] Matyjaszewski K, Spanswick J. Controlled/living radical polymerization. Mater. Today 2005;8:26–33.

[17] Liu FY, Seuring J, et al. A non-ionic thermophilic hydrogel with positive thermosensitivity in water and electrolyte solution. Macromol. Chem. Phys. 2014;215:1466–1472.

[18] Shantha KL, Harding DRK. Synthesis and evaluation of sucrose-containing polymeric hydrogels for oral drug delivery. J. Appl. Polym. Sci. 2002;84:2597–2604.

[19] Zhou M, Liu KY, et al. A facile preparation of pH-temperature dual stimuli-responsive supramolecular hydrogel and its controllable drug release. J. Appl. Polym. Sci. 2016;133:43279.

[20] Reddy NS, Rao KM, et al. Pectin/poly(acrylamide-co-acrylamidoglycolic acid) pH sensitive semi-IPN hydrogels: selective removal of Cu^{2+} and Ni^{2+}, modeling, and kinetic studies. Desalin. Water Treatment 2016;57:6503–6514.

[21] Kolb HC, Finn MG, et al. Click chemistry: diverse chemical function from a few good reactions. Angew. Chem. Int. Ed. 2001;40:2004–2021.

[22] Jirawutthiwongchai J, Krause A, et al. Chitosan-oxanorbornadiene: a convenient chitosan derivative for click chemistry without metal catalyst problem. ACS Macro Lett. 2013;2:177–180.

[23] Piluso S, Hiebl B, et al. Hyaluronic acid-based hydrogels crosslinked by copper-catalyzed azide-alkyne cycloaddition with tailorable mechanical properties. Int. J. Artif. Organs 2011;34:192–197.

[24] Seelbach RJ, Fransen P, et al. Multivalent dendrimers presenting spatially controlled clusters of binding epitopes in thermoresponsive hyaluronan hydrogels. Acta Biomater. 2014;10:4340–4350.

[25] Kaga S, Yapar S, et al. Photopatternable "clickable" hydrogels: "orthogonal" control over fabrication and functionalization. Macromolecules 2015;48:5106–5115.

[26] Truong VX, Ablett MP, et al. In situ-forming robust chitosan-poly(ethylene glycol) hydrogels prepared by copper-free azide-alkyne click reaction for tissue engineering. Biomater. Sci. 2014;2:167–175.

[27] Fan M, Ma Y, et al. Cytocompatible in situ forming chitosan/hyaluronan hydrogels via a metal-free click chemistry for soft tissue engineering. Acta Biomater. 2015;20:60–68.

[28] Codelli JA, Baskin JM, et al. Second-generation difluorinated cyclooctynes for copper-free click chemistry. J. Am. Chem. Soc. 2008;130:11486–11493.

[29] Uliniuc A, Popa M, et al. New approaches in hydrogel synthesis – click chemistry: a review. Cellul. Chem. Technol. 2012;46:1–11.

[30] Baskin JM, Prescher JA, et al. Copper-free click chemistry for dynamic in vivo imaging. Proc. Natl. Acad. Sci. U.S.A. 2007;104:16793–16797.

[31] DeForest CA, Sims EA, et al. Peptide-functionalized click hydrogels with independently tunable mechanics and chemical functionality for 3D cell culture. Chem. Mater. 2010;22:4783–4790.

[32] Yigit S, Sanyal R, et al. Fabrication and functionalization of hydrogels through "click" chemistry. Chemistry 2011;6:2648–2659.

[33] Zhang X, Malhotra S, et al. Micro- and nanogels with labile crosslinks – from synthesis to biomedical applications. Chem. Soc. Rev. 2015;44:1948–1973.

[34] McLemore R, Lee BH, et al. Surfactant effects on the kinetics of Michael-type addition reaction in reverse emulsion polymeric systems. J. Biomed. Mater. Res. B Appl. Biomater. 2009;89B:191–198.

[35] Peng JQ, Qi XL, et al. Octreotide-conjugated PAMAM for targeted delivery to somatostatin receptors over-expressed tumor cells. J. Drug Target. 2014;22:428–438.

[36] Bi XD, Liang AY, et al. Thiol-ene crosslinking polyamidoamine dendrimer-hyaluronic acid hydrogel system for biomedical applications. J. Biomater. Sci. Polym. Ed. 2016. DOI: 10.1080/09205063.09202016.01159473.

[37] Bi XD, Luckanagul JA, et al. Synthesis of PAMAM dendrimer-based fast cross-linking hydrogel for biofabrication. J. Biomater. Sci. Polym. Ed. 2015;26:669–682.

[38] Hoyle CE, Bowman CN. Thiol-ene click chemistry. Angew. Chem. Int. Ed. 2010;49:1540–1573.

[39] Hoyle CE, Lowe AB, et al. Thiol-click chemistry: a multifaceted toolbox for small molecule and polymer synthesis. Chem. Soc. Rev. 2010;39:1355–1387.

[40] Fairbanks BD, Schwartz MP, et al. A versatile synthetic extracellular matrix mimic via thiol-norbornene photopolymerization. Adv. Mater. 2009;21:5005–5010.

[41] Ganivada MN, Kumar P, et al. A unique polymeric gel by thiol-alkyne click chemistry. RSC Adv. 2015;5:50001–50004.

[42] Daniele MA, Adams AA, et al. Interpenetrating networks based on gelatin methacrylamide and PEG formed using concurrent thiol click chemistries for hydrogel tissue engineering scaffolds. Biomaterials 2014;35:1845–1856.

[43] Lin CC, Ki CS, et al. Thiol-norbornene photoclick hydrogels for tissue engineering applications. J. Appl. Polym. Sci. 2015;132:41563.

[44] Fan M, Ma Y, et al. Biodegradable hyaluronic acid hydrogels to control release of dexamethasone through aqueous Diels-Alder chemistry for adipose tissue engineering. Mater. Sci. Eng. C Mater. Biol. Appl. 2015;56:311–317.

[45] Fisher SA, Anandakumaran PN, et al. Tuning the microenvironment: click-crosslinked hyaluronic acid-based hydrogels provide a platform for studying breast cancer cell invasion. Adv. Funct. Mater. 2015;25:7163–7172.

[46] Desai RM, Koshy ST, et al. Versatile click alginate hydrogels crosslinked via tetrazine-norbornene chemistry. Biomaterials 2015;50:30–37.

[47] Jung S, Yi HM. An integrated approach for enhanced protein conjugation and capture with viral nanotemplates and hydrogel microparticle platforms via rapid bioorthogonal reactions. Langmuir 2014;30:7762–7770.

[48] Zhang H, Dicker KT, et al. Interfacial bioorthogonal cross-linking. ACS Macro Lett. 2014;3:727–731.

[49] Xiao YY, Zhang WZ, et al. Optimized preparation of thiolated hyaluronic acid hydrogel with controllable degree of substitution. J. Biomater. Tissue Eng. 2014;4:281–287.

[50] Shu XZ, Liu YC, et al. Disulfide-crosslinked hyaluronan-gelatin hydrogel films: a covalent mimic of the extracellular matrix for in vitro cell growth. Biomaterials 2003;24:3825–3834.

[51] Zhang YN, Avery RK, et al. A highly elastic and rapidly crosslinkable elastin-like polypeptide-based hydrogel for biomedical applications. Adv. Funct. Mater. 2015;25:4814–4826.

[52] Navath RS, Menjoge AR, et al. Injectable PAMAM dendrimer-PEG hydrogels for the treatment of genital infections: formulation and in vitro and in vivo evaluation. Mol. Pharm. 2011;8:1209–1223.

[53] Choh SY, Cross D, et al. Facile synthesis and characterization of disulfide-cross-linked hyaluronic acid hydrogels for protein delivery and cell encapsulation. Biomacromolecules 2011;12:1126–1136.

[54] Tomihata K, Ikada Y. Preparation of cross-linked hyaluronic acid films of low water content. Biomaterials 1997;18:189–195.

[55] Segura T, Anderson BC, et al. Crosslinked hyaluronic acid hydrogels: a strategy to functionalize and pattern. Biomaterials 2005;26:359–371.

[56] Binetti VR, Fussell GW, et al. Evaluation of two chemical crosslinking methods of poly(vinyl alcohol) hydrogels for injectable nucleus pulposus replacement. J. Appl. Polym. Sci. 2014;131:40843.

[57] Calvert P, Patra P, et al. Epoxy hydrogels as sensors and actuators – art. no. 65240M. In: BarCohen Y, editor. Electroactive Polymer Actuators and Devices; 2007. p. M5240–M5240.

[58] Nguyen QV, Huynh DP, et al. Injectable polymeric hydrogels for the delivery of therapeutic agents: a review. Eur. Polym. J. 2015;72:602–619.

[59] Balakrishnan B, Jayakrishnan A. Self-cross-linking biopolymers as injectable in situ forming biodegradable scaffolds. Biomaterials 2005;26:3941–3951.

[60] Ghosh S, Cabral JD, et al. Strong poly(ethylene oxide) based gel adhesives via oxime cross-linking. Acta Biomater. 2016;29:206–214.

[61] Oommen OP, Wang S, et al. Smart design of stable extracellular matrix mimetic hydrogel: synthesis, characterization, and in vitro and in vivo evaluation for tissue engineering. Adv. Funct. Mater. 2013;23:1273–1280.

[62] Vijayakumar V, Subramanian K. Diisocyanate mediated polyether modified gelatin drug carrier for controlled release. Saudi Pharmaceut. J. 2014;22:43–51.

[63] Dimida S, Demitri C, et al. Genipin-cross-linked chitosan-based hydrogels: reaction kinetics and structure-related characteristics. J. Appl. Polym. Sci. 2015;132:42256.

[64] Wang YC, Wu J, et al. Engineering nanoscopic hydrogels via photo-crosslinking salt-induced polymer assembly for targeted drug delivery. Chem. Commun. 2010;46:3520–3522.

[65] Mahalingam A, Jay JI, et al. Inhibition of the transport of HIV in vitro using a pH-responsive synthetic mucin-like polymer system. Biomaterials 2011;32:8343–8355.

[66] Wang HY, Heilshorn SC. Adaptable hydrogel networks with reversible linkages for tissue engineering. Adv. Mater. 2015;27:3717–3736.

[67] Azagarsamy MA, Anseth KS. Bioorthogonal click chemistry: an indispensable tool to create multifaceted cell culture scaffolds. ACS Macro Lett. 2013;2:5–9.

[68] DeForest CA, Polizzotti BD, et al. Sequential click reactions for synthesizing and patterning three-dimensional cell microenvironments. Nat. Mater. 2009;8:659–664.

[69] Truong VX, Ablett MP, et al. Simultaneous orthogonal dual-click approach to tough, in-situ-forming hydrogels for cell encapsulation. J. Am. Chem. Soc. 2015;137:1618–1622.

[70] Hall KK, Gattas-Asfura KM, et al. Microencapsulation of islets within alginate/poly(ethylene glycol) gels cross-linked via Staudinger ligation. Acta Biomater. 2011;7:614–624.

[71] He M, Li J, et al. Photodegradable supramolecular hydrogels with fluorescence turn-on reporter for photomodulation of cellular microenvironments. J. Am. Chem. Soc. 2013;135:18718–18721.

[72] Sletten EM, Bertozzi CR. A Bioorthogonal quadricyclane ligation. J. Am. Chem. Soc. 2011;133:17570–17573.

[73] Fairbanks BD, Scott TF, et al. Thiol-yne photopolymerizations: novel mechanism, kinetics, and step-growth formation of highly cross-linked networks. Macromolecules 2009;42:211–217.

[74] Wiggins KM, Brantley JN, et al. Polymer mechanochemistry: force enabled transformations. ACS Macro Lett. 2012;1:623–626.

[75] Teixeira LSM, Feijen J, et al. Enzyme-catalyzed crosslinkable hydrogels: emerging strategies for tissue engineering. Biomaterials 2012;33:1281–1290.

[76] Bae JW, Choi JH, et al. Horseradish peroxidase-catalysed in situ-forming hydrogels for tissue-engineering applications. J. Tissue Eng. Regen. Med. 2015;9:1225–1232.

[77] Zhang YM, Fan ZP, et al. Tough biohydrogels with interpenetrating network structure by bienzymatic crosslinking approach. Eur. Polym. J. 2015;72:717–725.

[78] Hammer N, Brandl FP, et al. Protein compatibility of selected cross-linking reactions for hydrogels. Macromol. Biosci. 2015;15:405–413.

[79] Drury JL, Mooney DJ. Hydrogels for tissue engineering: scaffold design variables and applications. Biomaterials 2003;24:4337–4351.

[80] Cui HT, Cui LG, et al. In situ electroactive and antioxidant supramolecular hydrogel based on cyclodextrin/copolymer inclusion for tissue engineering repair. Macromol. Biosci. 2014;14:440–450.

[81] Billiet T, Vandenhaute M, et al. A review of trends and limitations in hydrogel-rapid prototyping for tissue engineering. Biomaterials 2012;33:6020–6041.

[82] Schuurman W, Levett PA, et al. Gelatin-methacrylamide hydrogels as potential biomaterials for fabrication of tissue-engineered cartilage constructs. Macromol. Biosci. 2013;13:551–561.

[83] Hu YH, Wang HM, et al. Supramolecular hydrogels inspired by collagen for tissue engineering. Org. Biomol. Chem. 2010;8:3267–3271.

[84] Liu J, Zheng HY, et al. Hydrogels for engineering of perfusable vascular networks. Int. J. Mol. Sci. 2015;16:15997–16016.

[85] Chen YC, Lin RZ, et al. Functional human vascular network generated in photocrosslinkable gelatin methacrylate hydrogels. Adv. Funct. Mater. 2012;22:2027–2039.

[86] Pereira RF, Bartolo PJ. 3D bioprinting of photocrosslinkable hydrogel constructs. J. Appl. Polym. Sci. 2015;132:42458.

[87] Turksen K. Bioprinting in regenerative medicine preface. In: Turksen K, editor. Bioprinting in Regenerative Medicine. Totowa: Humana Press Inc.; 2015. p. 1–140.

[88] Tabriz AG, Hermida MA, et al. Three-dimensional bioprinting of complex cell laden alginate hydrogel structures. Biofabrication 2015;7:045012.

[89] Ma YF, Ji Y, et al. Bioprinting 3D cell-laden hydrogel microarray for screening human periodontal ligament stem cell response to extracellular matrix. Biofabrication 2015;7:044105.

[90] Lee W, Pinckney J, et al. Three-dimensional bioprinting of rat embryonic neural cells. Neuroreport 2009;20:798–803.

[91] Hsieh FY, Lin HH, et al. 3D bioprinting of neural stern cell-laden thermoresponsive biodegradable polyurethane hydrogel and potential in central nervous system repair. Biomaterials 2015;71:48–57.

[92] Hoare TR, Kohane DS. Hydrogels in drug delivery: progress and challenges. Polymer 2008;49:1993–2007.

[93] Mellott MB, Searcy K, et al. Release of protein from highly cross-linked hydrogels of poly(ethylene glycol) diacrylate fabricated by UV polymerization. Biomaterials 2001;22:929–941.

[94] Lin CC, Anseth KS. PEG Hydrogels for the controlled release of biomolecules in regenerative medicine. Pharm. Res. 2009;26:631–643.

[95] Amint M, Ahmad N, et al. Recent advances in the role of supramolecular hydrogels in drug delivery. Expert Opin. Drug Deliv. 2015;12:1149–1161.

[96] El Kechai N, Agnely F, et al. Recent advances in local drug delivery to the inner ear. Int. J. Pharm. 2015;494:83–101.

[97] Fan ZP, Zhang YM, et al. In situ injectable poly(-glutamic acid) based biohydrogel formed by enzymatic crosslinking. J. Appl. Polym. Sci. 2015;132:42301.

[98] Ray D, Gils PS, et al. Comparative delivery of diltiazem hydrochloride through synthesized polymer: hydrogel and hydrogel microspheres. J. Appl. Polym. Sci. 2010;116:959–968.

[99] Josef E, Barat K, et al. Composite hydrogels as a vehicle for releasing drugs with a wide range of hydrophobicities. Acta Biomater. 2013;9:8815–8822.

[100] Qiu Y, Park K. Environment-sensitive hydrogels for drug delivery. Adv. Drug Deliv. Rev. 2012;64:49–60.

[101] Li J, Hu WQ, et al. pH and glucose dually responsive injectable hydrogel prepared by in situ crosslinking of phenylboronic modified chitosan and oxidized dextran. J. Polym. Sci., Part A: Polym. Chem. 2015;53:1235–1244.

[102] Seidlits SK, Gower RM, et al. Hydrogels for lentiviral gene delivery. Expert Opin. Drug Deliv. 2013;10:499–509.

[103] Hu XF, Wang N, et al. Cyclodextrin-cross-linked diaminotriazine-based hydrogen bonding strengthened hydrogels for drug and reverse gene delivery. J. Biomater. Sci. Polym. Ed. 2013;24:1869–1882.

[104] Kidd ME, Shin S, et al. Fibrin hydrogels for lentiviral gene delivery in vitro and in vivo. J. Controlled Release 2012;157:80–85.

[105] Reyes-Ortega F, Cifuentes A, et al. Bioactive bilayered dressing for compromised epidermal tissue regeneration with sequential activity of complementary agents. Acta Biomater. 2015;23:103–115.

[106] Ajji Z, Othman I, et al. Production of hydrogel wound dressings using gamma radiation. Nucl. Instrum. Methods Phys. Res. B-Beam Interactions Mater. At. 2005;229:375–380.

[107] Lee PY, Li ZH, et al. Thermosensitive hydrogel as a Tgf-beta 1 gene delivery vehicle enhances diabetic wound healing. Pharm. Res. 2003;20:1995–2000.

[108] Hassan W, Dong YX, et al. Encapsulation and 3D culture of human adipose-derived stem cells in an in-situ crosslinked hybrid hydrogel composed of PEG-based hyper-branched copolymer and hyaluronic acid. Stem Cell. Res. Ther. 2013;4:32.

[109] Xu KM, Ge WW, et al. Bisphosphonate-containing supramolecular hydrogels for topical decorporation of uranium-contaminated wounds in mice. Int. J. Radiat Biol. 2008;84:353–362.

[110] Yang ZM, Liang GL, et al. d-Glucosamine-based supramolecular hydrogels to improve wound healing. Chem. Commun. 2007:843–845.

[111] Ng VWL, Chan JMW, et al. Antimicrobial hydrogels: a new weapon in the arsenal against multidrug-resistant infections. Adv. Drug Deliv. Rev. 2014;78:46–62.

[112] Mohamed NA, Abd El-Ghany NA, et al. Novel antimicrobial superporous cross-linked chitosan/pyromellitimide benzoyl thiourea hydrogels. Int. J. Biol. Macromol. 2016;82:589–598.

[113] Peng N, Wang Y, et al. Biocompatible cellulose-based superabsorbent hydrogels with antimicrobial activity. Carbohydr. Polym. 2016;137:59–64.

[114] Liu SQ, Yang C, et al. Antimicrobial and antifouling hydrogels formed in situ from polycarbonate and poly(ethylene glycol) via Michael addition. Adv. Mater. 2012;24:6484–6489.

[115] Salick DA, Kretsinger JK, et al. Inherent antibacterial activity of a peptide-based beta-hairpin hydrogel. J. Am. Chem. Soc. 2007;129:14793–14799.

[116] Abd El-Mohdy HL. Radiation synthesis of nanosilver/poly vinyl alcohol/cellulose acetate/gelatin hydrogels for wound dressing. J. Polymer Res. 2013;20:177.

[117] Gaharwar AK, Peppas NA, et al. Nanocomposite hydrogels for biomedical applications. Biotechnol. Bioeng. 2014;111:441–453.

[118] Han N, Johnson J, et al. Hydrogel-electrospun fiber composite materials for hydrophilic protein release. J. Controlled Release 2012;158:165–170.

[119] Abdel-Bar HM, Osman R, et al. Tunable biodegradable nanocomposite hydrogel for improved cisplatin efficacy on HCT-116 colorectal cancer cells and decreased toxicity in rats. Biomacromolecules 2016;17:407–414.

[120] Lee KY, Mooney DJ. Hydrogels for tissue engineering. Chem. Rev. 2001;101:1869–1879.

[121] Costa AMS, Mano JF. Extremely strong and tough hydrogels as prospective candidates for tissue repair – a review. Eur. Polym. J. 2015;72:344–364.

Chapter 7

Cellulose-Derivatives-Based Hydrogels as Vehicles for Dermal and Transdermal Drug Delivery

Lavinia Vlaia, Georgeta Coneac, Ioana Olariu,
Vicențiu Vlaia and Dumitru Lupuleasa

Additional information is available at the end of the chapter

http://dx.doi.org/10.5772/63953

Abstract

The use of water-soluble polymers of natural, semisynthetic, and synthetic origin for dermal and transdermal drug delivery systems is manifold. Among the most used biopolymers in the formulation of skin preparations, the cellulose ether derivatives as representatives of semisynthetic polymers distinguish through their specific physicochemical properties, by which the pharmacist can select the appropriate cellulose derivative for a particular purpose. The hydrogels containing cellulose derivatives as gelling agents are widely used as water-soluble ointment bases, because they usually associate the characteristics of both conventional and innovative hydrogels, including especially safety, biocompatibility, biodegradability, and a relatively easy way of preparation and low price. The present chapter describes the following issues: the physicochemical properties of water-soluble cellulose derivatives in relationship with their type and grade; physical and chemical properties of cellulose-derivatives-based hydrogels and their compatibility with other auxiliary substances commonly used in the formulation of pharmaceutical hydrogels; the development and manufacturing of these hydrogels on both small and large scales; the characterization of cellulose derivatives hydrogels as pharmaceutical dosage forms through different compendial and noncompendial methods; and well-recognized and novel applications of cellulose-derivatives-based hydrogels for dermal and transdermal drug delivery.

Keywords: cellulose derivative, hydrogel, dermal, drug delivery, gelation, viscosity

1. Introduction

Over the past decades, the delivery of drugs to and through the skin has gained an increased interest in research and in the pharmaceutical industry, mainly due to the fact that the skin is an easily accessible and painless route for drug administration, which in turn leads to increased patient compliance. Furthermore, dermal and transdermal drug delivery represents an attractive alternative to the conventional oral and parenteral route of administration, as it offers several advantages.

In present-day pharmaceutical and dermatological practice, most of the drug products applied to the skin for either localized or systemic effects are semisolid preparations, such as ointments, creams, and gels. Hydrogels, as conventional semisolid vehicles with hydrophilic properties, are highly valued in dermatology because they are transparent, completely water washable, greaseless, thixotropic, easily spreadable, suitable for the incorporation of lipophilic compounds or insoluble solids, and present good bioadhesive properties. In addition, innovative hydrogel formulations recently developed show other advantageous characteristics, namely biocompatibility, biodegradability, and sensitivity to various external stimuli. Biodegradability is nowadays a preferred and even a required property considering the need for environmentally friendly materials and technologies in different domains.

Hydrogels generally consist of two main components, the gelling agent responsible for the network formation and an aqueous liquid vehicle. Among the gelling agents extensively used in the pharmaceutical compounded and industrialized topical hydrogel formulations are cellulose derivatives (also referred to as cellulosics) such as methylcellulose, carboxymethylcellulose, and hydroxypropylmethylcellulose (HPMC). The cellulose-derivatives-based hydrogels are particularly attractive as ointment bases because they usually associate the characteristics of both conventional and innovative hydrogels. Moreover, the large availability in nature, nontoxicity, and the low cost of cellulose derivatives represent other important features, which recommend these polymers as first choice raw materials for the preparation of pharmaceutical hydrogels.

This chapter focuses on the current design, development, and applications of cellulose-derivatives-based hydrogels as semisolid pharmaceutical dosage forms intended for dermal and transdermal drug delivery.

2. Classification of cellulose derivatives

Cellulose is the most abundant naturally occurring biopolymer, found as the major component of annual plants and natural fibers (e.g., cotton, hardwoods and softwoods, linen, jute, and hemp) and also produced by some bacteria, fungi, and animals [1–5]. This glucose polymer is extensively used in pharmaceutical applications as it fulfills the two essential requirements: biocompatibility and biodegradability [6].

Plant-derived cellulose occurs as fibers, formed of macromolecules that contain hundreds of glucose molecules, determining the variability in chain length and molecular weight. The cellulose molecular weight can reach 1500 Da, each 40–50 glucose units being associated with longitudinal formations, named crystallites, which are oriented parallel to the longitudinal axis of the fiber and have 600–650 nm in length. In cellulose, as a linear polysaccharide polymer, the glucose monomers in pyranose form are linked to unbranched chains by β-1,4-glucosidic bonds, every glucose monomer being flipped related to the next one. Due to this structure, cellulose shows high crystallinity and rigidity and is practically insoluble in water and most organic solvents [7, 8].

In order to alter these disadvantageous properties, which limit its biomedical applications, chemical modification of cellulose, involving reactions of hydroxyl groups such as esterification or etherification, was performed. The extension of these reactions is expressed through the degree of substitution (DS), representing the average number of hydroxyl groups replaced by the substituents; the maximum value of DS is 3. For the cellulose derivatives used in pharmaceutical domain, the DS values correspond to pharmaceutical grades [8, 9].

Cellulose derivatives, named cellulosics, fall within the general class of hydrophilic colloids and have in common that they are hydrophilic, semisynthetic linear macromolecules obtained through chemical modification of cellulose. A more specific classification of cellulose derivatives can be made on the basis of other criteria, namely type of chemical modification of cellulose (etherification or esterification), electrolytic dissociation, and water solubility. Such classifications are useful because they facilitate the discussion of cellulose derivatives properties. Thus, based on the type of chemical modification of cellulose, cellulosics can be divided into two major groups:

- *polymers formed by hydroxyl etherification* with the appropriate alkyl halide of previously alkalinized cellulose. Principal examples of cellulose ethers derivatives used in pharmaceutical applications include methylcellulose (MC), ethylcellulose (EC), benzylcellulose (BC), hydroxyethylcellulose (HEC), hydroxypropylcellulose (HPC), hydroxypropylmethylcellulose (HPMC), hydroxyethylmethylcellulose (HEMC), and sodium carboxymethylcellulose (CMCNa);

- *polymers formed by hydroxyl esterification* with various organic acids, in the presence of a strong acid as catalyst. Some of the most important cellulose esters derivatives used in pharmaceutical domain are: cellulose acetate, cellulose nitrate, cellulose acetate phthalate, hydroxypropylmethylcellulose phthalate, and hydroxypropylmethylcellulose acetate succinate.

Table 1 presents the general chemical structures of cellulose ether derivatives and the formulas of the R groups in these biopolymers, which are further discussed in this chapter.

Cellulose derivatives may also be classified according to their electrolytic dissociation or charge as *nonionic (uncharged) polymers* that do not have an electric charge (i.e., MC, EC, HEC, HPC, HEMC, and HPMC) and *ionic (anionic and cationic) polymers* with electric charge. Among the above-mentioned cellulose derivatives, only NaCMC is an anionic or negatively charged polyelectrolyte at pH values above its isoelectric point, being sensitive to pH and ionic strength variations [9–11].

Cellulose ether derivatives	R groups
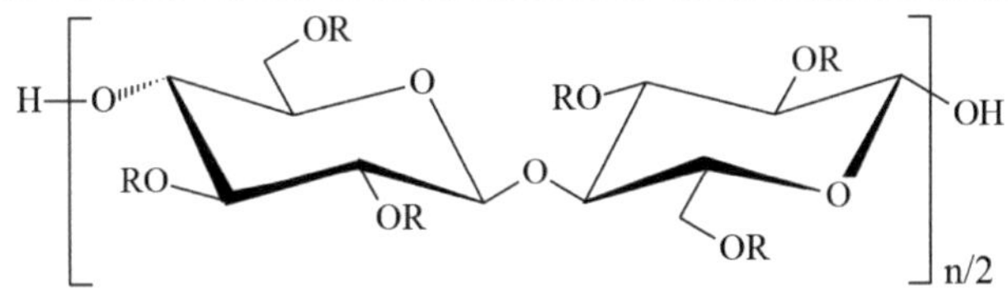	
Methylcellulose	H, CH_3
Ethylcellulose	H, CH_2CH_3
Hydroxyethylcellulose	H, $[CH_2CH_2O]_nH$
Hydroxypropylcellulose	H, $[CH_2CH(CH_3)O]_nH$
Hydroxyethylmethylcellulose	H, CH_3, $[CH_2CH_2O]_nH$
Hydroxypropylmethylcellulose	H, CH_3, $[CH_2CH(CH_3)O]_nH$
Carboxymethylcellulose	H,CH_2COONa

Table 1. Chemical structures of some important cellulose ether and ester derivatives.

In addition, the cellulose derivatives can also be categorized based on their water solubility into two groups: *water-soluble polymers*, including most of the cellulose ethers, and *water-insoluble polymers*, including the cellulose esters and both EC and BC from the group of cellulose ethers. It is to be mentioned that the water-insoluble cellulose derivatives are soluble in various organic solvents.

3. Water-soluble cellulose derivatives

In this section, it will be discussed in detail the physical and chemical properties of several water-soluble cellulose ether derivatives including methylcellulose (MC), hydroxyethylcellulose (HEC), hydroxypropylcellulose (HPC), hydroxypropylmethylcellulose (HPMC), and sodium carboxymethylcellulose (CMCNa), which are widely used as polymeric gelling agents in pharmaceutical hydrogels.

The commercial types of cellulose ethers for pharmaceutical applications are available in various grades, with different molecular weight, structural formula, and distribution of substituent groups and with different degree of substitution. These specific characteristics determine the physicochemical properties of cellulose ethers, such as solubility, viscosity in solution, surface activity and stability to biodegradation, heat, and hydrolytic and oxidative degradation. The solubility and the thermal gelation temperature of aqueous solutions of cellulose ethers are affected by their degree of substitution. But the viscosity of their solutions is directly related to their molecular weight and degree of polymerization. Therefore, the pharmacist has the possibility to select the appropriate cellulose derivative for a particular purpose [12].

3.1. Methylcellulose (MC)

Methylcellulose is included in various pharmacopoeias as *methylcellulose* (i.e., British Pharmacopoeia, BP; Japanese Pharmacopoeia, JP; European Pharmacopoeia, PhEur; United States Pharmacopoeia, USP; and International Pharmacopoeia) and is commercially available as different trade names *Benecel, Methocel, Metolose, Tylose* etc [13].

It is a nonionic, linear, and filiform macromolecule of cellulose, in which approximately 27–32% of the hydroxyl groups are in the form of methyl ether. At present, there are various grades of MC commercially available, having degrees of polymerization in the range of 50–1000, molecular weights (number average) in the range 10,000–220,000 Da, and a degree of substitution in the range 1.64–1.92 [13].

Methylcellulose occurs as a white, fibrous powder or granules, practically odorless and tasteless, and slightly hygroscopic. It swells and disperses slowly in cold water (1–5°C), forming a clear to opalescent, viscous colloidal dispersion, with pH 5.0–8.0. MC is practically insoluble in hot water (70°C), ethanol (95%), and glycerol. This methyl derivative of cellulose has the specific property of forming thermally reversible hydrogels on heating, being classified as a lower critical solution temperature polymer. In addition, other typical physicochemical properties of methylcellulose, which are to be considered for its selection as gelling agent in a pharmaceutical formulation, are presented in **Table 2**.

Specific property	Methylcellulose
Surface tension	MC has an important surface activity, decreasing the surface tension of water from 72.8 dyn/cm to 45–55 dyn/cm for a 0.1% (w/v) solution at 20°C
Stability	Solutions of methylcellulose are stable to alkalis and dilute acids at pH 3–11, at room temperature. At pH less than 3, acid-catalyzed hydrolysis of the glucose-glucose linkages occurs and the viscosity of methylcellulose solutions is reduced
Incompatibilities	MC is incompatible with various drugs (i.e., chlorocresol, mercuric chloride, phenol, resorcinol, tannic acid, silver nitrate, tetracaine), antimicrobial preservatives (i.e., cetylpyridinium chloride, methylparaben, propylparaben and butylparaben), salts of mineral acids, and strong oxidizing agents

Table 2. Several specific properties of methylcellulose important as selection criteria in the formulation stage of a pharmaceutical gel.

3.2. Hydroxyethylcellulose (HEC)

The grades of HEC used in pharmaceutical applications comply with the specifications defined in the *hydroxyethylcellulose* monograph included in different pharmacopoeia (i.e., BP, PhEur, USP, and International Pharmacopoeia). Commercially, HEC is available in different trade names such as *Cellosize HEC, Natrosol, Tylose H* etc.

Hydroxyethylcellulose, a partially substituted poly(hydroxyethyl) ether of cellulose, is a nonionic, hydrophilic polymer, with a linear and filiform chain, having a degree of substitution of minimum 1.5 (three hydroxyls substituted/two units). With a further increase in DS over this minimum, the water solubility of HEC is increased.

There are available several grades of HEC, which differ in viscosity (determined by molecular weight) and in degree of substitution; further, some grades are modified to improve their water dispersibility. Also, HEC may contain a suitable anticaking agent. HEC appears as a white, yellowish-white or grayish-white, odorless and tasteless, and hygroscopic powder. It dissolves readily in either hot or cold water, forming clear, smooth, uniform solutions, with pH 5.0–8.5. HEC is practically insoluble in acetone, ethanol (95% m/v), ether, and most other organic solvents, but it swells or is partially soluble in polar solvents, usually those that are miscible with water such as glycols, dimethyl formamide, dimethyl sulfoxide, and ethanol:water mixtures (70:30, 60:40, 30:70 by weight) [13, 14].

Since the aqueous formulations of HEC are sensitive to biological contamination, a water-soluble antimicrobial preservative should be added for a prolonged storage (i.e., sodium benzoate, sorbic acid, and methyl-propylparaben combinations).

In **Table 3**, other HEC typical properties of interest for its incorporation into a pharmaceutical formulation are presented.

Specific property	Hydroxyethylcellulose
Surface tension	HEC has an insignificant surface activity that lead to a negligible lowering of water surface tension, from 72.8 dyn/cm to 66.8 dyn/cm for a 0.1% (w/v) solution at 20°C
Stability	Although the aqueous solutions of hydroxyethylcellulose are stable over the pH range of 2–12, at room temperature, their greatest stability is achieved in the pH range of 6.5–8.0. The solutions are less stable below pH 3 and under highly alkaline conditions, due to acid hydrolysis and oxidative degradation respectively
Compatibilities	Due to its water solubility and nonionic character, HEC is soluble in many salt solutions that will not dissolve other water-soluble polymers
	For example, HEC dissolves in most 10% salt solutions and many 50% salt solutions, with several exceptions which are mentioned below. Further, HEC is compatible with a wide range of water-soluble materials, including other cellulosic hydrosoluble polymers, natural gums, and surfactants
Incompatibilities	HEC will precipitate in sodium carbonate and sodium sulfate 10% solutions, and in 50% (saturated) solutions of sulfates of bi- and trivalent metals (i.e., magnesium, zinc, and aluminum), di- and trisodium phosphate, ferric chloride, sodium nitrate, sodium sulfite. HEC is also incompatible with some quaternary disinfectants and partially compatible with the following water-soluble compounds: casein, gelatin, methylcellulose, polyvinylalcohol, and starch
Safety	Generally, HEC is considered as an essentially nontoxic and nonirritant material, being mainly used in ophthalmic and cutaneous pharmaceutical formulations

Table 3. Some specific properties of hydroxyethylcellulose as gelling agent.

3.3. Hydroxypropylcellulose (HPC)

According to Ph. Eur. and USP-NF, hydroxypropylcellulose intended for pharmaceutical use, is a partially substituted poly(hydroxypropyl) ether of cellulose, which may contain no more than 0.6% of silica or another suitable anticaking agent. HPC, also known as hyprolose and oxypropylated cellulose, is made by the Aqualon Division of Hercules Inc. and Nippon Soda Co., Ltd. (NISSO) under the brand names Klucel and NISSO HPC, respectively. The pharmaceutical grades of HPC are commercially available in several different viscosity types with molecular weight in the range of 50,000–1,250,000, depending on the degree of polymerization.

This ether of cellulose is a hydrophilic, nonionic, linear thread-like polymer where some of the hydroxyl groups of the cellulose have been hydroxypropylated, forming $-OCH_2CH(OH)CH_3$ groups. Moreover, the added hydroxypropyl groups can also be etherified during HPC preparation, leading to a value of moles of substitution (the number of moles of hydroxypropyl group per glucose ring) higher than 3. Hence, HPC with good water solubility must have a DS value of 2.5 and an MS value of approximately 4.

Hydroxypropylcellulose occurs as a white to slightly yellow-colored, odorless and tasteless powder. It feature a remarkable combination of properties: solubility in cold or hot polar organic solvents such as methanol (1:2), ethanol 95% (1:2.5), isopropyl alcohol (1:5), propyleneglycol (1:5); water solubility (1:2) below 38°C, forming a smooth, clear, colloidal solution; surface activity; aqueous thickening and stabilizing properties. HPC is insoluble in hot water, precipitates as a highly swollen floc in the temperature range 40–45°C, but this precipitation is completely reversible; thus, the gelation of HPC occurs on heating and yields to thermally reversible gels, similar to MC. It is also practically insoluble in aliphatic and aromatic hydrocarbons, glycerin, and oils [13, 15, 16].

In addition, other physical and chemical properties of pharmaceutical grades hydroxypropylcellulose are indicated in **Table 4**.

Specific property	Hydroxypropylcellulose
Surface tension	HPC is a surface active cellulose polymer, lowering the surface tension of water, from 72.8 dyn/cm to 46.3 dyn/cm for a 1% (w/v) solution, at 20°C
Stability	Aqueous solutions of hydroxypropylcellulose are stable in the pH range of 6–8, at room temperature. However, under highly acid or alkaline conditions, the degradation of polymer can occur, due to acid hydrolysis or alkaline oxidation
Compatibilities	The compatibility of hydroxypropylcellulose with inorganic salts varies according to the salt and its concentration
	For example, HPC dissolves in most 2–3% salt solutions and few 10% salt solutions. Further, HEC is compatible with a wide range of organic materials, including water-soluble as well as solvent-soluble resins, polymers (i.e., natural gums, semisynthetic, and synthetic polymers), surfactants, and organic liquids
Incompatibilities	In aqueous solutions, HPC will precipitate in the presence of 10% aluminum, ammonium or sodium sulfates, disodium phosphate, sodium acetate, carbonate, chloride, and thiosulfate. This polymer tends to be salted out in the presence of high concentrations of other dissolved compounds
Safety	Generally, HPC is considered as an essentially nontoxic and nonirritant material, being widely used as excipient in oral and topical pharmaceutical formulations. It does not exhibit skin irritation or skin sensitization. It is GRAS listed and is included in the FDA Inactive Ingredients Database (for oral solid dosage forms such as capsules and tablets, and also for topical and transdermal preparations)

Table 4. Some typical properties of hydroxypropylcellulose used in pharmaceutical applications.

3.4. Hydroxypropylmethylcellulose (HPMC)

It is also known as hypromellose, a nonproprietary name under which this cellulose derivative is found in different pharmacopoeias (BP, JP, PhEur, and USP) or their current editions. In addition, hydroxypropylmethylcellulose intended for use in pharmaceutical applications is produced by different manufacturers and commercialized under several trade names such as *Benecel MHPC, Methocel, Metolose, Pharmacoat, Tylopur, Tylose MO* [17–21].

Hypromellose is the propyleneglycol ether of methylcellulose, described by the PhEur as a partly *O*-methylated and *O*-(2-hydroxypropylated) cellulose. This nonionic, water-soluble polymer is available in several grades that differ in viscosity and extent of substitution. Different viscosity grades of HPMC are identified by an attached number indicating the apparent viscosity, in mPa s, of a 2% (w/w) aqueous solution at 20°C. The substitution type of hypromellose is specified in the pharmacopoeias as a four-digit number following the nonproprietary name, e.g., hypromellose 2208. The first two digits indicate the approximate percentage content of the methoxy group (OCH_3) and the second two digits indicate the approximate percentage content of the hydroxypropoxy group ($OCH_2CH(OH)CH_3$), calculated on a dried basis.

Methocel and *Metolose* grades	Compendial (JP, PhEur, USP) substitution type	Nominal viscosity (mPa s)
Methocel K3 Premium LV	2208	3
Methocel K100 Premium LVEP	2208	100
Methocel K4M Premium	2208	4000
Methocel K15M Premium	2208	15,000
Methocel K100M Premium	2208	100,000
Methocel E3 Premium LV	2910	3
Methocel E5 Premium LV	2910	5
Methocel E6 Premium LV	2910	6
Methocel E15 Premium LV	2910	15
Methocel E50 Premium LV	2910	50
Methocel E4M Premium	2910	4000
Methocel F50 Premium	2906	50
Methocel F4M Premium	2906	4000
Metolose 60S	2910	50, 4000, 10,000
Metolose 65SH	2906	50, 400, 1500, 4000
Metolose 90SH	2208	100, 400, 4000, 15,000

Table 5. The substitution type and typical viscosity values for 2% (w/v) aqueous solutions of Methocel (Dow Wolff Cellulosics) and Metolose (Shin-Etsu Chemical Co. Ltd.). Viscosities measured at 20°C [13, 17, 20, 21].

In the case of *Methocel* products of Dow Wolff Cellulosics, the substitution type is incorporated into the product name as an initial letter "*E*", "*F*," and "*K*," the number that follows stands for the viscosity (in mPa s) of that product measured at 2% concentration in water at 20°C; further, in referring the viscosity, the letter "*C*" is frequently used to represent 100 and the letter "*M*" is used to represent 1000 (**Table 5**). Also, several suffixes are used to identify special products: "*P*" denotes Methocel Premium grade products which means they are compliant with the USP, PhEur, and JP; "*LV*" stands for Low Viscosity products; "*CR*" refers to a Premium, Controlled Release grade (**Table 5**). On the other hand, the *Metolose SH* types of Shin-Etsu Chemical Co. Ltd. may be distinguished based on the degree of substitution by a number preceding the initial letters "*SH*" which identify the hypromellose products (**Table 5**) [13, 17, 20, 21].

The percentage content of methoxy and hydroxypropoxy groups affects both the molecular weight that is approximately 10,000–1,500,000, and the physicochemical properties of HPMC, such as solubility, surface activity, and thermal gelation [13]. HPMC appears as a white or creamy-white fibrous or granular powder, which is odorless and tasteless. It is soluble in cold water, forming a transparent, viscous, and surface active colloidal solution (the surface tension range from 42 dyn/cm to 64 dyn/cm). Certain types and grades of hypromellose are also soluble in various binary cosolvent systems such as ethanol/water, isopropanol/water, ethanol/dichloromethane, isopropanol/dichloromethane, providing a unique combination of solubility in organic solvents and water. However, hypromellose is practically insoluble in hot water, chloroform, ethanol (95%), and ether.

Another interesting characteristic of HPMC is the thermoreversible gelation behavior in aqueous media, which can be explained as follows: the aqueous solutions of hypromellose, obtained at room temperature, turn into gels when heated to their specific gel temperature (50–90°C); the resulting gels are completely reversible and liquefy upon cooling to room temperature, returning to their solution form. The thermoreversible gelation of hypromellose is particularly affected by several factors, such as its concentration in solvent media and the nature of additives. Thus, the additives that impart a solubilizing effect (i.e., ethanol, propyleneglycol, PEG 400) raise the gel point of HPMC, whereas the additives that exhibit a coagulant effect (i.e., glycerol, sorbitol, and most electrolytes) lower the gel temperature. In the past decade, numerous studies on thermoreversible gelation behavior of HPMC solutions have been carried out [22–26].

In addition, hypromellose possesses several other physicochemical characteristics, which are generally, similar to those of the above described cellulose ether derivatives of pharmaceutical grade. Some of these properties are presented below: HPMC in powder form is a stable material, although it is hygroscopic; aqueous solutions of HPMC are predisposed to microbial spoilage and require the addition of an antimicrobial preservative; in solution, hypromellose is stable in a wide pH range (3–11); HPMC is incompatible with some oxidizing agents, but exhibits a higher tolerance for salts in solution than MC; it is generally regarded as nontoxic an nonirritating excipient, being extensively used in topical pharmaceutical formulations and cosmetics; also, it is GRAS listed [13].

3.5. Carboxymethylcellulosesodium (CMCNa)

Among the two salt forms of carboxymethylcellulose (calcium and sodium) available for industrial use, sodium carboxymethylcellulose is commonly used for pharmaceutical preparations, including hydrogels. This cellulose ether derivative, also known as carmellose sodium, must comply with the compendial requirements of *carmellose sodium* or *carboxymethylcellulose sodium* monographs listed in BP, PhEur, JP, and respectively USP, which describe it as the sodium salt of polycarboxymethyl ether of cellulose, with a 6.6–10.8% sodium content. It is commercially available as various trade names such as *Akucell, Aqualon CMC, Aquasorb, Blanose, Tylose CB,* and *Walocel C.*

CMCNa is a hydrophilic, anionic polymer, with a linear, filiform chain, prepared by partial substitution of the two, three, and six hydroxyl groups of cellulose by carboxymethyl groups. The carmellose sodium pharmaceutical grades are available in a wide variety of types with regard to the degree of substitution, viscosity, and particle size. The value of DS varies in the range 0.6–1, affecting some of the polymer physicochemical properties. Hence, as the DS value is higher, the solubility in water and sodium content of CMCNa increase and a better polymer tolerance for other components in solution is achieved. The viscosity of different types of carmellose sodium depends on their polymerization degrees and molecular weights which are of 100–2000 and 90,000–700,000, respectively. Particle size has a pronounced effect on the ease of dispersing and dissolving CMCNa.

Sodium carmellose occurs as granular or fibrous, white to slightly off white, odorless, tasteless, and hygroscopic powder. It is slightly soluble in water at all temperatures, forming clear, viscous colloidal solutions, but is practically insoluble in most organic solvents such as ethanol (95%), methanol, acetone, ether, and toluene. However, it can be dissolved in aqueous mixtures if the content in water-miscible solvent is less than 40% (by weight). The powder of CMCNa presents a high chemical and microbiological purity and stability, and is not surface active. The aqueous solutions of carboxymethylcellulose sodium exhibit maximum viscosity and stability at pH 7–9, although they are stable over a broad pH range (2–10); at pH < 2, precipitation of CMC acid can occur, and at pH > 10 the viscosity of solutions decreases rapidly. Similar to HPMC, sodium carmellose exhibits thermoreversible gelation behavior in aqueous media. Also, its aqueous solutions are sensitive to microbiological attack and should contain a preservative for prolonged storage [13, 27, 28, 29].

Being a polyelectrolyte, CMCNa is sensible to pH and ionic strength variations. Therefore, its compatibility in solution with other components (**Table 6**) is another critical characteristic for the formulation of a pharmaceutical hydrogel.

In the past decade, cross-linked networks of CMC have been obtained and reported by applying chemically and physically cross-linking technologies. The chemically cross-linking method involves the use of bifunctional cross linkers such as epichlorhydrin [30], multifunctional carboxylic acids [31–34], ethyleneglycol diglycidyl ether [32], and polyethyleneglycol diglycidyl ether [35]. However, some of these reagents such as epichlorhydrin and ethyleneglycol diglycidyl ether produce large amounts of toxic byproducts under the cross-linking conditions, requiring their elimination through extensive washing, thus affecting the biocompatibility of the resulted hydrogel and the environmental safety of the production process. Considering these environmental and health safety risks, a physical cross-linking method, namely the radiation technology based on γ or electron-beam irradiation under relatively mild conditions, has attracted increased interest [36, 37]. This method presents some advantages: the addition of chemical reagents is not required, the side products are not present, and the simultaneously sterilization of the final hydrogel is possible.

Common excipients compatible with CMCNa in solution	Common excipients incompatible with CMCNa in solution
Most hydrophilic nonionic and anionic polymers and gums	Xanthan gum
	Gelatin, pectin and collagen forming complexes
Most 10% and 50% inorganic salts of monovalent cations, which form soluble salts of carboxymethylcellulose, with a prerequisite condition: the polymer should be dissolved in water before adding the salt	Most inorganic salts of bivalent (Ca^{2+}, Ba^{2+}, Mg^{2+}, Co^{2+}, Fe^{2+}, Mn^{2+}, and Zn^{2+}) and trivalent cations (Al^{3+}, Fe^{3+}), and also trace amounts of heavy metals cations (Ag^{+}, Cu^{2+}, Pb^{2+}, Zr^{2+}), which form precipitates

Table 6. Compatibility of carboxymethylcellulose sodium with other components in solution.

Although cross-linked CMC is a water-insoluble biopolymer, it is capable to absorb large amounts of water and swells to form superabsorbent hydrogels that exhibit superior mechanical properties and viscoelasticity compared with conventional sodium CMC-based hydrogels [38]. Due to this characteristic, the cross-linked CMC-based hydrogels were recently studied as potential wound dressing materials, as well as dermal and transdermal drug delivery systems [35, 38, 39].

Similar to the others cellulose ether derivatives, sodium carmellose is a safe excipient, being regarded as nontoxic and nonirritant. It is GRAS listed and included in different databases of inactive ingredients. Consequently, this polymer is extensively used in oral, topical, and some parenteral formulations, in cosmetics, toiletries, and food products [13].

4. Colloidal dissolution and gelation as processes involved in the formation of cellulose-derivatives-based hydrogels

To obtain a topical semisolid drug product, such as a medicated hydrogel, which meets the specific requirements, one of the major objectives of its formulation is the excipients selection and preparation of the base (hydrogel base) with semisolid consistency, jelly-like structure, and bioadhesive properties. Further, the hydrogel base must be stable, formed of compatible components, and therapeutically acceptable. Due to the specific jelly-like structure, hydrogels preserve their shape during the stockage, spread evenly, and adhere to the skin surface. In order to achieve this goal of hydrogel formulation, the knowledge of processes involved in the formation of hydrogels, such as colloidal dissolution and gelation of hydrophilic polymers, are of great importance as they are closely related with the physical and chemical properties of the gelling agent.

Based on the jelly-like structure, cellulose derivatives hydrogels are categorized as single-phase, reversible (or physical) gels composed of a network of organic macromolecules dissolved in water, without the existence of definite boundaries between the two components. However, due to the large size of the dissolved molecules, cellulose derivatives hydrogels are considered, on the microlevel, as biphasic colloidal systems, consisting of colloidal polymer and a liquid phase (water) [16, 40].

According to the formation mechanism of hydrogel, all cellulose derivatives described in the previous section are biopolymers that form hydrogels independently of pH, but in the case of MC, HPC, and HPMC the hydrogel formation is temperature dependent. Usually, the water-soluble cellulose derivatives form hydrogels in the concentrations of 1–10% (by weight), depending on the polymerization degree. The gelation of cellulose ether derivatives has been extensively studied for decades and numerous of these studies have been reviewed in several scientific journals and books [24, 39–47].

It is generally accepted that the gelation of water soluble cellulose ether derivatives is due to some physical processes such as molecular entanglements, hydrophobic associations between macromolecules and hydrogen bonding. The filiform macromolecules chains connect at both ends by forming some intermolecular bonds as a result of interaction between the existing functional groups. Also, entanglements develop from the interpenetration of random-coil flexible chains of polymers. Thus, they form a continuous three-dimensional network that occupies the entire system and entraps the entire amount of water. The formation of cellulose-derivatives-based hydrogels occurs in three stages. The first stage, identified as the diffusion of water molecules into the polymer network, is attributed to hydrogen bonding between the water molecules and the hydrophilic functional groups of macromolecules including carboxyl and hydroxyl groups. So, the solvent molecules will be oriented along the polymer chain, increasing the rigidity of the dispersed system. The second stage, correspond to relaxation of the polymer chains by hydration, when large amounts of water permeated into the polymer network are spontaneously absorbed by the macromolecules that swell and therefore greatly enhance the gel volume. In this stage of hydrogel formation, the interactions between polymer and water molecules through coordinate bonds lead to the formation of very stable complexes of hydration. Also, wetting determines the enhancement of macromolecule chains permeability and their stretching, accompanied by a more or less linear arrangement. The third stage, namely the polymer network expansion displayed through the increase of the gel volume, is due to water absorption and swelling of macromolecules. The stronger macromolecules hydration is the lower mobility of water molecules and the higher stability of the cross-linked gel network [9, 11, 16, 48].

The two main processes, namely colloidal dissolution and gelation, involved in the formation of cellulose-derivatives-based hydrogels, depend on several physicochemical factors (i.e., properties of cellulose-based polymers determined by chemical structure, macromolecular chain configuration, molecular weight, degree of substitution and also by the composition of the aqueous media, the pH, and the temperature) and thus influence to a great extent the formulation of a suitable pharmaceutical hydrogel. In general, lower temperature, higher concentration, and higher molecular weight of polymer promote gelation of cellulose ether derivatives, leading to firmer hydrogels. In most cases, the gelation of polymer is affected in the presence of relatively high concentrations of electrolytes, surfactants, sugar, or some natural gums, which reduce the polymer hydration and consequently the gelation temperature by the "salting out" phenomenon; the magnitude of this effect depends not only on the nature and concentration of the other dissolved components, but also on the substitution degree of the polymer. Generally, the cellulose ether derivatives with lower substitution degrees tend to exhibit a higher tolerance for the other dissolved components in the system. Due to the stability of aqueous solutions of water-soluble cellulose derivatives in a wide range of pH values (3–11), the gel formation is affected only under highly acid or alkaline conditions [9, 11, 16, 48].

Among the water soluble cellulose derivatives, MC, HPC, and HPMC are lower critical solution temperature polymers (**Table 7**) that form thermoreversible hydrogels [49, 50].

In order to understand and clarify the thermal gelation of these biopolymers, in the past two decades a large number of studies have been performed using different techniques. However, it was not reached a consensus in all cases, due the complexity of this process involving several different phenomena which can occur during the sol-gel transition.

Cellulose derivative	Trade name and type	Methoxy, DS	Methoxy, wt%	Hydroxypropoxy, MS	Hydroxypropoxy, wt%	Gelation temperature (°C)
MC	Benecel™ A	1.8	27.1–31.5	–	–	56
	Methocel A	1.8	30	–	–	50
	Metolose SM	1.8	Not specified	–	–	50–55
HPC	Klucel™ HPC	–	–	3.4–4.4	Not specified	40–45
	Nisso HPC	–	–	Not specified	53.4–77.5	40–45
HPMC	Benecel™K	Not specified	20.0–24.0	Not specified	7.0–12.0	80
	Benecel™E	Not specified	28.0–30.0	Not specified	7.0–12.0	63
	Methocel E	1.9	29.0	0.23	8.5	63
	Methocel F	1.8	28.0	0.13	5.0	63
	Methocel J	1.3	18.0	0.82	27	62
	Methocel K	1.4	22.0	0.21	8.1	85
	Metolose 60SH	1.9	28.0–30.0	0.25	7.0–12.0	65–66
	Metolose 65SH	1.8	27.0–30.0	0.15	4.0–7.5	61–65
	Metolose 90SH	1.4	19.0–24.0	0.20	4.0–12.0	75

Table 7. Thermal gelation temperatures of various grades and types of methyl- and/or hydroxypropyl-derivatives of cellulose.

In the case of the above-mentioned three cellulose derivatives (MC, HPC, and HPMC) is generally accepted that the formation of thermoreversible hydrogels occurs in two stages: (1) the formation of water clusters around the methyl or hydroxypropyl substituents of the polymer chains (the hydrophobic portions), which will be isolated one from another at low temperatures (below 50°C); (2) the phase separation accompanied by gelation at high temperatures (above 50°C).

For methylcellulose, the sol-gel transition is currently defined as the transformation from a clear solution to a turbid strong gel. Although the gelation of MC has been extensively studied by various analytical techniques (NMR, differential scanning calorimetry, IR attenuated total reflectance spectroscopy, small-angle neutron scattering, static and dynamic light scattering, and rheology) and the researchers have proposed different gelation mechanisms, the specific molecular interactions that govern this process is still unclear. However, the results of various studies, which were in good agreement one with another, indicated several factors and/or phenomena which play an important role in the gelation of MC [24, 51–64]:

- the first stage (pregel state) is mainly governed by hydrophobic interactions between highly methylated glucose regions, but also by bundles of residual native cellulose crystals, liquid crystal phases, and intermolecular bonding between unsubstituted hydroxyl groups along the polymer chains;
- the second stage (gel state), when the phase separation and gelation occur almost simultaneously, results from micellar interactions, formation of crystallites of trimethylated glucose rings, hydrophobic polymer-polymer interactions, and entangled physical cross links between chains leading to a three-dimensional network.

Recently, it was confirmed experimentally by rheological measurements that a nucleation and growth mechanism is also involved in the gelation of MC, as the gelation temperature depends on the heating rate [65].

The same group of researchers has also demonstrated by cryogenic transmission electron microscopy and small-angle neutron scattering that MC hydrogels have a heterogeneous fibrillar structure that is responsible for their turbidity [66, 67].

For HPMC, the formation of thermoreversible hydrogels is, similar to MC, a two-stage process. Conventionally, it is considered that the mechanism of HPMC gelation involves polymer reptation and dissociation of cellulosic bundles followed by exclusion of water (syneresis) from heavily methoxylated regions of the macromolecule. This syneresis allows hydrophobic interactions between the respective macromolecules accompanied with the formation of polymer clusters, which further associate in a three-dimensional network. This specific mechanism of gelation is supported by the results obtained from several experimental techniques including: differential scanning calorimetry, IR attenuated total reflectance spectroscopy, UV/VIS and fluorescence spectroscopy, polarized light microscopy, and oscillatory rheometry [22, 26, 68–71].

However, several studies have indicated some differences between MC and HPMC regarding thermal gelation properties and gel network structure. Generally, these differences are generated by the presence of hydroxypropyl substituents along the HPMC polymeric chains that hinder the gelation process by inhibiting intermolecular association. This observation is supported by the fact that HPMC has a higher gelation temperature and forms weaker thermosensitive gels compared to MC. Also, as the temperature rise to ca. 55°C, hydroxypropyl derivative of MC precipitates and causes only clouding of the solution, but no gelation, unlike MC [53, 72–74].

On the other hand, hydroxypropylcellulose being soluble in water at room temperature undergoes the gelation in these conditions. Increasing the temperature above the cloud point (40–45°C), the resulted hydrogel suddenly becomes opaque, loses its gel-like characteristics (evidenced by a marked reduction in/rapid decrease in viscosity) and exhibits a sudden shrinking in volume. These effects are due to separation of the polymer as a highly swollen precipitate at the above-mentioned temperature that corresponds to its phase transition temperature (or lower critical solution temperature). However, the polymer precipitation is reversible, because upon cooling the system below 40°C, redissolution of polymer and restoring of original viscosity take place [13, 15].

In contrast to MC and HPMC, only a few studies have investigated the gelation mechanism of HPC by NMR spectroscopy and thermal analysis [75, 76].

As the results from all these studies fitted well one with another, it was proposed the following plausible explanation of the HPC gelation mechanism. At room temperature, due to its high DS and uniform distribution of substituents along the polymeric chains, the HPC backbone has a little hydrophilic character that promotes a hydrophobic effect with water in solution, being extended, and mobilized. At elevated temperature, this hydrophobic structure breakdown and the cellulose backbones of the polymer chains coil into supramolecular helical structure. Although the cellulose backbones of the polymer chains are immobilized in this way, the hydroxypropyl substituents remain solvated projecting outward from the coiled backbone into the solvent and exhibiting a cilia-type motion. These cilia keep adjacent coils apart, leading to a highly dispersed solid phase (precipitate), with no gel-like properties.

Finally, the only one "smart" cellulose derivative, namely sodium carboxymethylcellulose, does not form thermoreversible hydrogels and is soluble in either hot or cold water, because it is an anionic polyelectrolyte. Consequently, the three-stage process of hydrogel formation, described above, is mainly determined in the presence of electrostatic charges attached to the polymer network, which enhance the polymer swelling capability in water by two mechanisms: (1) due to the electrostatic repulsion between the electric charges of the same sign, the macromolecule chains are forced to get more elongated than in a neutral network; (2) enhanced water penetration in the polymer network, due to the presence in the hydrogel of counterions to ensure the electrical neutrality [39]. Nevertheless, the gelation of sodium carboxymethylcellulose is influenced by pH and ionic strength variations (due to the presence of different inorganic salts). In addition, it is generally accepted that the firmness of NaCMC hydrogels increases with the increase in carboxymethyl substitution, molecular weight and polymer concentration [13, 28, 29].

5. Physical and chemical properties of cellulose derivatives hydrogels

Cellulose derivatives hydrogels are interesting as water-soluble ointment bases and topical drug delivery systems, because they present several advantageous characteristics: transparency (especially HPMC- and NaCMC-based hydrogels); high water content (80–95%), which is responsible for their favorable cooling effect and washability; nongreasy, being practically free from fats or fatty substances; porosity; nonocclusive; safety, as they are nontoxic and nonallergenic; are well tolerated by skin and different mucosa (oral, buccal, ophthalmic, nasal, auricular, vaginal, and rectal); are mucoadhesive and bioadhesive materials, adhering well to skin, mucosa and suppurative wounds, and providing intimate contact between the drug and the site of action; can be safely sterilized. They also may be formulated to ensure excellent spreading properties or to optimize drug delivery. Other important advantages of cellulose derivatives hydrogels include biocompatibility, biodegradability, smart stimuli-responsive behavior (in the case of NaCMC), compatibility with most active substances and auxiliary pharmaceutical components (excepting NaCMC which is an anionic polyelectrolyte). Also, the availability in nature and the low cost of cellulose derivatives make them more attractive as gelling agents [16, 45, 39, 77–81]. It is noteworthy that the cellulose derivatives hydrogels are considered biocompatible because they are similar to living tissues, with regard to physical properties, based on their water content, soft consistency, and low interfacial tension with water or biological fluids [39, 82].

The physical properties of cellulose derivatives hydrogels can be divided into two groups: transitional properties (including swelling behavior, sol-gel transition or gel point, and physical aging) and rheological properties (including rigidity, yield point, and rupture strength). These properties of cellulose derivatives hydrogels are directly related to their structure through their polymeric composition, because the selected cellulose derivative decisively influences the network structure as well as the final hydrogel properties [80].

5.1. Transitional properties

5.1.1. Swelling behavior

The capacity to absorb water or aqueous solutions and the permeability (the rate at which the liquid is absorbed into the hydrogel structure) are the most important features characterizing hydrogels, including those based on cellulose derivatives. Generally, the amount of the aqueous medium incorporated in a hydrogel is determined gravimetrically and is expressed as fractional hydration (W) or as swelling ratio (r):

$$W = (w_1 - w_0) / w_1 \text{ or } r = (w_1 - w_0) / w_0$$

where w_1 andw_0 are the weights of the swollen and dry gels, respectively. Also, the amount of water absorbed by a polymer can be described by the equilibrium swelling as the maximum degree of the polymer swelling.

Hydrogels of cellulose derivatives are considered highly swollen gels (containing large percentages of water), because of the high solubility of these polymers in water, high flexibility of the polymeric chains and a large free volume available between polymeric chains [40, 83–85].

However, the water content at the swelling equilibrium of cellulose derivatives hydrogels is influenced basically by the nature of the monomer that makes them up, by the type and density of the cross link (entanglements) and also, by other factors including temperature, pH, and the composition of the hydration medium (the presence of salts and nonsolvents). So, dynamic and equilibrium studies carried out by the Peppas group on hydrogel discs of HEC, HPC, HPMC K4M, and HPMC K100M [44] showed that the volume-swelling ratio and the swelling time (the time when hydrogel discs reached the equilibrium state, namely the gel thickness became essentially constant) of the polymer were dependent on their hydrophilicity and number of junctions or entanglements per original chain. Thus, the equilibrium state for HEC was reached after 240 h, for HPC and HPMC K100M after 260 h, and for HPMC K4M after 170 h, whereas the obtained volume-swelling ratios up to 190 h of swelling ranked the polymers as follows: HPC < HPMC K4M < HPMC K100M < HEC. The swelling kinetics of the studied HPMC polymers in the dynamic state was accurately described by the equation used to calculate the volume-swelling ratio, Q:

$$Q = V_s / V_d$$

where V_s is the volume of the swollen gel and V_d is the initial volume of the dry disc. At equilibrium, the swelling kinetics of HPMC polymers correlated well with the Flory-Rehner equation, describing the average molecular weight between two consecutive entanglements, $\overline{M_e}$:

$$\frac{1}{\overline{M_e}} = \frac{2}{\overline{M_n}} - \frac{\left(\frac{v}{V_1}\right)\left[ln\left(1 - v_{2,s}\right) + v_{2,s} + x_1 v_{2,s}^2\right]}{\left(v_{2,s}^{1/3} - \frac{v_{2,s}}{2}\right)}$$

where $\overline{M_n}$ is the number average molecular weight of the cellulose ether tested, v is its specific volume, V_1 is the molar volume of water, $v_{2,s}$ is the polymer fraction in the swollen gel at equilibrium ($v_{2,s} = 1/Q$), and x_1 is the Flory or polymer-water interaction parameter [44]. From detailed explanation given in this paper, further references are made only for HEC and HPC. In the swollen state, HEC formed the thicker hydrogel layer, signifying that the largest amount of water is physically and chemically trapped in the HEC hydrogel structure. This entrapment was attributed to the relatively high number of entanglements which holds the network structure together at equilibrium. In contrast, HPC formed the thinnest hydrogel layer, meaning that the interactions between the HPC chains are very strong and that water molecules occupy a smaller volume than in the case of HEC. This behavior is supported by the higher number of entanglements and the relatively high molar substitution for HPC, and also by the large volume occupied by hydroxypropyl substituents in the network structure of the HPC hydrogel [44].

As in the case of other hydrogels, the equilibrium degree of swelling influences different properties of cellulose derivatives hydrogels, including permeability, surface properties and surface mobility, optical properties, mechanical properties, and solute diffusion coefficient through these hydrogels. Consequently, the knowledge of the swelling characteristics of cellulose derivatives is of great importance in their pharmaceutical applications [80, 83].

5.1.2. Sol-gel transition (gel point)

Sol gel transition is an essential property of most cellulose derivatives hydrogels as they are thermoreversible and is dependent on several factors, including polymer concentration and temperature.

The critical gelling concentration is the concentration below which the polymer and the solvent form a sol rather than a macroscopic gel, under the current experimental conditions [86]. For thermoreversible hydrogels of cellulose derivatives (MC, HPC, and HPMC), the critical gelling concentration obviously depends on temperature because, above the melting temperature, this concentration is nominally infinite, requiring a maximum gelation temperature and a critical gelation time [87]. The critical gelling concentration of cellulose derivatives depends on the polymer-polymer and polymer-solvent interactions, their hydrophilic-lipophilic character and molecular weight, and the flexibility of the chain [16, 88, 89]. Moreover, certain additives such as electrolytes, sorbitol, sucrose, and solvents (glycerin, ethanol, propyleneglycol, and polyethyleneglycol 400) can have different effects on the gel point and critical gelling concentration of cellulose derivatives [23]. Thus, it was demonstrated that most electrolytes, as well as sorbitol, sucrose and glycerin depress the gel point, whereas other cosolvents (ethanol, propyleneglycol, and polyethyleneglycol 400) raise the gel point of various grades of methylcellulose [90]. In **Table 8**, the minimal concentration domains for the most used cellulose derivatives as gelling agents in pharmaceutical hydrogels are listed.

Cellulose derivative	Gelling concentrations (wt %)	Required additives
Carboxymethylcellulose	4–6	Na^+
	10–25	Na^+
Methylcellulose	2–4	
Hydroxypropylcellulose	8–10	
Hydroxypropylmethylcellulose	2–10	

Table 8. Gelling concentrations of cellulose derivatives commonly used in topical hydrogels [18].

5.1.3. *Physical aging*

As the structure of cellulose derivatives hydrogels have not reached the equilibrium, these gels physically age as they move toward respective equilibrium. It is important to consider this physical aging of cellulose derivatives hydrogels, because it is accompanied by changes in gels microstructure where noncovalent cross links are breaking and reforming [16].

5.2. Rheological properties

The mechanical properties of cellulose derivatives hydrogels as water-soluble ointment bases must be considered during the development of a topical medicated hydrogel, as they are very important not only for the establishment of the topical drug formulation and the manufacturing method, but also for packaging, storage, and application.

Due to their semisolid consistency, cellulose derivatives hydrogels are viscoelastic, semistiff gels that exhibit a pseudoplastic flow [91]. These rheological properties are strongly related to the concentration and average molecular weight of the polymer, the gel structure and interchain interactions, and entanglements. So, the hydrogels of methylcellulose and hydroxypropylcellulose of 3–6% concentration exhibit a pseudoplastic character, while those of sodium carboxymethylcellulose types of high molecular weight and low substitution, with a concentration of 5–6% exhibit thixotropy, in addition to pseudoplastic flow. The apparent viscosity or gel strength of cellulose derivative hydrogels increases with an increase in their effective cross-link density or in the concentration and molecular weight of the polymer. Also, a rise in temperature decrease the apparent viscosity of these hydrogels, but under normal conditions, this effect is reversible. In addition, as it was discussed in the previous section, changes in apparent viscosity and consequently in other rheological properties of cellulose derivatives hydrogels can occur when different components such as salts, surfactants, solvents or nonsolvents, and other compatible polymers are added in the formulation. The effects of these additives have been extensively studied in the past decades and are presented in each cellulose derivative technical book [15, 17, 28, 29, 92–94].

6. Preparation of cellulose-derivatives-based hydrogels as pharmaceutical dosage forms

Cellulose-derivative-based hydrogels are relatively easy to prepare. In pharmaceutical applications, these hydrogels can be formulated with or without a drug substance. Medicated dermal gels can contain, in addition to the cellulose derivative as gelling agent and active substances, antimicrobial preservatives (i.e., methylparaben and propylparaben, or chlorhexidinegluconate), stabilizers (i.e., edetate disodium), dispersing agents (i.e., alcohol and/or glycerol, propyleneglycol, sorbitol), and permeation enhancers [95–98]. Minimum gel-forming concentrations of cellulose derivatives are different, based on the type and the molecular weights of these derivatives, but the medium range is about 4–6% (w/v), as mentioned before (**Table 8**). The most widely used hydrophilic external phase, in the preparation of these gels, is purified water. If the addition of cosolvents as dispersing agents is necessary, care should be taken to avoid their evaporation or degradation during gel preparation.

In manufacturing of dermal pharmaceutical cellulose derivatives hydrogels on a small scale (as in extemporaneous compounding), but also on a large scale (industry), the obtaining of an uniform preparation depends in a great extent on several factors including order of mixing, processing conditions, duration of swelling, and removal of entrapped air.

Mixing the above-mentioned components with the gelling agent should be made considering their influence on the gelling process. If the rate and extent of swelling of the gelling agent is affected by these ingredients, they are mixed after the gel formation. If such interference does not occur, the drug and the other additives are mixed prior to the swelling process; in this case, the effects of swelling duration, mixing temperature, and other processing conditions on the physicochemical stability of the drug and additives are also to consider. In general, the following order of mixing is recommended: (1) the drug substances are dissolved or suspended in the hydrophilic phase necessary for the gel preparation; (2) the other additives are dissolved in the obtained solution or in a small amount of the hydrophilic phase respectively; (3) if necessary, the drug dispersion is mixed with the additives solution; (4) the powder of the gelling agent is added under light stirring in the obtained solution/dispersion and allowed to swell.

In the preparation of cellulose derivatives hydrogels, the temperature and pH of the dispersion are critical parameters, as the gelation mechanism of these polymers is temperature dependent and their optimum stability depends on pH. Thus, the macromolecules dispersion is recommended to be heated either before, or after the polymer swelling. In case of MC and HPMC, the heating is indicated before the swelling occurs, but in case of HPC, HEC, and NaCMC, which dissolve better in hot water, the heating of their molecular dispersions is performed after de particle swelling. Practically, the general preparation method of most cellulose derivatives hydrogels involve the dispersion of polymer powder in cold water by using mechanical mixing to form uniform lump-free dispersion, followed by heating to about 60–80°C of the obtained dispersion, which is then gradually cooled to room temperature to form a gel. Also, there are differences regarding the pH values of dispersion medium favorable for gel formation: NaCMC, MC, and HPMC form gels over a wide pH range (4–10), whereas HPC and HEC form gels at a pH of 6–8 and respectively at alkaline pH condition. Another critical parameter in the preparation of cellulose derivatives hydrogels is the duration of polymer swelling. Generally, a swelling duration of about 24–48 h helps in obtaining homogeneous gels, as the cellulose polymers require about 48 h for complete hydration.

Finally, removal of the entrapped air is also an important issue to consider in the manufacturing of cellulose derivatives hydrogels, because the presence of air bubbles in the gel, which is inevitable, affects their transparency. The incorporation of air bubbles into the gel can be minimized by positioning the propeller at the bottom of the mixing container. Further removal of air bubbles by different methods including long - term standing, low -temperature storage, sonication, or inclusion of silicon antifoaming agents is commonly used. In addition, in large scale production vacuum vessel deaerators are used to remove the entrapped air [95, 97–99].

Figure 1 shows the flow chart of the fabrication process of a cellulose derivative hydrogel.

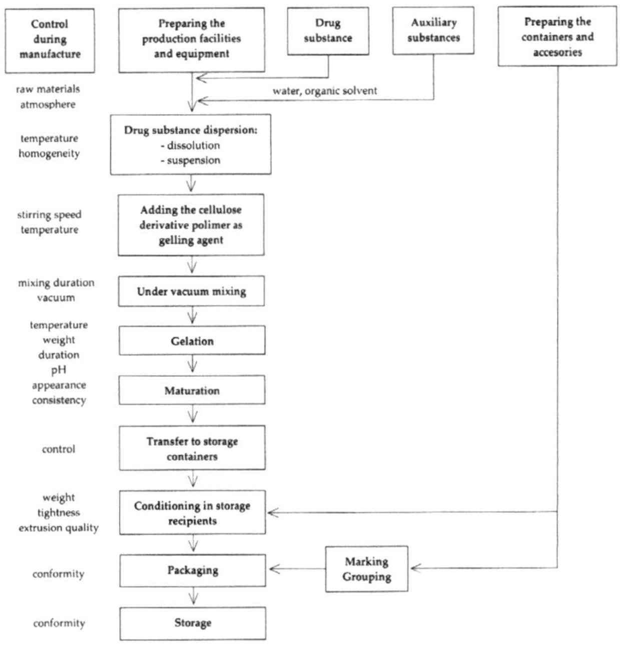

Figure 1. Flow chart of the manufacturing processes of a cellulose derivative hydrogel.

In small scale, the preparation of hydrogels by dispersing of polymer by hand or mechanically, in hot/cold water or in a nonsolvent, is carried out using simple equipment and utensils existing in the pharmacy laboratory, such as: mortars and pestles of porcelain or glass, beakers, magnetic stirrers, and different propeller mixers. In large scale production of pharmaceutical cellulose derivatives hydrogels, different mills, separators, mixers, deaerators, and shifters are used. **Figure 2** present two different processing machines for the hydrogels preparation: a single mixing kettle equipment with slow agitation to avoid air entrapment (a) and "one bowl" vacuum processing machine (b) which is designed to control the processing temperature, and to mix (using a counterrotating mixing system), homogenize (using a high-shear rotor/stator system) and remove the entrapped air (during the product recirculation) [95, 98].

7. Characterization of cellulose derivatives hydrogels as dermal pharmaceutical dosage forms

The physicochemical, microbial and biological characteristics of cellulose-derivatives-based hydrogels are evaluated by a variety of pharmacopoeial and nonpharmacopoeial tests that are carried out to assess the quality and performance of hydrogel formulations and to minimize the batch-to-batch variations. Generally, the US, European, Japanese, and many other pharmacopoeias recommend the following tests: appearance (transparency or clarity), homogeneity, particle size analysis, pH, minimum fill, rheological measurements (viscosity, consistency), stability, microbial screening, *in vitro* drug release testing and assay. Consequently, a range of experimental techniques are used, more often in tandem, to characterize these hydrogels, providing appropriate evidences for their quality and performance.

The transparency or "clarity" of hydrogels is determined by visual methods or is measured as light transmittance by spectroscopy. Their homogeneity is usually assessed by visual examination and the surface morphology by scanning electron microscopy, which analyze under an electron microscope the lyophilized hydrogel gold sputter coated [99, 100]. Optical microscopy, dynamic light scattering and laser diffraction are currently used to determine the size, shape and granulometric distribution of particles of the suspended active substances, as these particles characteristics can influence several properties of a medicated hydrogel such as the rheological properties, stability, and therapeutic activity. Since these properties of cellulose derivatives hydrogel may be also affected by their pH, this parameter must be considered in their quality control. The pH is measured by potentiometry, using either special pH electrodes designed for viscous gels, or conventional pH electrodes for water diluted gels [101]. The content uniformity assessment of hydrogel product can be supported by the minimum-fill test that is performed to compare the weight of product filled into each container labeled with their weight.

Rheological analysis of cellulose derivatives hydrogels is very important for their characterization and quality control, providing valuable information for their formulation, manufacturing, use, and therapeutic activity. Rheological properties of cellulose derivatives hydrogels are determined by mechanical techniques using more often small-deformation and penetrometric measurements. Small-deformation measurements can be performed under both continuous shear and dynamic oscillatory conditions, being used to determine the viscoelastic properties of these gels. Continuous shear instruments (i.e., rotational viscosimeter, cone-

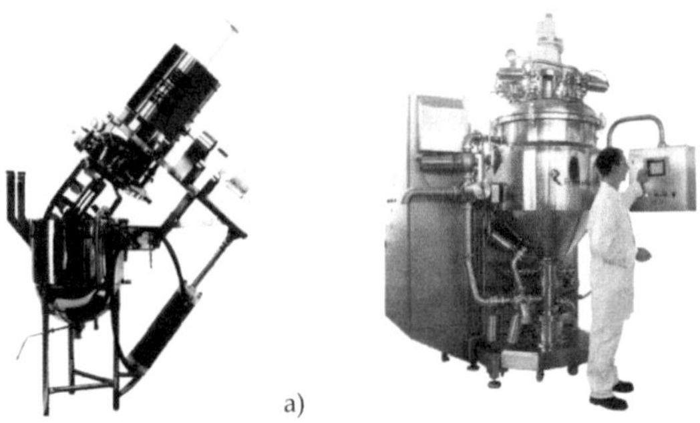

a) b)

Figure 2. Processing machines for the hydrogels preparation: (a) Stainless steel jacketed kettle with agitator (courtesy of Lee Industries, Inc., Philipsburg, Pennsylvania); (b) vacuum processing machine used for the preparation of gels (courtesy of Fryma Koruma, Rheinfelden, Switzerland).

and-plate viscosimeter) measures the apparent viscosity, yield stress, and shear modulus *G*(or modulus of rigidity), and generates a complete rheogram for a particular hydrogel, allowing to identify its flow behavior. Unlike continuous shear measurements, dynamic oscillatory testing, using rheometers, does not modify the hydrogel microstructure and measures another two important parameters, namely the storage modulus G' and the loss modulus G'', reflecting the hydrogel viscoelasticity [23, 102–105]. The consistency determination is a specific test for semisolids, including the hydrogels and is performed by penetrometry, according to pharmacopoeial specifications [101]. This compendial method measures the hardness of hydrogels, but the other rheological characters of these systems reflected by consistency (i.e., rigidity, spreadability, elasticity, and adhesiveness) can be also measured by different noncompedial techniques, using modern instruments such as texture analyzers [106]. Further, as it was mentioned in Section 5, rheological measurements are powerful experimental techniques for probing both the sol-gel transition and swelling behavior of cellulose derivatives hydrogels.

Due to their high water content, cellulose derivatives hydrogels are susceptible to physical, chemical, and microbial stability alterations. Syneresis, a commonly observed phenomenon of physical instability of these hydrogels, can be determined by water loss from the gel network after a heating-cooling cycle. In the case of medicated cellulose derivatives hydrogels, the chemical stability of active substances into the gel is evaluated by accelerated aging studies, which use different analytical methods to evidence the possible drug degradation in the hydrogel, under extreme conditions of temperature, moisture and light [95].

Although is a noncompendial test, *in vitro* drug release testing is of high interest as it can reflect the combined effects of several formulation parameters (i.e., the type of cellulose derivative as gelling agent, the drug concentration, the pH, and viscosity of hydrogel) on the *in vitro* drug release and permeation parameters. In the last decade, continuous efforts in international harmonization have been made to standardize the methodology and the study protocols of *in vitro* drug release from dermal products [107, 108].

8. Applications of cellulose-derivatives-based hydrogels for dermal and transdermal drug delivery

Cellulose-derivatives-based hydrogels, as pharmaceutical dosage forms, have been developed with their end use mainly in topical drug delivery [109]. These hydrogels are widely used as water-soluble ointment bases usually for dermal and, in a lesser extent, for transdermal delivery of various categories of drugs including nonsteroidal anti-inflammatory agents, antifungals, antibiotics, anesthetics, analgesics, antiallergics, antiseptics, keratolytics, revulsive agents, β-blockers etc. From the formulations intended for dermal delivery, the drug has to pass through the *stratum corneum*, the outermost layer of the skin to reach the subjacent skin layers. Therefore, medicated dermal formulations ensure the drug localization in the skin layers. Instead, from transdermal formulations the drug is transported in the skin dermis and then enters into the systemic circulation [110]. However, the excellent barrier properties of the *stratum corneum* limit the drugs penetration through the skin in both dermal and transdermal drug delivery. Therefore, several strategies have been used to enhance and control the drug transport across the skin, and to increase the number of the delivered drugs. These strategies involve both chemical methods, based on the use of penetration enhancers that temporarily increase the skin permeability, and physical methods, in which a driving force is provided to act on the drug [110].

8.1. Dermal drug delivery

Dermal medicated hydrogels based on cellulose derivatives are intended to treat different acute or chronic, mild or moderate skin conditions (i.e., eczema, dermatitis, psoriasis, acne, warts, inflammations, and allergies).

Medicated gel formulations based on cellulose derivatives were obtained by dispersing the drug directly into the hydrogel vehicle or by entrapping the drug in colloidal carriers (micro- and nanoemulsions, liposomes, niosomes etc.) acting as percutaneous enhancers. In the past decades, the favorable effect of cellulose derivatives hydrogels on dermal drug delivery has been demonstrated by a large number of published original scientific papers, academic reviews, monographs and books focused on preparation, characterization and applications of cellulose derivative hydrogels [85, 89, 111–117].

Due to their specific properties, some of gel-forming cellulose derivatives, such as NaCMC and MC are used for the development of wound dressings, usually in combination with other hydrophilic polymers and propyleneglycol, which act as humectant and preservative. These hydrogels proved to be effective as treatment for burned tissues and nonhealing diabetic ulcers [16, 39, 118, 119].

In **Table 9** are included the main studies that have been conducted in the past decade on cellulose-derivatives-based hydrogels as vehicles for dermal delivery of different drugs.

This section provides a survey of the relevant results of the recent reports published in the last decade.

Kouchak and Handali [121] studied the effect of some skin penetration enhancers, including sodium tauroglycocholate, lauric acid, and ethanol on the *in vitro* permeation of aminophylline from 3% HPMC-based hydrogels through snake skin. Sodium tauroglycholate and ethanol at concentration of 100 μg/ml and 60%, respectively, produced a 6-fold increase of the permeation parameters (flux and permeation coefficient), being the optimum enhancers for aminophylline, intended to be used as anticellulitic agent.

From the class of antifungal agents, clotrimazole, bifonazole, and fluconazole were selected as model drugs that were solubilized in emulsion or microemulsion systems, which were then loaded in a cellulose derivative-based hydrogel in order to improve the viscosity of the emulsion/microemulsion and make it suitable for cutaneous application. Shahin et al. [125] studied several jojoba oil-based emulgel formulations as potential clotrimazole delivery vehicles. Jojoba oil was used as lipophilic phase, Span 60 and Brij 35 were used as surfactants, propyleneglycol was selected as humectant and for its aesthetic benefits, and hydroxypropylmethylcellulose and/or Carbopol 934P were chosen as gelling agents; triethanolamine was used to neutralize the formula containing carbomer to the pH range 5.5–6.5. It was found that the concentration and type of the gelling agent significantly influenced the viscosity and the consistency of the examined systems. Also, the *in vitro* clotrimazole release data showed an inverse correlation between the concentration of the gelling agent and the extent of the released drug. The formulation containing the combination of two gelling agents, HPMC and Carbopol 934P, in concentrations of 1% and 0.2%, respectively, showed stable rheological properties (shear thinning behavior with little thixotropy), high extent of drug release and superior antifungal activity in comparison to a commercially available preparation [125].

Drug	Cellulose derivative as gelling agent	Pharmaceutical dosage form	References
Alaptide	HEC, HPC, MC	Hydrogel	[120]
Aminophylline	HPMC	Hydrogel	[121]
Bifonazole	HPMC	Microemulsion-loaded hydrogel	[122]
Clarithromycin	HPMC	Emulgel	[123]
Chlorphenesin	HPMC	Emulgel	[124]
Clotrimazole	HPMC	Emulgel	[125]
Diclofenac sodium	CMCNa	Hydrogel	[126]
Etoricoxib	HPMC	Hydrogel	[127]
Fluconazole	CMC	Emulgel, microemulsion-loaded hydrogel	[128]
	HPMC	Hydrogel	[129]
Ketorolac tromethamine	HPMC	Hydrogel	[130]
Lidocaine	HPC	Hydrogel	[131]
Lidocaine hydrochloride			
Meloxicam	HPC	Hydrogel	[132, 133]
Mepivacaine		Hydrogel	[134]
Piroxicam	HPMC	Emulgel	[135]
	MC, HPMC, CMC	Microemulsion-loaded hydrogel	[136]
Propranolol hydrochloride	HPMC	Hydrogel	[137]

Table 9. Overview of cellulose-derivatives-based hydrogels as vehicles for cutaneous delivery of drugs.

In order to improve the solubility and skin permeability of bifonazole, Sabale et al. [122] investigated several oil-in-water microemulsion-loaded hydrogels composed of bifonazole (1%) water, oleic acid as oil, Tween 80/isopropyl alcohol as surfactant/cosurfactant mixture, and two grades of HPMC (K15M and K100M) in concentrations of 1, 1.5, and 2%. After optimization of the microemulsion formulation by 3^2 factorial design and of the polymer concentration for preparation of microemulsion-loaded hydrogel based on viscosity, the composition of the optimized microemulsion-loaded hydrogel was bifonazole (1%), oleic acid (6.25%), Tween 80/isopropyl alcohol (55%, 3:1), water (38.75%), and HPMC K100M (2%).

The results of the *ex vivo* permeation study through excised rat skin indicated that bifonazole presented a permeability of 84% within 10 h and a sustained release from the optimized formulation, due to the presence of the gelling agent; the drug release was accurately described by the zero order model. Also, the developed preparation exhibited a good stability over a period of 3 months, no skin irritancy and an antifungal activity comparable with a marketed bifonazole cream [122].

Salerno et al. [128] studied the *in vitro* fluconazole release from different topical dosage forms, aiming to determine a formulation with the capacity to deliver the whole active compound and maintain it within the skin, in order to be considered as a useful formulation, either for topical mycosis treatment or as adjuvant in a combined therapy for Cutaneous Leishmaniasis. Sodium carboxymethylcellulose was used for emulgels and microemulsion-loaded hydrogels as gelling agent, and propyleneglycol and diethyleneglycol monoethyl ether (Transcutol P®) were used for each dosage form as solvent for the drug and also as penetration enhancers. The microemulsion-loaded hydrogel containing Transcutol P® delivered the whole applied dose of fluconazole, and showed the highest ability to penetrate pig skin (a four times greater of total amount drug released than that from lipogels) and an important ability to keep the drug within the skin layers; also this formulation proved to be the most effective in regard of the *in vitro* antifungal activity. Further, the authors concluded that in the case of CMCNa-based dosage forms (emulgels and microemulsion-loaded hydrogels), the viscosity was not the main parameter governing the fluconazole release, as these systems showed similar viscosity, but different rates of the released drug [128]. Other study investigated the *in vitro* release of fluconazole from different hydrophilic gels, containing HPMC, chitosan or poloxamer 407 as gelling agents, propyleneglycol as cosolvent and various penetration enhancers (glycerol, PEG 400, Tween 80, and cetrimide). Among the studied hydrogels, those based on HPMC produced the higher percentages of fluconazole released after 6 h, either in the presence or absence of glycerol (66.66 and 71.65%, respectively), glycerol was found to be the most effective release enhancer [129].

Nonsteroidal anti-inflammatory agents, such as diclofenac sodium, oxicams (piroxicam, meloxicam), and ethoricoxib, were used as model drugs in several studies investigating the effects of the structure and the components of the vehicle on drug release and penetration from cellulose-derivative-based hydrogels [126, 127, 130, 132, 133, 135, 136]. A base-gel formulation consisting of 2.5% hydroxypropylcellulose (Klucel), propyleneglycol, ethanol, and water (1:1:1) was selected to investigate the effect of four penetration enhancers (dimethylsulfoxide, Tween 20, oleic acid, and menthol) on *in vitro* permeability of meloxicam through IPM-saturated cellulose membranes and human cadaver skin [132]. Permeation studies through human cadaver skin showed that the highest flux value (2.43 ± 0.47 μg/cm^2/h), with a corresponding enhancement ratio of 27.5, was obtained from HPC-based hydrogel containing 5% menthol as penetration enhancer. According to the authors, the developed meloxicam gel formulation consisting of 2.5% Klucel® gel, 0.3% meloxicam, 5% menthol, and the mixture of propyleneglycol, ethanol, and water (1:1:1) offers the possibility to deliver through the skin therapeutically effective amounts of meloxicam [132]. The potential of 2.5% hydroxypropylcellulose hydrogel, containing a combination of enhancers, to produce a required flux of meloxicam to maintain therapeutic concentration, was confirmed by another study performed by Chang et al. [133]. The effects of a combination of four penetration enhancers (ethanol, propyleneglycol, menthol, and azone) acting by different mechanisms, on the penetration of meloxicam sodium from HPC-based hydrogels through rat skin, was investigated by the response surface methodology. Also, the uniform design technique was applied to prepare systematical model formulations, which were composed of four formulation factors (the content of ethanol, propyleneglycol, menthol and azone); the penetration rate (flux) was chosen as the response. The obtained results demonstrated, on one hand, that the optimal meloxicam sodium hydrogel formulation can be designed using this response surface methodology, and on the other hand, that menthol influenced, in the greatest extent, the skin permeation of meloxicam sodium, followed by azone, ethanol, and propyleneglycol respectively. The optimal transdermal formulation, containing 1% meloxicam, 2.5% HPC, 37.1% ethanol, 15.4% propyleneglycol, 2.9% menthol, 4.3% azone, and water, had appropriate flux value (467.6 ± 89.3 μg/cm^2/h) which met the required flux value of meloxicam (400 μg/h) for maintaining a therapeutic concentration. Further, the *in vivo* absorption study showed that meloxicam could be determined at 1 h after 2.3 cm^2 topical administration and reached steady-state concentration in about 12 h; the bioavailability of the optimal meloxicam sodium gel was about 50.1% [133]. In another study, aiming to determine the optimal base that promote the *in vitro* release of piroxicam, several gel bases consisting of different polymers in various concentrations (3% MC, 2% CMC, 3% HPMC, 0.5% Carbopols 934 and 940, and 20% Pluronic F-127) were evaluated [136]. In each of these gel bases, piroxicam (0.5%) was incorporated as such and in the form of a microemulsion, consisting of oleic acid as oil, Tween 80 as surfactant and propyleneglycol as cosurfactant. The developed gel formulations were evaluated for their rheological properties, stability and *in vitro* piroxicam release through an artificial membrane. Comparison of the *in vitro* release results showed that the 3% MC and 3% HPMC gel bases loaded with piroxicam-microemulsion released the higher amounts of drug after 180 min (97% and 94%, respectively). However, considering also the rheological properties and shelf life, the above mentioned HPMC gel-base loaded with microemulsion was proposed as the most suitable vehicle for topical delivery of piroxicam [136].

Prakash et al. studied the *in vitro* release of ethoricoxib through rat epidermis and human cadaver skin from several hydroethanolic gels based on different polymers as gelling agents (Carbopol, HPMC K4M, MC, and HPC) in the presence of various permeation enhancers such as DMSO, lemongrass oil, menthe oil, and oleic acid [127]. The results of *ex vivo* permeation study revealed that hydroethanolic gel containing 2% HPMC with 2% lemongrass oil produced the highest cumulative amounts of ethoricoxib permeated at 6 h (99.28%), similarly as the studied marketed product (99.89%). Also, this formulation showed comparable anti-inflammatory activity with the marketed product.

In our research work, a hydroethanolic gel based on 2% HPMC and 60% ethanol was used as vehicle for a hydrophilic model drug (propranolol hydrochloride) in the presence of some terpenes as penetration enhancers (menthol, camphor, eucalyptol, thymol, and α-bisabolol) at 5% concentration [137]. The results of our *in vitro* permeation study through pig ear skin, indicated that eucalyptol and α-bisabolol were the most effective terpenes in enhancing the skin transport of propranolol hydrochloride from HPMC-based hydroethanolic gels. Thus, we considered that the gel formulations consisting of 3% propranolol hydrochloride, 2% HPMC, 60% ethanol, and 5% terpene (eucalyptol or α-bisabolol) should be an alternative to oral dosage forms of this drug, recently reported as an effective treatment for infantile haemangioma [137].

These studies, but also many other, showed that the release properties of topical formulations and consequently of the base, are mainly controlled by the drug thermodynamic activity, particle size and diffusion through the preparation. Generally, the diffusion coefficient of a solute in a base is inversely related to the viscosity of the continuous phase. Consequently, the drug release decreases with the increase of viscosity, which is due to the increase in gelling agent concentration. Also, it was suggested that high polymer concentrations increase the resistance to diffusion in a greater extent than expected, because the drug particles are trapped by the polymer macromolecules and they are much closed to the respective entities. Moreover, at high polymer concentrations the density of chain structures increases, thus limiting the drug movement area. Therefore, in the case of gels, including those based on cellulose derivatives, it is considered more appropriate to relate the diffusion coefficient of drug particles in solution to the gel microviscosity, which controls the movement of the particles [138, 139].

8.2. Transdermal drug delivery

Nowadays, transdermal drug delivery is considered an attractive alternative to the conventional drug delivery methods (oral administration and injection) used for the systemic delivery of drugs. Hydrogels, including cellulose-derivatives-based hydrogels, were proposed and studied as transdermal drug delivery systems due to their several benefits including ease of application and delivery, sustained and controlled drug release, reduced systemic side effects, bypass of hepatic first pass effects and potential to provide a better feeling for the skin compared with conventional ointments and patches [140]. A recent research paper evidenced that a 1% HPMC-based hydrogel formulation produced higher permeation rates of diltiazem hydrochloride through rabbit skin (1569.5 $\mu g/cm^2$) than other studied gel formulations (organogels and bigels). Moreover, *in vivo* study of diltiazem hydrochloride as transdermally applied antihypertensive agent revealed that the HPMC-based hydrogel formulation produced a faster and sustained antihypertensive effect compared with other studied gels. The superiority of diltiazem hydrochloride hydrogel based on 1% HPMC in terms of drug permeation through the skin was attributed to high water content that facilitates the release of the hydrophilic drug from the gel-base and increases epidermal cell hydration, thus enhancing the drug diffusion across *stratum corneum* [141].

However, numerous drugs such as hydrophilic, high molecular weight, and charged active substances are not able to penetrate the skin, because of their structure and physicochemical properties. Therefore, in the recent past, physical penetration enhancement techniques including ionophoresis, sonophoresis, electroporation, and laser irradiation gained attention in transdermal drug delivery research. The therapeutic effects of above-mentioned electrically assisted techniques, used alone as well in combination or further, in conjunction with chemical enhancers, for transdermal drug delivery were intensively investigated [110, 142–147]. Furthermore, hydrogels proved to be suitable formulations for assisted transdermal delivery by ionophoresis, sonophoresis, and electroporation due to their advantageous characteristics such as ease of loading into the device, suitability with the electrode design, good flexibility and fitting with the skin contour, strength, transparency, stability, and high electrical conductivity, which is attributed to their high water content [116, 148, 149].

A number of studies reported the successfully transdermal delivery of different drugs from cellulose-derivatives-based hydrogels using ionophoresis alone and combined with other enhancement physical or chemical techniques. Tavakoli et al. [150] investigated the transdermal iontophoretic delivery of celecoxib from several gel formulations, containing different gelling agents (sodium alginate, sodium carboxymethyl cellulose, hydroxypropyl methylcellulose, and Carbopol 934P). Among the studied gel-bases, the hydrogel containing 4% HPMC K4M was considered the optimal formulation for the iontophoretic studies, as it showed higher spreadability and ability to retain on the skin, and released the highest percent of celecoxib after 5 h (41.5%). The findings of the *ex vivo* studies showed that iontophoretic transport of celecoxib from HPMC K4M-based hydrogels through rat skin was 2-fold higher than the passive flux [150].

Recently, the feasibility of using a 2% HEC-based hydrogel as gel-base for successful iontophoretic transdermal delivery of the E-selectin antagonist CGP69669A, a sialyl Lewisx-glycomimetic with potential activity against inflammatory skin diseases was reported. Although the *in vitro* drug permeation through porcine and human skin from the hydrogel formulation was lower than from aqueous solution, the skin deposition (more relevant for the local treatment of dermatological conditions) was 3-fold higher, which was attributed to the occlusive effect of the gel layer on the skin surface, increasing the degree of dermal hydration and consequently promoting the permeation of the hydrophilic CGP69669A [151].

Nandy et al. studied the efficacy of iontophoresis used alone and in combination with chemical enhancers (L-menthol and Tween 20) on transdermal delivery of atenolol from an aqueous solution and several hydrogel formulations based on 3% sodium carboxymethly cellulose or 3% methylcellulose, through excised abdominal rat skin. The obtained results demonstrated that cellulose-derivatives-based hydrogels were more suitable than a solution as transdermal iontophoretic delivery systems, ensuring a sustained release of atenolol. Also, compared with passive delivery, iontophoresis increased the atenolol transport from all studied gel formulations through rat skin. Moreover, the synergistic effects of iontophoresis and chemical enhancers were revealed. Considering this and the superiority of L-menthol as penetration enhancer compared to Tween 20, the NaCMC- and MC-based hydrogel formulations, containing 1.5% atenolol and 2% L-menthol combined with iontophoresis were considered as the best drug delivery systems to achieve the desired drug level [152]. The advantageous combination of iontophoresis and penetration enhancers for transdermal delivery was also recently evidenced in the case of diclofenac sodium formulated as hydroxyethylcellulose-based hydrogels containing different terpenes. The diclofenac sodium hydrogel containing geraniol produced an iontophoretic flux 5.16-fold higher than the passive control, being considered as optimal formulation [153]. Another research report revealed the feasibility of using carboxymethylcellulose-based hydrogels for transdermal delivery of buprenorphine under the application of iontophoresis or electroporation, separately or together [154].

8.3. Commercial products

A list of examples of pharmaceutical gel products, based on cellulose derivatives with their corresponding therapeutic activity, is shown in **Table 10**. The pharmaceutical gel products containing HPC offer the advantage of the polymer compatibility with high percentages of alcohol.

Commercial gel and manufacturer	Cellulose derivative	Drug and therapeutic activity
Persa-Gel 10 (Otho Derm)	Hydroxypropyl methylcellulose	Benzoyl peroxide; antibacterial
Diclac gel (Hexal AG)	Hydroxypropyl methylcellulose	Diclofenac sodium; antiinflammatory, analgesic
Xylocaine jelly (Astra)	Hydroxypropyl methylcellulose	Lidocaine HCl; anesthetic
ArthriCare triple-medicated gel (Commerce)	Hydroxypropyl methylcellulose	Methyl salicylate, menthol; analgesic, revulsive
Erygel (Herbert)	Hydroxypropylcellulose	Erythromycin; antibiotic for acnee
Retin-A gel (Ortho)	Hydroxypropylcellulose	Retinoic acid; antiacne
Compound W gel (Whitehall)	Hydroxypropylcellulose	Salicylic acid; keratolytic
DuoPlant gel (Schering-Plough)	Hydroxypropylcellulose	Salicylic acid; keratolytic
Hydrisalic gel (Pedinol)	Hydroxypropylcellulose	Salicylic acid; keratolytic
Keralyt gel (Summers)	Hydroxypropylcellulose	Salicylic acid; keratolytic
Fungicure-Tolnaftate gel (Alva-Amco Pharmacal Companies, Inc.)	Hydroxypropylcellulose	Tolnaftate; antifungal
Naftin (Merz Pharmaceuticals, LLC)	Hydroxyethylcellulose	Naftifine; antifungal
Nurofen gel (Reckitt Benckiser Healthcare Int. Ltd.)	Hydroxyethylcellulose	Ibuprofen; antiinflammatory, analgesic
IntraSite™ Gel (Smith and Nephew)	Sodium carboxymethyl cellulose	Wound dressing
GranuGel™ (ConvaTec)	Sodium carboxymethyl cellulose	Wound dressing
Purilon Gel™ (ColoPlast)	Sodium carboxymethyl cellulose	Wound dressing
Aquacel Ag™ (ConvaTec)	Sodium carboxymethyl cellulose	Silver ions; wound dressing

http://dx.doi.org/10.5772/63953

Commercial gel and manufacturer	Cellulose derivative	Drug and therapeutic activity
Silvercel™ (Johnson & Johnson)	Sodium carboxymethyl cellulose	Silver ions; wound dressing

Table 10. Some commercially available gels containing cellulose derivatives.

9. Conclusion

Due to their typical properties, several water-soluble cellulose derivatives, namely methylcellulose, hydroxyethylcellulose, hydroxypropylcellulose, hydroxypropylmethylcellulose, and sodium carboxymethylcellulose, have gained increasing interest for pharmaceutical applications, being currently used as gelling agents in the development of topical drug delivery systems. Among the specific properties, which support the extensively use of water-soluble cellulose derivatives as thickening agents in dermal gel formulations, it can be mentioned the diversity of cellulose derivatives, with broad compatibility with many active substances and wide range of viscosities correlated with variable polymer concentrations. In the presence of water, these hydrophilic cellulose derivatives form reversible hydrogels composed of networks of dissolved macromolecules, which possess specific properties in terms of swelling behavior, gel point and sensitivity to external stimuli. The cellulose-derivatives-based hydrogels are widely used as water-soluble ointment bases for dermal and transdermal drug delivery, due to their unique characteristics such as transparency, high water content, spreadability, good skin tolerability, bioadhesiveness, availability, and also biocompatibility and biodegradability, which make them similar with living tissues. In addition, these hydrogels with or without drug substances are relatively easy to prepare either on a small scale (as in extemporaneous compounding) and on a large scale (industry). Although numerous medicated and nonmedicated hydrogels based on cellulose derivatives have been developed, studied and even patented, only few have reach the pharmaceutical market. Therefore, more studies on this topic are expected so as to fully explore the potential of cellulose derivatives hydrogels as dermal and transdermal drug delivery systems.

Author details

Lavinia Vlaia[1*], Georgeta Coneac[1], Ioana Olariu[1], Vicenţiu Vlaia[1] and Dumitru Lupuleasa[2]

*Address all correspondence to: vlaia.lavinia@umft.ro

1 Faculty of Pharmacy, "Victor Babeş" University of Medicine and Pharmacy, Timişoara, Romania

2 Faculty of Pharmacy, "Carol Davila" University of Medicine and Pharmacy, Bucureşti, Romania

References

[1] Nevell TP, Zeronian SH. Cellulose Chemistry and Its Applications. New York: Halsted Press, John Wiley; 1985. p. 22–26.

[2] Kimura S, Itoh T. Evidence for the role of the glomerulocyte in cellulose synthesis in the tunicate, Metandrocarpauedai. Protoplasma. 1995;186:24–33.

[3] Lin CC, Aronson JM. Chitin and cellulose in cell walls of oomycete, *Apodachlya* sp. Arch Mikrobiol. 1970; 72:111–114.

[4] Römling U. Molecular biology of cellulose production in bacteria. Res Microbiol. 2002; 153:205–212.

[5] Somerville C. Cellulose synthesis in higher plants. Annu Rev Cell Dev Biol. 2006;22:53–78.

[6] Prabhu DM, Li WJ. The most abundant and natural source: cellulose and its derivatives and their applications. In: Mondal IH, editor. Cellulose and Cellulose Derivatives: Synthesis, Modification and Applications. New York: Nova Science Publishers Inc.; 2015. p. 3–16.

[7] Varshney VK, Naithani S. Chemical functionalization of cellulose derived from nonconventional sources. In: Kalia S, editor. Cellulose Fibers: Bio- and Nano-Polymer Composites. Heidelberg: Springer-Verlag Berlin; 2011. p. 43–60.

[8] Kamel S, Ali N, Jahangir K, Shah SM, El-Gendy AA. Pharmaceutical significance of cellulose: a review. eXPRESS Polym Lett. 2008;2(11):758–778.

[9] Popovici I. Dispersii liofile. Dispersii hidrofile. In: Popovici I, Lupuleasa D, editors. Tehnologie Farmaceutică, vol. 2. Iaşi: Polirom; 2008. p. 43–89.

[10] Tosh B. Esterification and etherification of cellulose: synthesis and application of cellulose derivatives. In: Mondal IH, editor. Cellulose and Cellulose Derivatives: Synthesis, Modification and Applications. New York: Nova Science Publishers Inc.; 2015. p. 259–298.

[11] Bowman BJ, Ofner III CM, Schott H. Colloidal dispersions. In: Troy D, editor. Remington: The Science and Practice of Pharmacy, 21st ed. Philadelphia: Lippincott Williams & Wilkins; 2005. p. 293–318.

[12] Bajerová M, Gajdziok J, Dvořáčková K, Masteiková R, Kollár P. Semisynthetic cellulose derivatives as the base of hydrophilic gel systems. Čes Slov Farm. 2008; 57(2):63–69.

[13] Rowe RC, Sheskey PJ, Quinn ME, editors. Handbook of Pharmaceutical Excipients. London: Pharmaceutical Press; 2009. p. 118–121, 311–314, 317–322; 438–441.

[14] Aqualon, NATROSOL®Hydroxyethylcellulose A Nonionic Water-Soluble Polymer. Physical and Chemical Properties. Wilmington; Ashland Aqualon Functional Ingredients. Technical literature; 1999.

[15] Ashland Aqualon Functional Ingredients. Technical literature: Klucel Hydroxypropylcellulose Physical and Chemical Properties; 2012.

[16] Ofner III CM, Klech-Gelotte CM. Gels and Jellies.In: Swarbrick J, editors. Encyclopedia of Pharmaceutical Technology, 3rd ed. vol. 3. New York: Informa Healthcare; 2007. p. 1875–1890; 1885–1886.

[17] Dow Chemical Company. Methocel Cellulose Ethers Technical Book. USA; 2002.

[18] Dow Chemical Company. Using METHOCEL Cellulose Ethers for Controlled Release of Drugs in Hydrophilic Matrix Systems. USA; 2000.

[19] Grover JA. Methylcellulose and derivatives. In: Whistler RL, BeMiller JN, editors. Industrial Gums: Polysaccharides and Their Derivatives, 3rd ed. California: Academic Press Inc.; 1993. p. 475–500.

[20] ShinEtsu: Metolose. Water-Soluble Cellulose Ethers. Japan; 2005

[21] Thakur VK, Thakur MK. Handbook of Polymers for Pharmaceutical Technologies, Structure and Chemistry. Beverly: Scrivener Publishing LLC; 2015.

[22] Silva SM, Pinto FV, Antunes FE, Miguel MG, Sousa JJ, Pais AA. Aggregation and gelation in hydroxypropylmethyl cellulose aqueous solutions. J Colloid Interface Sci. 2008;327(2):333–340.

[23] Joshi SC. Sol-gel behavior of hydroxypropyl methylcellulose (HPMC) in ionic media including drug release. Materials. 2011;4:1861–1905.

[24] Banks SR, Sammon C, Melia CD, Timmins P. Monitoring the thermal gelation of cellulose ethers in situ using attenuated total reflectance Fourier transform infrared spectroscopy. Appl Spectrosc. 2005;59(4):452–459.

[25] Acevedo A, Takhistov P, de la Rosa CP, Florián V. Thermalgelation of aqueous hydroxypropylmethylcellulose solutions with SDS and hydrophobic drug particles. Carbohydr Polym. 2014;102:74–90.

[26] Yoo YJ, Um IC. Examination of thermo-gelation behavior of HPMC and HEMC aqueous solutions using rheology. Korea-Australia Rheol J. 2013;25(2):67–75.

[27] Dow Chemical Company. Pharma & Food Solutions. WALOCEL™ C. USP/EP Carboxymethyl Cellulose Sodium for Pharmaceutical Applications; 2014.

[28] Aqualon Division of Hercules Inc. Booklet 250-10H. Aqualon® Sodium carboxymethylcellulose. Physical and Chemical Properties. Wilmington; 1999.

[29] Hennink WE, Van Nostrum CF. Novel crosslinking methods to design hydrogels. Adv Drug Deliv Rev. 2002;54(1):13–36.

[30] Chang C, Duan B, Cai J, Zhang L. Superabsorbent hydrogels based on cellulose for smart swelling and controllable delivery. Eur Polym J. 2010;46:92–100.

[31] Kono H, Onishi K, Nakamura T. Characterization and bisphenol A adsorption capacity of cyclodextrin-carboxymethylcellulose-based hydrogels. Carbohydr Polym. 2013;98:784–792.

[32] Lin OH, Kumar RN, Rozman HD, Noor MAM. Grafting of sodium carboxymethyl cellulose (CMC) with glycidylmethacrylate and development of UV curable coatings from CMC-g-GMA induced by cationic photoinitiators. Carbohydr Polym. 2005;69:57–69.

[33] Demitri C, Del Sole R, Scalera F, Sannino A, Vasapollo G, Maffezzoli A, Ambrosio L, Nicolais L. Novel superabsorbent cellulose-based hydrogels crosslinked with citric acid. J App Polym Sci. 2008;110:2453–2460.

[34] Akar E, Antinis KA, Seki Y. Preparation of pH and ionic-strength responsive biodegradable fumaric acid crosslinked carboxymethylcellulose. Carbohydr Polym. 2012;90:1634–1641.

[35] Kono H. Characterization and properties of carboxymethyl cellulose hydrogels crosslinked by polyethyleneglycol. Carbohydr Polym. 2014;106:84–93.

[36] Wach RA, Rokita B, Bartoszek N, Katsumura Y, Ulanski P, Rosiak JM. Hydroxyl radical-induced crosslinking and radiation-initiated hydrogel formation in dilute aqueous solutions of carboxymethylcellulose. Carbohydr Polym. 2014;112:412–415.

[37] Wach RA, Mitomo H, Nagasawa N, Yoshii F. Radiation crosslinking of carboxymethylcellulose of various degree of substitution at high concentration in aqueous solutions of natural pH. Radiat Phys Chem. 2003;68:771–779.

[38] Barbucci L, Leone G, Vecchiullo A. Novel carboxymethylcellulose-based microporous hydrogels suitable for drug delivery. J Biomater Sci Polym Ed. 2004;15(5):607–619.

[39] Sannino A, Demitri C, Madaghiele M. Biodegradable cellulose-based hydrogels: design and applications. Materials. 2009;2:353–373.

[40] Gulrez SKH, Al-Assaf S, Phillips GO. Hydrogels: methods of preparation, characterization and applications. In: Carpi A, editor. Progress in Molecular and Environmental Bioengineering – From Analysis and Modeling to Technology Applications. Rijeka: InTech; 2011. p. 117–150.

[41] Desbrières J, Hirrien M, Rinaudo M. A calorimetric study of methylcellulose gelation. Carbohydr Polym. 1998;37(2):145–152.

[42] Hirrien M, Chevillard C, Desbrières J, Axelos MAV, Rinaudo M. Thermogelation of methylcelluloses: new evidence for understanding the gelation mechanism. Polymer. 1998;39(25):6251–6259.

[43] Burchard W. Solubility and solution structure of cellulose derivatives. Cellulose. 2003;10(3):213–225.

[44] Baumgartner S, Kristl J, Peppas NA. Network structure of cellulose ethers used in pharmaceutical applications during swelling and at equilibrium. Pharm Res. 2002;19(8):1084–1090.

[45] Augustine R, Rajendran R, Cvelbar U, Mozetic M, George A. Biopolymers for health, food and cosmetic applications. In: Thomas S, Durand D, Chassenieux C, Jyotishkumar P, editors. Handbook of Biopolymer-Based Materials: From Blends and Composites to Gels and Complex Networks. Weinheim: Wiley-VCH; 2013. p. 801–819.

[46] Omidian H, Park K. Hydrogels. In: Siepman J, Siegel R, Rathbone M, editors. Fundamentals and Applications of Controlled Release Drug Delivery. New York: Springer; 2012. p. 75–106.

[47] Hatakeyama H, Hatakeyama T. Interaction between cellulosic polysaccharides and water. In: Nishinari K, editor. Hydrocolloids. Part I. Philadelphia: Elsevier Science BV; 2000. p. 261–270.

[48] Durand A, Hervé M, Hourdet D. Thermogelation in aqueous polymer solutions. In: McCormick CL, editor. Stimuli-Responsive Water Soluble and Amphiphilic Polymers. ACS Symposium Series, vol. 780. Washington, DC: American Chemical Society; 2009. p. 181–207.

[49] Yokota S, Matsuyama K, Kitaoka T, Wariishi H. Thermally responsive wettability of self-assembled methylcellulose nanolayers. Appl Surf Sci. 2007;253:5149–5154.

[50] Jain S, Sandhu S, Malvi R, Gupta B. Cellulose derivatives as thermoresponsive polymers: an overview. J Appl Pharm Sci. 2013;3(12):139–144.

[51] Kometani N, Tanabe M, Su L, Yang K, Nishinari K. In situ observations of thermoreversible gelation and phase separation of agarose and methylcellulose solutions under high pressure. J Phys Chem B. 2015;119(22):6878–6883.

[52] Budslov DK, Sushko NI, Tretinnikov ON. Study on thermal gelation of methylcellulose in water using FTIR-ATR spectroscopy. Appl Spectrosc. 2008;75(4):514–518.

[53] Haque A, Morris ER. Thermogelation of mehylcellulose. Part I: molecular structures and processes. Carbohydr Polym. 1993;22:161–173.

[54] Takahasi M, Shimazaki M, Yamamoto J. Thermoreversible gelation and phase separation in aqueous methylcellulose solutions. J Polym Sci B. 2001;39:91–100.

[55] Li L. Thermal gelation of methylcellulose in water: scaling and thermoreversibility. Macromolecules. 2002;35:5990–5998.

[56] Li L, Thangamathesvaran PM, Yue CY, Tam KC, Hu X, Lam Y C. Gel Network structure of methylcellulose in water. Langmuir. 2001;17(26):8062–8068.

[57] Sarkar NJ. Thermal gelation properties of methyl and hydroxypropyl methylcellulose. Appl Polym Sci. 1979;24:1073–1087.

[58] Joshi SC, Su JC, Liang CM, Lam YC. Gelation of methylcellulose hydrogels under isothermal conditions. J Appl Polym Sci. 2008;107(4): 2101–2108.

[59] Kato T, Yokoyama M, Takahashi A. Melting temperatures of thermally reversible gels. Colloid Polym Sci. 1978;256:15–21.

[60] Kobayashi K, Huang C, Lodge TP. Thermoreversible gelation of aqueous methylcellulose solutions. Macromolecules. 1999;32:7070–7077.

[61] Sekiguchi Y, Sawatari C, Kondo T. A gelation mechanism depending on hydrogen bonding formation in regioselectively substituted O-methylcelluloses. Carbohydr Polym. 2003;53:145–153.

[62] Chatterjee T, Nakatani AI, Adden R, Brackhagen M, Redwine D, Shen H, Li Y, Wilson T, Sammler RL. Structure and properties of aqueous methylcellulose gels by small-angle neutron scattering. Biomacromolecules. 2012;13(10):3355–3369.

[63] Thirumala S, Gimble JM, Devireddy RV. Methylcellulose based thermally reversible hydrogel system for tissue engineering applications. Cells. 2013;3:460–475.

[64] Balaghi S, Edelby Y, Senge B. Evaluation of thermal gelation behavior of different cellulose ether polymers by rheology. AIP Conf Proc. 2014;1593:755–761

[65] Arvidson SA, Lott JR, McAllister JW, Zhang J, Bates FS, Lodge TP, Sammler RL, Li Y, Brackhagen M. Interplay of phase separation and thermoreversible gelation in aqueous methylcellulose solutions. Macromolecules. 2013;46(1):300–309.

[66] Lott JR, McAllister JW, Arvidson SA, Bates FS, Lodge TP. Fibrillar structure of methylcellulose hydrogels. Biomacromolecules. 2013;14:2484–2488.

[67] Lott JR, McAllister JW, Wasbrough M, Sammler RL, Bates FS, Lodge TP. Fibrillar structure in aqueous methylcellulose solutions and gels. Macromolecules. 2013; 46(24): 9760–9771.

[68] Ford JL. Thermal analysis of hydroxypropylmethylcellulose and methylcellulose: powders, gels and matrix tablets. Int J Pharm. 1999;179:209–228.

[69] Sarkar N. Kinetics of thermal gelation of methylcellulose and hydroxypropylmethylcellulose in aqueous solutions. Carbohydr Polym. 1995;26(3):195–203.

[70] Xu XM, Song YM, Ping Q., Wang Y, Liu XYJ. Effect of ionic strength on the temperature-dependent behavior of hydroxypropyl methylcellulose solution and matrix tablet behavior. Appl Polym Sci. 2006;102 (4):4066–4074.

[71] Sammon C, Bajwa G, Timmins P, Melia CD. The application of attenuated total reflectance Fourier transform infrared spectroscopy to monitor the concentration and

state of water in solutions of a thermally responsive cellulose ether during gelation. Polymer. 2006;47:577–584.

[72] Haque A, Richardson RK, Morris ER, Gidley MJ, Caswell DC. Thermogelation of methylcellulose. Part II: Effect of hydroxypropyl substituents. Carbohydr Polym. 1993;22:175–186.

[73] Fairclough JPA, Yu HAO, Kelly O, Ryan AJ, Sammler RL, Radler M. The interplay between gelation and phase separation in aqueous solutions of methylcellulose and hydroxypropyl methylcellulose. Langmuir. 2012;28(28):10551–10557.

[74] Bodvik R, Dedinaite A, Karlson L, Bergström M, Bäverbäck P, Pedersen J.S, Edwards K, Karlsson G, Varga I, Claesson PM. Aggregation and network formation of aqueous methylcellulose and hydroxypropulmethylcellulose solutions. Colloid Surf A: Physicochem Eng Aspects. 2010;354:162–171.

[75] Robitaille L, Turcotte N, Fortin S, Charlet G. Calorimetric study of aqueous solutions of hydroxypropyl cellulose. Macromolecules. 1991;24(9):2413–2418.

[76] Ibbett RN, Philp K, Price DM. ^{13}C NMR studies of the thermal behavior of aqueous solutions of cellulose ethers. Polymer. 1992;33(19):4087–4094.

[77] Fernandes EM. Bionanocomposites from lignocellulosic resources: properties, applications and future trends for their use in the biomedical field. Prog Polym Sci. 2013;38(10–11):1415–1441.

[78] Hoare TR, Kohane DS. Hydrogels in drug delivery: progress and challenges. Polymer. 2008;49:1993–2007.

[79] Patel A, Mequanint K. Hydrogel biomaterials. In: Fazel-Rezai R, editor. Biomedical Engineering – Frontiers and Challenges. New York: InTech; 2011. p. 275–296.

[80] Blanco MD, Olmo RM, Teijón JM. Hydrogels In: Swarbrick J, editor. Encyclopedia of Pharmaceutical Technology, 3rd ed., vol. 3. New York: Informa Healthcare;2007. p. 2021–2039.

[81] Ahmed EM. Hydrogel: preparation, characterization, and applications. A review. J Adv Res. 2015;6(2):105–121.

[82] Miyamoto T, Takahashi S, Ito H, Inagachi H, Nioshiki Y. Tissue biocompatibility of cellulose and its derivatives. J Biomed Mater Res. 1989;23(1):125–133.

[83] Peppas N. Hydrogels. In: Ratner BD, Hoffman AS, Schoen FJ, Lemons JE, editors. Biomaterial Science. An Introduction to Materials in Medicine. 2nd ed. New York: Elsevier Academic Press; 2004. p. 100–107.

[84] Peppas NA, Mikos AG. Preparation methods and structure of hydrogels. In: Peppas NA, editor. Hydrogels in Medicine and Pharmacy, vol. 1. Boca Raton: CRC Press; 1986. p. 1–27.

[85] Hoffman AS. Hydrogels for biomedical applications. Adv Drug Deliv Rev. 2002;43: 3–12.

[86] Clark AH, Ross-Murphy SB. Structural and mechanical properties of biopolymer gels. Adv Polym Sci. 1987;83:57–192.

[87] Nijenhius K. Thermoreversible networks. Viscoelastic properties and structure of gels. Adv Polym Sci. 1997;130:169–193.

[88] Florence AT, Attwood D. Physicochemical Principles of Pharmacy. 5th ed. London: Pharmaceutical Press; 2011. p. 292–294.

[89] Peppas NA, Bures P, Leobandung W, Ichikawa H. Hydrogels in pharmaceutical formulations. Eur J Pharm Biopharm. 2000;50:27–46.

[90] Levy G, Schwarz TW. The effect of certain additives on the gel point of methylcellulose. J Am Pharm Assoc. 1958;47(1):44–46.

[91] Vennat B, Lardy F, Arvouet-Grand A, Pourrat A. Comparative texturometric analysis of hydrogels based on cellulose derivatives, carraghenates, and alginates: evaluation of adhesiveness. Drug Dev Ind Pharm. 1998;24(1):27–35.

[92] Michailova V, Titeva St, Kotsilkova R, Krusteva E, Minkov E. Influence of aqueous medium on viscoelastic properties of carboxymethylcellulose sodium, hydroxypropylmethyl cellulose, and thermally pre-gelatinized starch gels. Colloids Surf A. 1999;149(1–3):515–520.

[93] Mata J, Patel J, Jain M, Ghosh G, Bahadur P. Interaction of cationic surfactants with carboxymethylcellulose in aqueous media. J Colloid Interf Sci. 2006;297:797–804.

[94] Kolodziejska J. The effect of saccharic alcohols on rheological parameters of dental anti-inflammatory gels and on pharmaceutical availability of sodium ibuprofen. Acta Pol Pharm. 2006;63(2):127–133.

[95] Mahalingam R, Li X, Jasti BR. Semisolid dosages: ointments, creams, and gels. In: Gad SC, editor. Pharmaceutical Manufacturing Handbook. Production and Processes. New Jersey: Wiley-Interscience, John Wiley & Sons Inc.; 2008. p. 291–306.

[96] Betageri G, Prabhu S. Semisolid preparations. In: Swarbrick J, editor. Encyclopedia of Pharmaceutical Technology, 3rd ed., vol. 3. New York: Informa Healthcare;. 2007. p. 3257–3274.

[97] Popa EA, Popovici I, Braha SL. Forme farmaceutice bioadezive. In: Popovici I, Lupuleasa D, editors. Tehnologie Farmaceutică, vol. 2. Iaşi: Polirom; 2008. p. 715–791.

[98] Allen LV Jr, Ansel HC. Ointments, creams, and gels. Semisolid dosage forms and transdermal delivery systems. In: Ansel's Pharmaceutical Dosage Forms and Drug Delivery Systems. 10th ed. Philadelphia: Wolters Kluwer Health; 2014. p. 316–342.

[99] El Fray M, Pilaszkiewicz A, Swieszkowski W, Kurzydlowski KJ. Morphology assessment of chemically modified cryostructured poly(vinyl alcohol) hydrogel. Eur Polym J. 2007;43:2035–2040.

[100] Aouada FA, de Moura MR, Fernandes PRG, Rubira AF, Muniz EC. Optical and morphological characterization of polyacrylamide hydrogel and liquid crystal systems. Eur Polym J. 2005;41:2134–2141.

[101] Directorate for the Quality of Medicines of the Council of Europe. European Pharmacopoeia. 8th ed. Strasbourg Cedex: Council of Europe; 2014. p. 24, 28, 299–301.

[102] Lin SY, Amidon GL, Weiner ND, Goldberg AH. Viscoelasticity of cellulose polymers and mucociliary transport on frog palates. Int J Pharm. 1993;95:57–65.

[103] Rudraraju VS, Wyandt CM. Rheological characterization of microcrystalline cellulose/sodium carboxymethyl cellulose hydrogels using a controlled stress rheometer: part I. Int J Pharm. 2005;292(1–2):53–61.

[104] Dolz-Panas M, Roldan-Garcia C, Herraez-Dominiquez JV, Belda-Maximino R. Thixotropy of different concentrations of microcrystalline cellulose:sodium carboxymethyl cellulose gels. J Pharm Sci. 1991;80:75–79.

[105] Hoare T, Zuralowski D, Langer R, Kohane DS. Rheological blends for drug delivery. I. Characterization in vitro. J Biomed Mater Res A. 2010;92:575–585.

[106] Jones DS, Woolfson AD, Djokic J. Texture profile analysis of bioadhesive polymeric semisolids: mechanical characterization and investigation of interactions between formulation components. J Appl Polym Sci. 1996:61(12):2229–2234.

[107] OECD Guidance document for the conduct of skin absorption studies. OECD series on testing and assessment. No. 28. Paris: Environment Directorate;, 2004. p. 12–24.

[108] FDA Guidance for Industry SUPAC-SS. Nonsterile Semisolid Dosage Forms, Scale-Up and Postapproval Changes: Chemistry, Manufacturing, and Controls. In Vitro Release Testing and In Vivo Bioequivalence Documentation. U.S. Department of Health and Human Services, Food and Drug Administration, Center for Drug Evaluation and Research; June 1997.

[109] Shokri J, Adibkia K. Application of cellulose and cellulose derivatives in pharmaceutical industries. In: van de Ven T, Godbout L, editors. Cellulose – Medical, Pharmaceutical and Electronic Applications. Rijeka: InTech; 2013. p. 47–66.

[110] Trommer H, Neubert RHH. Overcoming the stratum corneum: the modulation of skin penetration. Skin Pharmacol Physiol. 2006;19:106–121.

[111] Ibrahim MM, Hafez SA, Mahdy MM. Organogels, hydrogels and bigels as transdermal delivery systems for diltiazem hydrochloride. Asian J Pharm Sci. 2013;8(1):48–57.

[112] Vlaia L, Olariu I, Coneac G, Vlaia V, Popoiu C, Stanciulescu C, Mut AM, Szabadai Z, Lupuleasa D. Percutaneous penetration enhancement of propranolol hydrochloride

from HPMC-based hydroethanolic gels containing terpenes. Farmacia. 2014;62(5):991–1008.

[113] Pillai AB, Nair JV, Gupta NK, Gupta S. Microemulsion-loaded hydrogel formulation of butenafine hydrochloride for improved topical delivery. Arch Dermatol Res. 2015;307(7):625–633.

[114] Csóka G, Pásztor E, Marton S, Zelkó R, Antal I, Klebovich I. Formulation of thermoresponsive transdermal therapeutic systems using cellulose derivatives. Acta Pharm Hung. 2007;77(2):102–107.

[115] Valenta C, Auner BG. The use of polymers for dermal and transdermal delivery. Eur J Pharm Biopharm. 2004;58(2):279–289.

[116] Caló E, Khutoryanskiy VV. Biomedical applications of hydrogels: a review of patents and commercial products. Eur Polym J. 2015;65:252–267.

[117] Trombino S, Cassano R. Cellulose and its derivatives for pharmaceutical and biomedical applications. In: Mondal IH, editor. Cellulose and Cellulose Derivatives: Synthesis, Modification and Applications. NewYork: Nova Science Publishers; 2015. p. 405–419.

[118] Zheng WJ, Gao J, Wei Z, Zhou J, Chen YM. Facile fabrication of self-healing carboxymethylcellulose hydrogels. Eur Polym J. 2015;72:514–522.

[119] Aly FU. Preparation and evaluation of novel topical gel preparations for wound healing in diabetics. Int J Pharm Pharm Sci. 2012;4(4):76–77.

[120] Sklenář Z, Vitková Z, Herdová P, Horáčková K, Šimunková V. Formulation and release of alaptide from cellulose-based hydrogels. Acta Vet Brno. 2012;81:301–306.

[121] Kouchak M, Handali S. Effects of various penetration enhancers on penetration of aminophylline through shed snake skin. Jundishapur J Nat Pharm Prod. 2014;9(1):24–29.

[122] Sabale V, Vora S. Formulation and evaluation of microemulsion-based hydrogel for topical delivery. Int J Pharm Invest. 2012;2(3):140–149.

[123] Baibhav J, Singh G, Rana AC, Saini S. Development and characterization of clarithromycin emulgel for topical delivery. Int J Drug Dev Res. 2012;4(3):310–323.

[124] Mohamed MI. Optimization of chlorphenesin emulgel formulation. AAPS J. 2004;6(3): 81–87.

[125] Shahin M, Hady SA, Hammad M, Mortada N. Novel jojoba oil-based emulsion gel formulations for clotrimazole delivery. AAPS PharmSciTech. 2011;12(1):239–247.

[126] Gupta A, Mishra AK, Singh AK, Gupta V, Bansal P. Formulation and evaluation of topical gel of diclofenac sodium using different polymers. Drug Invent Today. 2010;2(5):250–253.

[127] Prakash PR, Rao NGR, Soujanya C. Formulation, evaluation and antiinflammatory activity of topical etoricoxib gel. Asian J Pharm Clin Res. 2010;3(2):126–129.

[128] Salerno C, Carlucci AM, Bregni C. Study of in vitro drug release and percutaneous absorption of fluconazole from topical dosage forms. AAPS PharmSciTech. 2010;11(2): 986–993.

[129] Abdel-Mottaleb MMA, Mortada ND, Elshamy AA, Awad GAS. Preparation and evaluation of fluconazole gels. Egypt J Biomed. Sci. 2007;23:35–41.

[130] Hosny KM, Tayeb MM, Fallatah OM, Mahmoud AA, Mandoura MS, Al-Sawahli MM. Preparation and evaluation of ketorolac tromethamine hydrogel. Int J Pharm Sci Rev Res. 2013;20(2):269–274.

[131] Sawant PD, Luu D, Ye R, Buchta R. Drug release from hydroethanolic gels. Effect of drug's lipophilicity (log *P*), polymer-drug interactions and solvent lipophilicity. Int J Pharm. 2010;396(1–2):45–52.

[132] Jantharaprapap R, Stagni G. Effects of penetration enhancers on *in vitro* permeability of meloxicam gels. Int J Pharm. 2007;343:26–33.

[133] Chang JS, Huang YB, Hou SS, Wang RJ, Wu PC, Tsai YH. Formulation optimization of meloxicam sodium gel using response surface methodology. Int J Pharm. 2007;338:48–54.

[134] Cho CW, Choi JS, Shin SC. Enhanced local anesthetic action of mepivacaine from the bioadhesive gels. Pak J Pharm Sci. 2011;24(1):87–93.

[135] Shokri J, Azarmi Sh, Fasihi Z, Hallaj-Nezhadi S, Nokhodchi A, Javadzadeh Y. Effects of various penetration enhancers on percutaneous absorption of piroxicam from emulgels. Res Pharm Sci. 2012;7(4): 225–234.

[136] Abd-Allah FI, Dawaba HM, Ahmed AMS. Preparation, characterization, and stability studies of piroxicam loaded microemulsions in topical formulations. Drug Discov Ther. 2010;4(4):267–275.

[137] Vlaia L, Olariu I, Coneac G, Vlaia V, Popoiu C, Stănciulescu C, Muţ AM, Szabadai Z, Lupuleasa D. Percutaneous penetration enhancement of propranolol hydrochloride from HPMC-based hydroethanolic gels containing terpenes. Farmacia. 2014;62(5):991–1008.

[138] Al-Khamis KI, Davis SS, Hadgraft J. Microviscosity and drug release from topical gel formulations. Pharm Res. 1986;3(4):214–217.

[139] Davis SS, Khanderia M. The influence of the rheological properties of the vehicle on the release of drugs from semi-solid ointments. Proceedings 1st International Conference on Pharmaceutica Technology, APGI, Paris; 1977. p. I: 30–47.

[140] Simões S, Figueiras A, Veiga F. Modular hydrogels for drug delivery. J Biomater Nanobiotechnol. 2012;3:185–199.

[141] Ibrahim MM, Hafez SA, Mahdy MM. Organogels, hydrogels and bigels as transdermal delivery systems for diltiazem hydrochloride. AJPS. 2013;8:48–57.

[142] Prausnitz MR, Langer R. Transdermal drug delivery. Nat Biotechnol. 2008;26(11):1261–1268.

[143] Azagury A, Khoury L, Enden G, Kost J. Ultrasound mediated transdermal drug delivery. Adv Drug Deliv Rev. 2014;72:127–43.

[144] Wang Y, Thakur R, Fan Q, Michniak B. Transdermal iontophoresis: combination strategies to improve transdermal iontophoretic drug delivery. Eur J Pharm Biopharm. 2005;60(2):179–191.

[145] Charoo NA, Rahman Z, Repka MA, Murthy SN. Electroporation: an avenue for transdermal drug delivery. Curr Drug Deliv. 2010;7(2):125–136.

[146] Khan A, Yasir M, Asif M, Chauhan I, Singh AP, Sharma R, Singh P, Rai S. Iontophoretic drug delivery: history and applications. JAPS. 2011;1(3):11–24.

[147] Escobar-Chávez JJ, Bonilla-Martínez D, Villegas-González MA, Rodríguez-Cruz IM, Domínguez-Delgado CL. The use of sonophoresis in the administration of drugs throughout the skin. J Pharm Pharm Sci. 2009;12(1):88–115.

[148] Vikram K, Kiran B, Rahul T. Enhancement of iontophoretic transport of diphenhydramine hydrochloride thermosensitive gel by optimization of pH, polymer concentration, electrode design and pulse rate. Pharm Sci Tech. 2007;8:E1–E6.

[149] Stamatialis DF, Rolevink HH, Gironès M, Nymeijer DC, Koops GH. *In vitro* evaluation of a hydroxypropyl cellulose gel system for transdermal delivery of timolol. Curr Drug Deliv. 2004;1:313–319.

[150] Tavakoli N, Minaiyan M, Heshmatipour M, Musavinasab R. Transdermal iontophoretic delivery of celecoxib from gel formulation. Res Pharm Sci. 2015;10(5):419–428.

[151] Gratieri T, Wagner B, Kalaria D, Ernst B, Yogeshvar N, Kalia YN. Cutaneous iontophoretic delivery of CGP69669A, a sialyl Lewisx-mimetic, *in vitro*. Exp Dermatol. 2012;21:226–228.

[152] Nandy BC, Gupta RN, Rai VK, Das S, Tyagi LK, Roy S, Meena KC. Transdermal iontophoretic delivery of atenolol in combination with penetration enhancers: optimization and evaluation on solution and gels. Int J Pharm Sci Drug Res. 2009;1(2):91–99.

[153] Arunkumar S, Shivakumar HN, Desai BG, Ashok P. Effect of gel properties on transdermal iontophoretic delivery of diclofenac sodium. e-Polymers. 2016;16(1):25–32.

[154] Fang JY, Sung KC, Wang JJ, Chu CC, Chen KT. The effects of iontophoresis and electroporation on transdermal delivery of buprenorphine from solutions and hydrogels. J Pharm Pharmacol. 2002;54(10):1329–1337.

Chapter 8

An Engineering Point of View on the Use of the Hydrogels for Pharmaceutical and Biomedical Applications

Gaetano Lamberti, Anna Angela Barba,
Sara Cascone, Annalisa Dalmoro and Diego Caccavo

Additional information is available at the end of the chapter

http://dx.doi.org/10.5772/64299

Abstract

In this chapter, the modern uses of hydrogels in pharmaceutical and biomedical applications are revised following an engineering point of view, i.e. focusing the attention on material properties and process conditions. The chapter discusses the applications following the increase in scale-size. First, the nanoscale systems, i.e. hydrogel nanoparticles (HNPs), are analysed in terms of preparative approaches (polymerization methods and uses of preformed polymers) and with a brief mention of the future trends in the field. Secondly, systems based on hydrogel microparticles (HMPs) are examined following the same scheme (polymerization methods, uses of preformed polymers, a mention of novel and future trends). Thirdly, and last but not the least, the hydrogel-based drug delivery systems (macroscopic HB-DDSs) are presented, focusing in particular on tablets made of hydrogels, discussing the characterization methods and on the modelling approaches used to describe their behaviour. Other macroscopic systems are also discussed in brief. Even if the vastness of the field makes its discussion impossible in a single chapter, the presented material can be a good starting point to study the uses of hydrogels in pharmaceutical and biomedical sciences.

Keywords: nano-hydrogels, micro-hydrogels, macro-hydrogels, drug release, modelling

1. Introduction

Hydrogels are hydrophilic polymer networks, which are able to absorb and retain large amounts of water. Networks can be composed of homopolymers or copolymers, and their network structure and physical integrity are due to the presence of cross-links of chemical (tie-points, junctions) or physical (entanglements, crystallites) nature. Based on the stability/strength of these cross-links, hydrogels can either withstand exposure to water or they can degrade and dissolve in water, after a given exposure time [1]. Because of the wide spectrum of chemical and mechanical properties as well as their excellent biocompatibility, hydrogels have been extensively investigated for pharmaceutical and biomedical applications [2–6]. In particular, hydrogels are the main component in controlled drug delivery systems (DDSs), as described in reviews by Kamath and Park [7], Peppas [8], Peppas et al. [4] and Hoare and Kohane [9]. For some drugs difficult to be administered, such as the protein, the hydrogels are one of the most important ways to delivery [10–15]. Hydrogels are also interesting because they can be prepared *in situ* [16–20], and their degradation times can be tailored by the used building blocks, the chemical nature of cross-links and the cross-link density [21]. Importantly, because of their high water content and soft nature, hydrogels are well tolerated by cells and tissues and therefore they possess a good biocompatibility [22]. Hydrogels are also under investigation in the form of nanogels as delivery systems for NABDs (nucleic acid-based drugs, such as pDNA, siRNA and mRNA) [23,24].

This chapter is intended to give an overview of the hydrogels in pharmaceutical applications, with a particular—even if not exclusive—focus on the research activities carried out at the University of Salerno by Transport Phenomena and Processes Group (http://grup-potpp.unisa.it). The chapter is organized following the size of the described systems: first of all, the applications of nanoscale hydrogels are discussed, then the micro-scale hydrogels are analysed and, last but not the least, macroscopic drug delivery systems based on hydrogels are presented. Of course, since this is a huge topic, the chapter cannot be exhaustive but it has to be considered as a guiding path to explore the active researches in the field.

2. Nano systems

Over the past few years, hydrogel nanoparticles (HNPs) or nanogels have been investigated as carrier systems for site-specific and/or time-controlled drug delivery. The reason for this interest is because of the combination of the features of a hydrogel, i.e. hydrophilicity, high water content and swelling ability, useful for controlled release [25] with those of a nanoparticle, especially the small size [26]. Both synthetic and natural polymers can be used in hydrogel nanoparticles production for their different surface properties or bulk erosion rates to control the release rate of active molecules. The former (synthetic polymers) include especially block copolymers consisting of two or more segments of simple polymers (blocks) joined in some arrangement, such as the biodegradable and biocompatible poly(D,L lactic acid) (PLA), poly(glycolic acid) (PGA) and their copolymer poly(D,L-lactic-co-glycolic acid) (PLGA)polyvinyl alcohol and polyethylene oxide [27]. The latter (natural polymers) are characterized by a variety of functional groups, a wide range of molecular weights and variable chemical composition. Among them, polysaccharides are the more often used: they are carbohydrate-based polymers formed of repeating units (monosaccharides) joined together by glycosidic bonds and can be of algal (alginate), plant (cellulose, starch) and animal (chitosan) sources [28]. Therefore, hydrogel nanoparticles are ideal drug-delivery systems due to their excellent drug loading capacity, high stability, biologic consistence, flexibility, versatility, biocompatibility and response to a wide variety of environmental stimuli (such as ionic strength, pH and temperature) [29]. The techniques for preparing hydrogel nanoparticles can be divided into two main categories: methods exploiting the direct polymerization of monomers (based on chemical cross-links) and methods based on the use of preformed polymers (based on physical interactions) [30].

2.1. Polymerization methods

The preparation of HNPs via monomer polymerization includes two simultaneous steps, polymerization and formation of nanostructures, that can be accomplished by interfacial polymerization, by emulsion polymerization or by controlled radical polymerization [31]. For example, an inverse emulsion polymerization method was used for poly(ethylene glycol) (PEG) cross-linked acrylic nanoparticles for the controlled release of curcumin, a hydrophobic molecule that inhibits proliferation and induces apoptotic cell death in numerous cell lines established from malignancies such as leukaemia, breast, lung, prostate and colon tumours [32]. In particular, emulsification was obtained by dispersing the aqueous phase, made of 10% acrylic acid, 5% sodium hydroxide and 15% water, in a continuous lipophilic phase consisting of liquid paraffin (68%) and emulsifiers (2%): Span 80 and Tween 80 (75:25 ratio). To the mixture, first a PEG diacrylate (1%) as cross-linker and then the initiator, ammonium persulphate, were added. The polymerization was performed at 60°C for 6 h. The obtained particles were centrifuged, washed and finally freeze-dried. Curcumin was loaded in polymer nanoparticles after polymerization: a nanoparticle/water solution was placed in contact with a solution of curcumin dissolved in chloroform under constant stirring (by vortex) and sonication. The curcumin-loaded nanoparticles were then lyophilized. They showed a higher entrapment efficiency and lower particle size with the decrease of the cross-linking degree, and a release rate dependent on the pH of the releasing medium, in particular, a high swelling at pH 7.4 than at pH 2.2. However, a higher cross-linking degree caused a reduction of the HNPs swelling at pH 7.4, thus a more controlled increase in mesh size that allowed a lower initial burst release followed by a sustained release. Moreover, curcumin HNPs showed *in vitro* a cellular uptake similar to their free counterpart confirming their ability to overcome the barrier of curcumin's aqueous dispersibility, facilitating the *in vivo* administration. Again, they showed more cytotoxicity and apoptotic effects towards cancer cells than free curcumin, thanks to both nanoparticulate form and the use of hydrophilic polymers for minimizing the opsonization and prolonging the *in vivo* circulation of HNPs. A different polymerization technique was the PRINT (particle replication in non-wetting templates) particle fabrication technique, which is characterized by a higher degree of difficulty, but by complete, orthogonal control over particle characteristics, and an easiness of scaling up [33]. Ma et al. proposed a PRINT particle fabrication technique for the production of hydrogel nanoparticles conjugated with siRNA (short interfering RNA, highly used for gene therapy) [34]. The same technique was used by Kai et al. for the nano-encapsulation of cisplatin, a cytotoxic drug used in therapy against a wide variety of cancers [33]. In both cases, a pre-particle solution was prepared by dissolving different percentages of reactive monomers in methanol, where the reactive monomers were: a cure-site monomer, the oligomeric poly(ethylene glycol) (PEG,

MM: 700 g/mol) with terminal acryloxy functionality; a hydrophilic monomer, tetraethylene glycol monoacrylate (HP4A); an amine containing monomer, 2-aminoethyl methacrylate hydrochloride (AEM), for providing the amine functionality needed to conjugate PEG onto the surface of the PRINT particles; a photoinitiator, diphenyl (2,4,6-trimethylbenzoyl)-phosphine oxide (TPO). A thin film was drawn onto polyethylene terephthalate (PET), laminated to the patterned side of the mould and delaminated at the laminator nip roll. Then particles were cured by passing the filled mould through a UV-LED; a polyvinyl alcohol harvesting sheet was hot laminated to the filled mould and cooled to room temperature and particles were removed from the mould by splitting the harvesting sheet from the mould. Particles were then harvested by dissolving the polyvinyl alcohol in water and passed through a filter. After centrifuge, the cisplatin-containing particles were also subjected to PEGylation and succinylation. Incubation of the active molecules for obtaining the conjugation with the HNPs followed. siRNA-conjugated HNPs were characterized by a loading efficiency of up to 29% and high transfection efficiency *in vitro* by efficiently controlling hydrogel composition, surface modification and siRNA loading ratio. For cisplatin HNPs, an inverse correlation between PEG density (conformation that surface-bound PEG chains achieve: 'mushroom conformation', with low density PEG coverage, i.e. PEG chains are not fully extended away from the nanoparticle surface, and 'brush conformation' with the PEG chains extending away from the nanoparticle surface, resulting in a thick layer) and drug loading, and an improved exposure in the blood (higher total amount of drug reaching circulation) and tumour accumulation (due to both nanodimensions and PEG presence), with concurrent renal protection were observed.

2.2. Preformed polymers: physical and chemical interactions

Chemically cross-linked nanogels need the presence of reactive sites along the polymer chains. Some polymers possess inherent reactive sites, such as polyethyleneimine (PEI) with a high density of amine functional groups, or natural polymers containing amine or carboxylic acid groups. Chemical cross-linking provides nanogels with a great structural stability owing to irreversible inter- or intra-molecular covalent bonding formations. The chemical networks in the core inhibit the simple diffusion of hydrophobic drug molecules, giving enhanced encapsulation efficiency. However, some drawbacks can arise—sometimes the desired functionality requires complex chemical modifications of cross-linkers causing the loss of nanogel versatility; also, small chemicals for the cross-linking reaction can be toxic for medical uses [35].

2.3. Novel and future trends

Physically cross-linked HNPs, usually obtained in aqueous media with mild conditions, can be modified in size by changing some parameters, such as the pH value, ionic strength and temperature [29]. In particular, amphiphilic polymers can self-assemble into micelles with core–shell structure: the hydrophilic outer shell allows a prolonged circulation, i.e. it is a barrier against recognition and opsonization, and the hydrophobic core increases the drug loading via hydrophobic or electrostatic interactions [36]. Physical interactions are of two typologies:

amphiphilic association, based on Van der Waals links, including hydrogen bond and hydrophobic interaction, and electrostatic interaction. Used preformed polymers can be consequently divided into amphiphilic or triblock copolymers that form self-aggregates in water by undergoing intra- and/or inter-molecular association between hydrophobic moieties due to the minimization of interfacial free energy, and polymers modified with the aim to have reactive sites to form physical cross-links via electrostatic interaction [37]. To obtain physical cross-linking by self-association, hydrophobic polymers chains, such as poly(lactic acid), were grafted on a backbone of dextran in order to obtain water-soluble biodegradable graft copolymers of dextrose and PLA able to form nanometric aggregates in an aqueous solution [38]. The nanogel was prepared by the solvent exchange method: the copolymer was first dissolved in dimethyl sulphoxide (solvent for both PLA and dextran); then distilled water was added at a rate of 1 drop every 10 s under high shear stirring until a water content of about 30–50 wt% was reached; the resulting solution was dialyzed for organic solvent removal, after that the solution turned slightly opaque because of an aggregate formation with the hydrophobic PLA cores and the hydrophilic dextran skeleton. The mean diameters of the aggregates ranged from 16 to 73 nm with narrow size distribution. Moreover, copolymers showed low critical aggregation concentration values, further decreasing by increasing the amount of hydrophobic PLA, indicating a strong tendency of the copolymers towards formation of stable nanogels.

Recent trends in preparation of HNPs exploit the use of a hydrogel core in liposome nanoparticles to obtain lipogels. Liposomes are vesicular structures consisting of an aqueous core enclosed in one or more phospholipid layers which possess low intrinsic toxicity and immunogenicity, capability to incorporate hydrophilic and hydrophobic drugs and good biocompatibility [39]. Wang et al. demonstrated for the first time the possibility to form a poly(acrylic acid) (PAA) hydrogel core in liposomes for the encapsulation of an anticancer drug 17-DMAPG, a geldanamycin (GA) derivative [40]. In particular, a PAA hydrogel core was formed inside liposomes (already produced by the thin film hydration method) through UV-initiated activation and polymerization of acrylic acid (AA) and N,N'-methylenebis(acrylamide) (BA). An optimized pH gradient and electrostatic/hydrophobic interactions between cationic drug and anionic gel in the liposomal core were used to obtain about 90% of loading efficiency of the active molecule and its sustained release independently of the external solution pH, confirming that the lipid bilayer was intact in the presence of the gel core. Moreover, lipogels did not exert cytotoxicity to cells *in vitro* neither at the highest concentration tested (about 0.4 mg/ml of material).

These few examples showed how polymeric hydrogel nanoparticles are particularly attractive, thanks to their easiness of production, affordability and ability of incorporating a variety of active molecules, including proteins, peptides and oligosaccharides, anti-tumour agents and vaccines. However, the development of these systems still require a more comprehensive characterization, especially about the structure of the HNPs, the application of novel materials with targeting and environmental sensitivity, the toxicology effect and the interaction with cells and tissues *in vivo*, before their full potential can be exploited.

3. Micro systems

When hydrogels are in the form of macroscopic links confined to micrometric dimensions, they are termed as microgels or hydrogel microparticles (HMPs), which are different in size from the sub-micrometrc HNPs (nanogels) [41]. The size of the particles is a crucial factor in determining their properties: the specific surface area of a spherical particle, as well as the diffusion rate, is inversely proportional to the diameter, and the time required for a stimulus to reach the centre of a gel particle is proportional to the square of the particle's diameter. Thus, a smaller particle is a much quicker and more serviceable tool; however, if the particles are too small, they sometimes can cause problems such as difficulty in handling, difficulty in recovery, unavoidable dissipation and so on. Due to their size and environmentally sensitive nature, microhydrogels can be used as smart micro-containers, micro-reactors, building block for intelligent materials and for optically functionalized devices [42]. Microgels show unique advantages in comparison with other polymer systems: in particular, a fast response rate to external stimuli; suitability for subcutaneous administration; high biocompatibility for the large amount of water in the swollen state; a large surface area for multivalent bioconjugation; an internal network for the incorporation of several active molecules, thus large drug loading capacities; and adjustable chemical and mechanical properties, and soft architecture enabling them to flatten onto vascular surfaces, thus simultaneously anchoring in multiple points [43]. As already seen for HNPs, hydrogel microparticles can be formed from both natural and synthetic polymers and produced by essentially two methods: particle-forming polymerization, including inverse emulsion polymerization and precipitation polymerization, and molecular assembly of existing polymer molecules in aqueous solutions, aided by the external stimuli, such as temperature, pH and/or the presence of polivalent ions [44]. In particular, microgels of synthetic polymers are usually obtained starting from a monomer, such as acrylic acid, methyl methacrylate, acrylamide and ethylene glycol. Some examples are pH-sensitive hydrogel P(MAA-co-EGMA) microparticles obtained by dispersion photopolymerization by using MAA and ethylene glycol (EG) as monomers [45] and pH-sensitive microhydrogels prepared from N-vinylcaprolactam and methacrylic acid monomers by free radical polymerization [46]. Instead, biopolymer microgels are obtained either by the polymer itself, for example, alginate or chitosan that is gelled in a microparticle, or by a macro-gel, which is mechanically comminuted for producing anisotropic and irregularly shaped microgels, of high interest for tissue replacement applications [47], for biosensor, diagnostics and targeted drug delivery [48]. Methods for producing HMPs are shown in **Table 1**.

Particle properties which can be manipulated to obtain good entrapment and release properties are: the degree of crosslinking; particle size and size distribution; polymer type (controlling their responsiveness to the environment pH, temperature or enzymatic response); particle surface charge; and particle shape from spherical to elongated rods and fibres. HMPs can also be designed to shrink or swell by creating an imbalance in the osmotic pressure between their inside and outside via alteration of pH, temperature or ions in the suspension [49]. Moreover, existing microhydrogels can be modified to improve their properties or functions by using post-treatments such as surface modification, that is, surface grafting or surface graft polymerization, and composite formation by inclusion in HMPs of functional nanoparticles, such as colloidal metals, semiconductor nanocrystals and magnetic nanoparticles.

Polymerization methods	Molecular assembly
Dispersion photopolymerization: It involves the mixing of monomers with the addition of cross-linker, UV-initiator and dispersion stabilizer followed by ultraviolet light irradiation	**Emulsion technique:** Uniform-sized droplets (obtained by microfluidics or membrane method) are dispersed into a continuous phase resulting in oil in water, water in oil or multiple emulsions. After the formation of emulsion, the cross-linking agent is incorporated to harden the final product.
Free radical precipiation polymerization: It is a heterogeneous polymerization that involves a continuous phase in which monomer and initiator are completely soluble. When the reaction starts, the yielded polymer remains insoluble and precipitates.	**Physical gelation:** Uniform-sized droplets (obtained by atomization by different ways, such as twin nozzle or ultrasonic atomization) of biopolymer form a network structure through non-covalent bonds, such as hydrogen bonding, electrostatic (ionic) interactions and hydrophobic interactions, under specific conditions (for example, temperature or pH).
Free radical polymerization: It involves a repeated addition of free radicals building blocks and requires polymer, monomer, initiator, and cross-linker. In most cases, the HMPs are obtained by passing hydrogel through a sieve of desired size, using an anti-solvent, or by stirring at a high rate.	**Chemical gelation:** Uniform-sized droplets (obtained by atomization by different ways, such as twin nozzle or ultrasonic atomization) of biopolymer form a network structure via covalent bonds with a chemical cross-linking agent.
Inverse emulsion polymerization: It involves the water in oil polymerization method: water-soluble droplets are uniformly dispersed by mechanical stirring in a continuous organic phase with the help of oil-soluble surfactants. Polymerization is initiated in aqueous droplets upon addition of a radical initiator.	**Macro-gelation:** It involves the comminution of a macro-gel during its gelation in microgel by shearing the gel to form uniform droplets or, for proteins, by interrupting the aggregation process of random coil proteins with shearing or denaturing the protein, followed by cooling to form a network gel structure.
Ionic gelation method: A solution containing the monomer–cross-linker and a solution with the polymer and an initiator are mixed, and the reaction is carried out at a given temperature for a specific time.	

Table 1. Methods for producing HMPs.

3.1. Polymerization methods

As shown in **Table 1**, the production of HMPs via polymerization can be accomplished by several methods, that is, photopolymerization, precipitation polymerization, free radical polymerization, inverse emulsion polymerization and ionic gelation polymerization. Han et al. produced hollow-structured poly(vinyl amine) (PVAm) hydrogel particles, which are able to encapsulate macromolecules, by using the *in situ* hydrolysis/cross-linking reaction on poly(N-vinylformamide) (PNVF) particles, previously obtained by using dispersion polymerization [50]. PNVF particles were produced by performing the polymerization at 70°C for 24 h of a mixture of 19 g of *N*-vinyl formamide and 1 g of *N,N'*-methylene-bisacrylamide (MBA) in 170 g of methanol containing an initiator, 2,2'-azobis(isobutyronitrile) (AIBN) and a stabilizer, poly(2-ethyl-2-oxazoline) (PETOZO). After polymerization, all unreacted monomers and additives were completely removed by repeated centrifugation with an ethanol. The PNVF particles (about 1 g) were then re-dispersed in an ethanol solution containing glutaraldehyde, GA (76 ml, conc. GA: 0.1 mol/l). Then, 40 g of a 2 N sodium hydroxide aqueous solution was slowly added and the reaction was carried out at 80°C for 12 h. After washing away by-products, PVAm hydrogel particles were obtained. They were generated by competition between the partial disconnection of the cross-linked network in the centre part of PNVF particles and the secondary cross-linking induced by additional reaction of glutaraldehyde with primary amine groups in the PVAm chains located at the periphery of the particles (the secondary cross-linking depended on the glutaraldehyde diffusion from the continuous phase). Particles were characterized by an average diameter of 2.3 μm and showed a discrete spherical hollow capsule structure. As the cross-linked hydrogel shell is formed by the diffusion-limited secondary cross-linking, the thickness of the shell was constant at approximately 250 nm (the cross-linking density could be tuned by changing the concentration of glutaraldehyde). Except the cases using a high amount of glutaraldehyde, all the particles were found to be non-toxic to the cells, that is, it is the glutaraldehyde moiety that affects the cell viability, and not the cationic moieties in PVAm chains. Moreover, regardless of the cross-linking density of the hydrogel phase, the HMPs were very adhesive to both the normal human keratinocyte cells and the normal human dermal fibroblast cells. Thus, the carrier system seemed to be applicable for macromolecule delivery into the cell.

Thermo-sensitive microgels useful as drug carriers were produced by precipitation copolymerization of N-isopropylacrylamide (NIPAM) and N-hydroxyethylacrylamide (HEAM) with various concentrations of a cross-linker in the presence of an anionic surfactant, sodium dodecylsulphate (SDS) [43]. In particular, the poly(NIPAM-co-HEAM) HMPs were synthesized by free radical precipitation polymerization: a water solution of NIPAM, HEAM, N,N'-methylene-bis-acrylamide (BIS) and SDS was first heated to 70°C under a gentle stream of nitrogen, after 1 h a solution of potassium persulphate (KPS) was inserted to start the reaction, that proceeded for 8 h. After cooling the product to room temperature, the obtained HMPs were purified by dialysis and then freeze-dried. The monomer ratio was defined by the knowledge that the linear copolymer poly(NIPAM-co-HEAM) at 10:2 molar ratio in the initial reaction mixture has a lower critical solution temperature (LCST) close to the physiological temperature of 37°C. The volume phase transition temperature (VPTT, a characteristic of all thermoresponsive gels) was evaluated by different methods and it approached very close to the human body temperature, thus confirming the applicability of these HMPs for biomedical and biological applications (requiring a sharp phase transition around the physiological temperature). In effect, the *in vitro* release rates of a model drug, propranolol, was strongly influenced by the temperature: below the VPTT, when HMPs were in a swollen state and no steric or hydrophobic interactions occurred, the drug was rapidly released, whereas above the VPTT, shrunk HMPs produce a reduced release rate.

3.2. Molecular assembly

Self-assembly methods are characterized by the use of amphiphilic polymers that spontaneously form self-aggregates in water (in this way the uses of solvents and harsh reaction conditions are avoided) by undergoing intra- and/or inter-molecular association between hydrophobic moieties due to the minimization of interfacial free energy. This process allow the production of both particles with a shell-core structure and good stability, depending on hydrophobic/hydrophilic constituents and polyelectrolytes complexes between two polymers having different charges, for example, polysaccharides of complementary charge. Moreover, polyelectrolytes can aggregate into microparticles by cross-linking with substances with opposite charge via electrostatic interaction by the ionotropic gelation [30]. For example, chitosan undergoes ionic gelation due to the complexation with oppositely charged species, such as tripolyphosphate (TPP). Barba et al. deeply studied the release kinetics and the influence of the structure of chitosan-TPP complexed particles on transport phenomena [51]. In particular, vitamin B12-loaded chitosan microparticles were produced by dripping or nebulizing via ultrasonic atomization in a water solution of 1% w/w chitosan (containing 0.2% w/w of B12) in the cross-linking solution of 1% w/w TPP (with defined reticulation time and stirring), as shown in **Figure 1**. Then particles were filtered, washed and finally stabilized by convective drying at 50°C. The concentration of chitosan was chosen taking into account that low values brought low encapsulation efficiency and poor particles consistence, instead, at

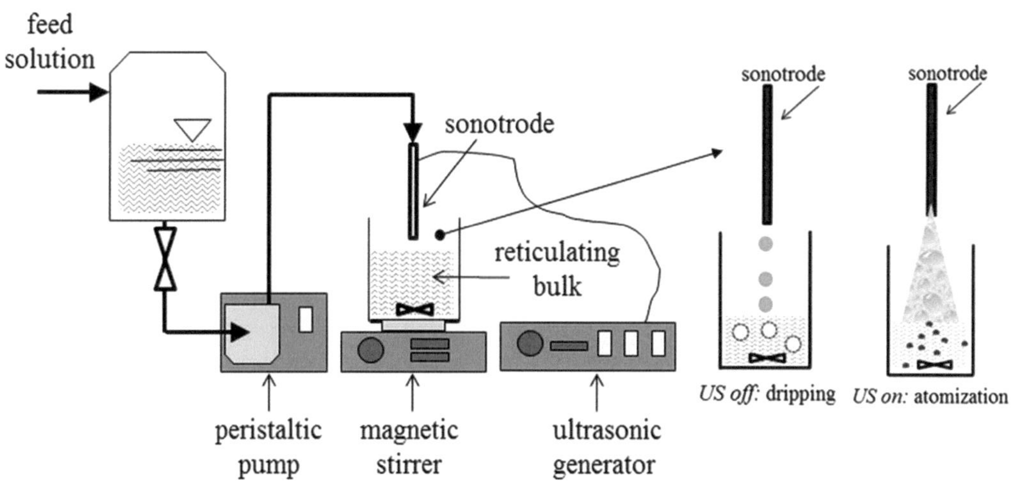

Figure 1. Sketch of the homemade apparatus assembled and used to produce chitosan and alginate cross-linked HMPs. From Barba et al. [51], Copyright © (2012) Croatian Society of Chemical Engineers.

large values, highly viscous solutions that are difficult to process are obtained. In a similar manner, the concentration of TPP was chosen, taking into account that chitosan cross-linked structure obtained at pH 3 had a higher density (and lower porosity) then those reticulated at pH 9. The milli- and micrometric particles (from dripping and ultrasonic atomization, respectively) presented an average diameter of 3.75 mm and 613 μm, respectively. High losses of B12 in the reticulating bulk, giving a low encapsulation efficiency, and a very fast B12 release rate were observed for both the particles: after 30 min more than of 70% of active molecule was transferred in the dissolution bulk. These phenomena suggested that diffusion was the dominant transport phenomenon, thus polymeric network mesh-size (PNMS) was placed under investigation. A simple model of the drug release was proposed, considering a drug-containing spherical particle inserted in a dissolution medium, to obtain the evolution of the drug concentration with time, in function of known or estimated parameters, such as number of particles, particles radius, volume of dissolution medium and initial drug loading concentration, and of the unknown parameter B12 diffusivity:

$$C_A^O(t) = N\frac{4\pi a^2}{V_2}\frac{D}{\delta}C_{A0}^I\tau\left[1-\exp\left(-\frac{t}{\tau}\right)\right] \tag{1}$$

In Eq. (1) the time constant τ is:

$$\tau = \left[\left(4\pi a^2\right)\frac{D}{\delta}\left(\frac{1}{NV_1}+\frac{1}{V_2}\right)\right]^{-1} \tag{2}$$

The fitting between experimental data and model data allowed to calculate the B12 diffusivity through the polymeric shell resulted in larger milli-metric particles because the kinetics (and the extent) of the reticulation may be affected by the size of the produced droplets and ultrasound can have an effect on gel network. These values were inserted in a relationship obtained starting from the volume free theory (proposed by Lustig and Peppas and tested by Amsden):

$$\frac{D}{D_0} = \left(1-\frac{r_s}{\xi}\right)\exp\left(-Y\frac{\varphi}{1-\varphi}\right) \tag{3}$$

In this way, they were able to calculate the mesh sizes of two kind of particles (2.3 nm for milli-particles and 1.2 nm for microparticles), that were shown to be larger than the B12 Stokes radius (0.86 nm, estimated by www.chemspider.com); thus, they were able to demonstrate that the hydrogel structure had to be modified (by physical protocols, for example, by microwaves curing, or by the use of reactive chemical agents) in order to keep the active molecule inside.

The same relationship was already used for estimating the release of a little molecule, theophylline (TP), from calcium chloride-cross-linked alginate HMPs [52]. Production of particles was performed in the same plant used for chitosan (**Figure 1**). An alginate/TP aqueous solution was atomized into a beaker containing 80 ml of a stirred calcium chloride solution (concentration 8.9 g/l) in order to have a rigid network deriving from the coordination of alginate molecules by bivalent positive ions. Once reticulated, the alginate particles were separated by centrifugation or by filtration and dried. Also in this case, the mesh size of

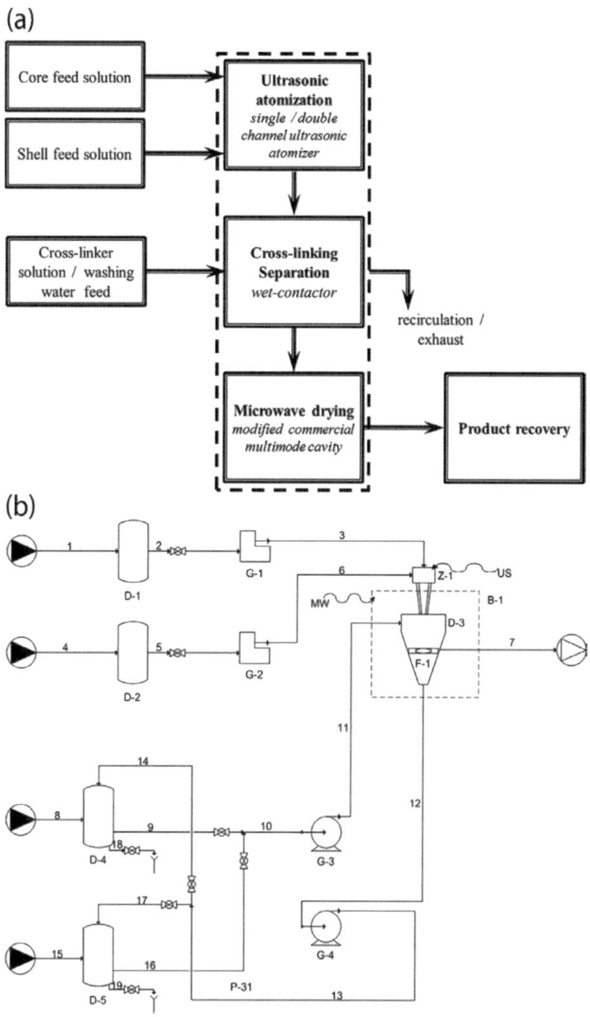

Figure 2. (A) Schematic diagram of the microparticles production. (B) Layout of the microencapsulation single-pot semi-continuous bench scale apparatus (Z-1, ultrasonic double channel atomizer; D-3, wet collector; B-1, microwave oven cavity; G-1/G-2, peristaltic pumps-core and shell feeding channels; G-3/G-4, centrifugal pumps-reticulation solution feeding and suspension recirculation; D-1/D-2, core and shell feeding tanks; D-4/D-5, reticulation solution tank, rinsewater tank; and F-1, filter; box with dashed line: MW cavity). Reprinted with permission from Dalmoro et al. [54], Copyright © (2014) American Chemical Society.

polymer network was not able to retain the small molecule: from the equation of the release kinetic by diffusion from a sphere, using the estimated TP diffusivity in sodium alginate membranes and the particles size (about 100 μm), a 99.9% release of the encapsulated drug after 2–3 s in a water medium (thus the cross-linking bulk) was estimated. To improve the entrapment of a model drug in cross-linked alginate particles, an external protection could be the right choice, thus shell-core particles were produced and compared to matrix (only core) system. Dalmoro et al. produced both shell-core and matrix beads of alginate encapsulating vitamin B12: the shell water solution (alginate 1.5% w/w) and the core water solution (alginate 1.5% w/w and B12 0.525 g/l and some drops of Tween 80) were separately pumped, under controlled conditions, into a stainless steel coaxial double-channel device (a variant of a typical drop generation systems, such as a syringe used to produce large drops) and dropped into a stirred water calcium solution (0.89% w/v) to promote alginate chains reticulation. For matrix particles production, only core channel was fed. After a given time of exposure to calcium ions, they were filtered, and washed by distilled water. Dissolution tests of B12-loaded particles showed that the presence of an additional thin layer of reticulated alginate (the shell-core structure) reduced losses during production, giving a loading B12 ratio of about 70% for shell-core systems, higher than that of matrix systems, of about 45% (same value theoretically estimated from the equation of the release by diffusion from a sphere), and allowed a delayed B12 release [53]. An evolution of shell-core dripping was the use of a double-channel ultrasonic nozzle for the production of shell-core HMPs. Dalmoro et al. proposed the design of a novel process to produce microparticles with a shell-core structure in a bench scale apparatus purposely realized, involving the coupling of atomization assisted by ultrasonic energy and microwave heating for stabilization (**Figure 2**) [54].

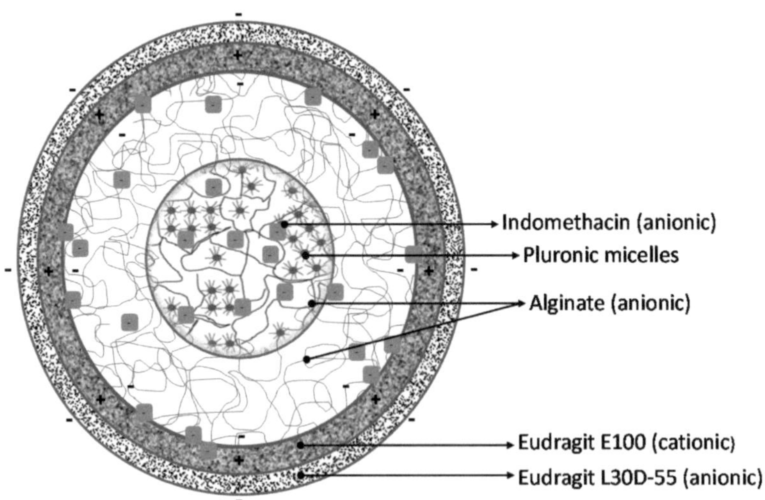

Figure 3. Structure of the HMPs produced by ultrasonic-atomization—two stage polyelectrolyte complexation. From Dalmoro et al. [59], Copyright © (2016).

A core feed aqueous solution, composed of 1% active principle, α-tocopherol, 1.5% alginate and a mesh size reducer, Tween 80 (as previously described in [53]), together with the shell feed solution, made of an aqueous solution of 1.5% alginate, were sent to the coaxial ultrasonic atomizer for the nebulization in the $CaCl_2$ cross-linking aqueous solution. Then they were filtered and stabilized in the same apparatus and finally recovered. Also in this case, the matrix system was obtained by atomizing only the core solution, containing the active molecule and alginate, in the inner channel leaving the external one empty. Both the kinds of HMPs (shell-core and matrix) showed an encapsulation efficiency of around 100% and globally good enteric release properties because α-tocopherol was negligibly released at simulated stomach pH 1, and completely released only at intestine simulated pH 6.8. Again, shell-core HMPs were preferable to matrix ones, confirming the functional role of the shell structure because they showed a better gastro-resistance, that is, a smaller amount of α-tocopherol released at acidic pH. Moreover, microwave treatments, used for stabilization, proved to be useful in controlled drug release since they caused a little delay in α-tocopherol release, especially for shell-core HMPs. The described production process not only gave alginate HMPs with good drug release and entrapment properties, but shown several other advantages [55], that is, it was simple, able to operate at room conditions and in absence of solvents, low energy consumption, thanks to the use of ultrasound compared to conventional techniques of atomization [56], with easily predictable and tuneable features (size) of the micro-droplets as function of process parameters [56,57]. The process proved to be versatile allowing the encapsulation of a molecule, ergocalciferol (vitamin D2), which is less lipophilic than α-tocopherol [58]. The alginate concentration and flow rates of shell and core solution were kept unchanged. The variations needed to encapsulate D2 were: vitamin D2 was first dissolved in ethanol and then emulsified with the aqueous core feeding solution in order to achieve a concentration of 0.2% (w/v) of D2; Tween 80 in core solution as stabilizer was replaced by Pluronic F127 for its amphiphilic properties, that is, for its ability to link to alginate with the hydrophilic moiety and to D2 with the hydrophobic one. Optimization of manufacturing parameters allowed to obtain high encapsulation efficiency (nearly 92%); good gastro-resistance properties (less than 10% release at pH of 1.0, nearly 100% release at pH of 6.8 and 240 min of dissolution); limited ergocalciferol degradation after 5 months of storage, reduction of D2 degradation during microwave-assisted stabilization process compared to the convective one. Another modification of the process, that is, the coupling of the ultrasonic atomization technique with the polyelectrolyte complexation, allowed to produce enteric shell-core HMPs encapsulating an anionic gastrolesive drug, indomethacin (**Figure 3**) [59]. Again, both the solutions, core and shell feed (with unchanged features), were sent to the coaxial ultrasonic atomizer where they were nebulized and placed in contact with the cross-linking solution. In this case, the cationic Eudragit® E 100 (E100) was used as a new complexing agent for the anionic alginate and also to interact with the anionic drug to raise encapsulation efficiency. However, E100 dissolves at pH < 5, thus particles obtained after cross-linking were filtered and put in a solution of the anionic Eudragit® L30D-55 (L30D) copolymer that interacted with the cationic E100 and formed an external gastro-resistant layer on the fine droplets (fresh shell-core microparticles). HMPs were then centrifuged and freeze-dried. The interaction of the cationic E100 with alginate first, and then E100 with L30D-55, obviously caused the formation of larger particles than the previous cases where there was the interaction of alginate with bivalent ions. Also, HMPs obtained by the double complexation method showed high encapsulation efficiency (about 74%) and good enteric release properties.

3.3. Novel and future trends

The methods used for HMPs production are numerous and their suitability depends on the polymer carrier as well as the functionality, particle size and particle size distribution required. Some methods are accessible at industrial scale, although methods producing HMPs with a mono-disperse size distribution are currently best suited to laboratory-scale manufacturing only [49]. Moreover, HMPs are sometimes very sensitive to the presence of contaminants, thus this feature must be improved for enlarging their application field. Therefore, the research must aim at the development of novel innovative matrices with controlling mechanical viscoelastic properties, versatile swelling performance, size and internal organization, triggered degradation behaviour and the enhancement of their biological interactions with body components. It will be also interesting to explore 'softness' for developing biomimetic particles (i.e. artificial cells), shape changing colloids or particles which evolve or move in time based on local chemical cues from each other or other entities (e.g. cells). Fundamental will be also the studies about the mechanisms of passive (traditional methods, i.e. polymerization or gelation) and driven (innovative methods, such as lithographic processing where hydrogel colloids are templated in a highly controlled and reproducible manner) self-assembly in HMPs. There are also still many opportunities not yet realized in the food industry, such as the possibility to use them for food preservation (thus, compatibility with foods), satiety control, encapsulation of phytonutrients and prebiotics and texture control for healthier food formulations.

4. Macro systems

4.1. Compressed tablets

Among the compressed systems used in oral administrations, the hydrogel-based tablets and matrices are the most diffused and, consequently, studied. The process of drug release from a hydrogel-based matrix is very complex and different transport phenomena have to be taken into account. For these reasons, during the years several experimental techniques and modelling approaches have been developed to describe the whole release process, starting from simplified systems (i.e. simple geometry, reduced number of components or 1D mass transport allowed), to more complex systems (i.e. complex geometry, pharmaceutical forms composed of several units, each with a specific behaviour).

4.1.1. Experimental approaches

The first step to characterize a hydrogel matrix behaviour is to measure the swelling phenomenon, which can be quantified by water uptake measurement [60], by evaluation of the swelling ratio [61] or by rheological analyses [62]. Each of these measurements can be performed using several experimental techniques, ranging from the simple gravimetric technique to evaluate the weight gain after a certain dissolution time [60], to a colorimetric technique combined with the image analysis to identify the position of the swelling, the diffusion and the erosion fronts [63]. In general, to describe the swelling phenomenon, the use of a combination of experimental techniques, each with its pros and cons [64], is widely diffused [65] due to the plurality of the aspects involved.

From an engineering point of view, the experimental measurements that characterize the hydrogel behaviour inside a compressed pharmaceutical form can be divided into two groups: (i) the macroscopic measurements, which involve the changes of the whole matrix, how the entire system evolves during the time; and (ii) the microscopic measurements, which involve the changes inside the matrix, how a single part of the system is affected by the external conditions during the time.

Typical examples of macroscopic measurements on a tablet are the evaluation of the total water uptake of the polymer eroded and the drug released after a certain dissolution time. The water uptake and the polymer eroded can be easily obtained by a gravimetric technique, which allows to determine their amounts in the whole matrix during the dissolution time. Concerning the drug eventually contained into the matrix, using an analytical technique, such as a spectrometric or a HPLC method, the amount of drug released can be measured and, since the initial amount of drug in the tablet is usually a known quantity, the drug amount residual into the matrix can be calculated. Thus, by the use of simple techniques all the macroscopic profiles can be measured [66].

In order to understand which parameters and phenomena influence the hydrogel behaviour, a microscopic analysis can be of aid in evaluating what happens inside the polymeric matrix. As described above, when the water diffuses inside the matrix, the polymer starts to swell, leading to the formation of a gel layer, which modulates the rate of drug release. For this reason, evaluate the water distribution inside the matrix is a key topic in the pharmaceutical applications and it is the basis of a deeper knowledge of the swelling process, necessary to understand the tablet behaviour. One of the most used techniques to evaluate the water distribution inside a polymer matrix (i.e. for a cylindrical tablet along the radius of the matrix) is the image analysis technique. The basic principle of this method is that, during the hydration of a tablet inside a dissolution medium, a camera records several grey-scale pictures of the tablet, which has a different grey level, depending on the hydration level [67] (i.e. in the glassy white core of the tablet, where the water has not penetrated, the image remains white, whereas in the hydrated gel the image is darker). Based on this non-destructive technique, after a proper calibration, it is possible to determine the water mass fraction along the radius of a swelling matrix made of pure hydrogel in which the water uptake is allowed only in the radial direction [67]. This last hypothesis is fundamental to maintain the water content uniform in a given section of the matrix. Moreover, with the aid of the image analysis, the movements of the erosion, swelling and diffusion fronts can be evaluated [68]. The main drawback of this technique is that if a third component (i.e. a drug) is added to the polymer (and water) in the tablet, it may interfere with the light intensity and the correlation may fail.

Focused on the development of a non-invasive method to study the behaviour of hydrophilic matrices, a technique based on the nuclear magnetic resonance imaging (NMRI) to produce a two-dimensional map of the density of nuclei inside a complex object has been developed [69]. This technique is very useful to characterize the water content and the diffusion phenomena in a swollen hydrogel matrix, and the evolution of the matrix diameter during the swelling [70]. The extent of the swelling can be identified evaluating the position of the interface between the dry core of the tablet and the swollen region; the erosion front can be identified at the interface between the dissolution medium and the swollen region. By the coupling of a NMR micro-imaging analysis and of an *ad hoc* experimental apparatus [71], it is possible to study the water uptake and the swelling behaviour of extended release tablets of hydrogels [72]. Once a NMR image has been taken, at each pixel of the image can be associated a numerical value that can be put in relation with the proton T2 relaxation (due to the water protons). Once the proton relaxation has been associated with the amount of water inside the tablet with a calibration [73], the hydration level of the matrix during the dissolution can be evaluated both in the axial and in the radial direction, which is a great improvement compared with the simple image analysis technique.

A laborious but effective technique to determine all the mass fractions inside the matrix is the gravimetric technique [66], which can be used both for simplified systems, for example, a cylinder in which the water penetrates only by the radial surface, and in more complex ones, in which the water uptake is allowed both from radial and axial surface. After a certain dissolution time, the tablet is withdrawn from the dissolution medium and, using several punches of different diameters, a series of radial sections and a central core can be obtained. In each of these sections, measuring the water content (by weighting the hydrated tablet and drying until constant weight, and weighting once again), the drug content (by dissolving the section content and assaying the drug concentration in the solution) and the polymer content (by difference between the initial weight of the section material and the water and drug weights), it is possible to obtain the components' mass fractions. Repeating these evaluations for several dissolution times, it is possible not only to determine what happens during the time to the matrix, but even the mass evolution along the radius. These data are particularly useful to identify the most relevant transport phenomena involved in the dissolution and to build a predictive tool of the hydrogel-based matrices behaviour. The described method is very labour intensive and time consuming, but allows to gain a lot of information about the matrix behaviour. However, the main drawback of this technique is the fact that it is destructive for the samples. A combined approach could be used to compare the image analysis results with the gravimetric ones [74].

A recent approach to evaluate the degree of hydration of a hydrogel matrix is the use of texture analysis, which is a non-destructive method based on indentation or compression tests on the swollen matrix to evaluate the gel layer position and strength. It is based on the fact that the resistance of the swollen matrix to the penetration (in the indentation tests) is related to the water content. Indeed, the higher the water concentration inside the polymer matrix, the larger the amount of gel formed and the lower the resistance opposed by the matrix. Moreover, due to the difference between the strength of the glassy core and the gel layer, the position of the swelling front and the thickness of the gel layer formed can be easily evaluated during the swelling by measuring the force opposed to the penetration from the slope of the force-displacement diagram. The use of this technique alone does not allow to quantify the water amount into the matrix, but it gives qualitative indication. For this reason, a method based on the coupling of the texture analysis and the gravimetric technique has been developed to relate the water content to the penetration work resulting from an indentation test [75] for a simple system, in which the water uptake is allowed only by lateral surface. Then, this technique has been improved to correlate the slope of the force-displacement diagram to the water amount even for more complicate system in which the matrix swells both in radial and axial directions [76].

4.1.2. Modelling approaches

Once the mechanisms which affect the drug release from a hydrogel-based compressed system have been clarified and quantified using both the macro- and microscopic approaches, the final goal of the engineering applied to the study of such pharmaceutical systems is to develop a mathematical model. This model should be able to describe the observed phenomena and to predict the matrix behaviour, with the aim to reduce time and costs required by the development of novel dosage forms. During the years, several modelling approaches have been proposed, starting from the empirical to the mechanistic ones.

First of all, the empirical models have been used to describe the release profile, relating the fractional drug release from a thin film to the square root of time by Higuchi [77], an approach later generalized relating the release to a general *n*th-power of time, where '*n*' is a function of the drug transport regime [78]: generally, for a thin film, it is 0.5 for Fickian diffusive process (Higuchi's equation), 1 for the Case-II transport (swelling-controlled drug release) and takes intermediate values for anomalous transport. The hypotheses on which the model was founded are constant diffusivities and negligible swelling, thus it is not particularly suitable for hydrogel-based systems and, despite it is widely used to describe the experimental results, its use should be avoided to derive phenomenological interpretation.

Then, a more complex modelling approach, which takes into account the simultaneous presence of the solvent, polymer matrix and drug during the dissolution process, is the mechanistic approach. To proper describe the complex phenomena which take place during the dissolution, the mass balance equations, coupled with the momentum conservation equation, have to be used. Often these models obtain the constitutive equations from the mixing and elastic free energy of the hydrogel, following the theory of Flory-Huggins and the affine network theory, respectively [1]. Being the gel made of cross-linked polymer and water, the degree of hydration, which in turns is influenced by the gel elasticity, is crucial to the drug release purpose [79]. Indeed, the amount of water modifies the stretching of the polymer chains and tunes the mobility of the active ingredient. Another factor, which affects sensibly the drug release from such matrices, is the tortuosity of the diffusion path inside the hydrogel [2]. The tortuosity is an aspect usually difficult to be estimated and it depends on the pore size into the matrix, on the pore size distribution and it is influenced by the composition and cross-link density of the hydrogel polymer network. A recent trend in pharmaceutical application is to produce superporous hydrogels, which are characterized by fast swelling and large swelling ratios [80].

The mechanistic models could be divided into two groups: (i) the multicomponent mixture models, in which the hydrogel is considered as a single phase constituted by several components; and (ii) the multiphasic models, in which the hydrogel is considered as constituted by different phases.

The *multiphasic* models have the advantages of using different phases for the solid (polymer) and liquid (solvent plus dissolved components) parts, that makes them theoretically easier to extend to very complex systems such as drug delivery matrices [81]. However, this approach has several drawbacks like the high numbers of non-linear partial differential equations (PDEs), that make the system numerically solvable under very limitative conditions, and the scarce physical meaning of some parameters. Instead, the *multicomponent* approach, with more rigorous thermodynamic bases, limits the numbers of PDEs and constrains the variables to physical quantities. However, the theoretical complexity, until now, has led to simplified approaches which take into account only the mass transport equations. One of the first multicomponent models for drug delivery systems was based on purely diffusive mass transport equations by Siepmann et al. [82]. Considering the matrix composed by water, polymer and drug, the water and drug mass transport equations were solved, coupled with their boundary conditions, in a 2D axisymmetric domain, under several hypotheses: (i) no volume contraction upon mixing, (ii) fast drug dissolution compared to drug diffusion, (iii) perfect sink condition for the drug, (iv) strong dependence of the diffusivities (of water and drug) on the hydration level and (v) affine deformations:

$$\begin{cases} \dfrac{\partial \rho_i}{\partial t} = \dfrac{1}{r}\dfrac{\partial}{\partial r}\left(D_i \dfrac{\partial \rho_i}{\partial r}\right) + \dfrac{\partial}{\partial z}\left(D_i \dfrac{\partial \rho_i}{\partial z}\right) \qquad i = 1,3 \\ @t = 0, \forall r, \forall z, \rho_i = \rho_{i,0} \\ @r = R(t), \forall t, \forall z \;\; \rho_i = \rho_{i,eq} \\ @z = Z(t), \forall t, \forall r \;\; \rho_i = \rho_{i,eq} \\ @r = 0, \forall t, \forall z \;\; \dfrac{\partial \rho_i}{\partial r} = 0 \\ @z = 0, \forall t, \forall r \;\; \dfrac{\partial \rho_i}{\partial z} = 0 \end{cases} \tag{4}$$

where ρ_i is the mass concentration of the ith species, $\rho_{i,eq}$ is the interface mass concentration of the ith species and D_i the diffusion coefficient of the ith species. $R(t)$ and $Z(t)$ represent the radius and the thickness, respectively, of the erosion boundaries. The diffusivity has been described by a Fujita-type equation [83] and the polymer mass release during the time (erosion) has been evaluated by the product of an erosion constant and the erosion surface area, which changes during the time due to the swelling:

$$\begin{cases} \dfrac{dm_2}{dt} = -k_{er} A_{er}(t) \\ @t = 0\, m_2 = m_{2,0} \end{cases} \tag{5}$$

where m_2 and $m_{2,0}$ are the polymer mass and the initial polymer mass, k_{er} is an erosion constant and A_{er} is the erosion surface, that is the surface exposed to the external medium. The model is able to describe the evolution of water, polymer and drug masses during the time for a given system. Based on this work, a 1D model to describe the hydration of pure HPMC tablet confined between glass slabs has been developed [67]. The system swelling has been described considering the variation of the total mass due to both the water inlet and the polymer erosion. The efficacy of this model has been proved by the authors comparing its results and the experimental data of water mass fraction profiles along the radial direction obtained by analysing the normalized light intensity of swollen tablet pictures for several dissolution times. The model has been later improved [84] to describe the drug release kinetics from different shaped matrices. These models are very useful to describe the behaviour of geometrically simple systems, but the hypothesis of affine deformation is a big limitation for more complex systems. For this reason, recently a 2D axisymmetric model was proposed [85] to overcome the affine deformation hypothesis. In this model, the transport equations for water and drug have been solved for the respective mass fractions with a finite element method, along with the proper initial and boundary conditions, and the constitutive equation for the system density has been obtained considering the summability of the specific volumes of the single species. The novelty of this model consists in the description of the swelling phenomenon: the inlet water flux has been divided into two components, the first one responsible for the tablet swelling ($j_{1,swe}$) and proportional to a swelling constant (k_{swe}), and the latter responsible for the inner layer hydration($j_{1,diff}$). The deformation rate of the hydrogel has been defined by a water mass balance on a boundary element ($A\delta$), where simultaneously the swelling and the erosion phenomena (the first leads to a volume increasing and the latter to a volume decreasing) happen:

$$\begin{gathered} \rho \frac{d}{dt}\left(A\delta\omega_{1,eq}\right) = \mathbf{A}\cdot \boldsymbol{j}_{1,diff} - \mathbf{A}\cdot \boldsymbol{j}_1 \\ v_{swe} = \frac{d\delta}{dt} = -\frac{\boldsymbol{j}_{1,swe}}{\rho\,\omega_{1,eq}} = -\frac{k_{swe}\boldsymbol{j}_{1,diff}}{\rho\,\omega_{1,eq}} \\ v_{eros} = -k_{eros} \end{gathered} \tag{6}$$

The model has been proved to work in comparison with experimental macroscopic results. Recently, a mechanistic model based on water diffusion in concentrated systems has been proposed [86,87]. This model is based on the assumption that: (i) the swelling is due to the water uptake and to the translocation of polymer, thus the polymer flow at the interface between the tablet and the external medium causes the swelling and the shape change, without polymer release; and (ii) the erosion phenomenon is due to interactions between the tablet surface and external fluid. Due to this new approach, the model has been found able to describe not only the macroscopic behaviour of a matrix, but even what happens inside the hydrogel matrix, and it is in good agreement with the experimental data.

4.2. Hydrogels for biomedical applications

During the years, biomedical applications of the hydrogels in the tissue engineering are gaining increasing interest, which aims to create biological body parts to replace harvested tissues and organs. In this application, the polymer plays a major role in the determination of the cell adhesion, in the formation and growth of a new 3D structured tissue in the human body; and its selection is governed by the physical, the mass transport properties and the biological interaction requirements. For these reasons, and due to their properties, the hydrogels have been explored as systems suitable for tissue engineering [88]. Despite the critical parameter in the selection of the hydrogel to be used in tissue engineering, the biocompatibility, the mechanism of gelling [89], the mechanical properties, the controlled degradation, the interaction with cells [90] and the gel structure are the most important design parameters to be taken into account [91]. That is, alginate hydrogels, already well exploited for biomedical applications like wound dressing and drug release, are receiving particular attention in tissue engineering applications, thanks to the property of forming macro-porous anisotropic structure with ordered capillaries [92]. This 3D structure can mimic the primary microstructural elements of some tissues (nerve, bone, microvasculature, etc.) and facilitate the cells growth processes. Recently, with a combined experimental and modelling approach, the variables that affect the capillaries formation and length, like ion and alginate sol concentration, have been investigated [93].

A consolidated biomedical application of the hydrogels is the development of patches based on water swellable polyacrylates for long-term transdermal drug delivery. One of the most relevant side effects of a conventional transdermal delivery system is the skin occlusion due to the long-term application. In fact, the conventional patches were based on hydrophobic polymer matrices, and therefore intrinsically occlusive; this drawback can be easily overcame with non-occlusive water swellable matrices as hydrogels. The ability of these polymers to exchange water with the skin increases their suitability for the transdermal applications [94]. Of great relevance for the hydrogels used in the development of transdermal patches are the mechanical properties, due to the high mechanical stresses experienced by these systems. In fact, the force exerted to promote the adhesion on skin surface and the continuous mechanical stimuli connected with the patient movements are a key parameter to be taken into account during the development of these delivery systems [95]. Among the various patches types, the muco-adhesive buccal patches are largely diffused due to large permeability of the mucus membranes that allow rapid uptake of a drug into the systemic circulation. Bio-adhesion, which is a phenomenon taking place between a biological material and another, usually a polymer, is the fundamental property in the designing of these systems [96]. The main factor promoting bio-adhesion is the presence of carboxyl and hydroxyl groups on the polymer backbone, to form hydrogen bonds. It was seen that anionic polymers form stronger bonds with respect to neutral or cation polymers. Good polymer wettability is also required to expose the bond-forming sites. The polymer molecular weight should be kept within an optimum range so that the dry system can quickly uptake water (upper limit for the Mw) to form a gel, that should not be too weak (lower limit for the Mw) [97]. However, in mucoadhesive applications, although the formation of H-bonds remains crucial, the modification of the mucus layer beneath the mucoadhesive material becomes important. This step, needed to overcome the anti-adherent properties of mucus, can be attributed to the macromolecular interpenetration effect or to the dehydration effect [98]. To better understand the mechanisms that lead to the adhesion phenomenon and to develop a mathematical model able to describe the mechanical behaviour and the water uptake of the system, following the engineering point of view, the adhesion phenomenon has been studied between a water-rich agarose system (which simulates the biological membrane), and a Carbopol tablet (which simulates the patch) [99].

Author details

Gaetano Lamberti[1*], Anna Angela Barba[2], Sara Cascone[1], Annalisa Dalmoro[1,2] and Diego Caccavo[1]

*Address all correspondence to: glamberti@unisa.it

1 Department of Industrial Engineering, University of Salerno, Fisciano, SA, Italy

2 Department of Pharmacy, University of Salerno, Fisciano, SA, Italy

References

[1] Caccavo D, Cascone S, Lamberti G, Barba A A, Larsson A: Swellable hydrogel-based systems for controlled drug delivery. In: Sezer AD, editor. Smart Drug Delivery System: InTech Open, Rijeka, Croatia; 2016. p. 237–303. DOI: 10.5772/61792

[2] Hoffman A S: Hydrogels for biomedical applications. Advanced Drug Delivery Reviews. 2012; 64: 18–23.

[3] Peppas N A. Hydrogels in Medicine and Pharmacy. CRC Press, Boca Raton, FL; 1987.

[4] Peppas N, Bures P, Leobandung W, Ichikawa H: Hydrogels in pharmaceutical formulations. European Journal of Pharmaceutics and Biopharmaceutics. 2000; 50: 27–46.

[5] Peppas N A, Hilt J Z, Khademhosseini A, Langer R: Hydrogels in biology and medicine: from molecular principles to bionanotechnology. Advanced Materials. 2006; 18: 1345.

[6] Cabral J, Moratti S C: Hydrogels for biomedical applications. Future Medicinal Chemistry. 2011; 3: 1877–1888.

[7] Kamath K R, Park K: Biodegradable hydrogels in drug delivery. Advanced Drug Delivery Reviews. 1993; 11: 59–84.

[8] Peppas N A: Hydrogels and drug delivery. Current Opinion in Colloid & Interface Science. 1997; 2: 531–537.

[9] Hoare T R, Kohane D S: Hydrogels in drug delivery: progress and challenges. Polymer. 2008; 49: 1993–2007.

[10] Vermonden T, Censi R, Hennink W E: Hydrogels for protein delivery. Chemical Reviews. 2012; 112: 2853–2888.

[11] Censi R, Di Martino P, Vermonden T, Hennink W E: Hydrogels for protein delivery in tissue engineering. Journal of Controlled Release. 2012; 161: 680–692.

[12] Van Tomme S R, Hennink W E: Biodegradable dextran hydrogels for protein delivery applications. Expert Review of Medical Devices. 2007; 4: 147–164.

[13] Franssen O, Vandervennet L, Roders P, Hennink W: Degradable dextran hydrogels: controlled release of a model protein from cylinders and microspheres. Journal of Controlled Release. 1999; 60: 211–221.

[14] Hennink W, Franssen O, van Dijk-Wolthuis W, Talsma H: Dextran hydrogels for the controlled release of proteins. Journal of Controlled Release. 1997; 48: 107–114.

[15] Hennink W, Talsma H, Borchert J, De Smedt S, Demeester J: Controlled release of proteins from dextran hydrogels. Journal of Controlled Release. 1996; 39: 47–55.

[16] Boere K W M, Soliman B G, Rijkers D T S, Hennink W E, Vermonden T: Thermoresponsive injectable hydrogels cross-linked by native chemical ligation. Macromolecules. 2014; 47: 2430–2438. 10.1021/ma5000927

[17] Censi R, Fieten P J, di Martino P, Hennink W E, Vermonden T: In situ forming hydrogels by tandem thermal gelling and michael addition reaction between thermosensitive triblock copolymers and thiolated hyaluronan. Macromolecules. 2010; 43: 5771–5778. 10.1021/ma100606a

[18] Van Tomme S R, Storm G, Hennink W E: In situ gelling hydrogels for pharmaceutical and biomedical applications. International Journal of Pharmaceutics. 2008; 355: 1–18.

[19] Van Tomme S R, van Steenbergen M J, De Smedt S C, van Nostrum C F, Hennink W E: Self-gelling hydrogels based on oppositely charged dextran microspheres. Biomaterials. 2005; 26: 2129–2135.

[20] Hennink W E, van Nostrum C F: Novel crosslinking methods to design hydrogels. Advanced Drug Delivery Reviews. 2002; 54: 13–36. http://dx.doi.org/10.1016/S0169-409X(01)00240-X

[21] De Jong S, Van Eerdenbrugh B, van Nostrum C V, Kettenes-Van Den Bosch J, Hennink W: Physically crosslinked dextran hydrogels by stereocomplex formation of lactic acid oligomers: degradation and protein release behavior. Journal of Controlled Release. 2001; 71: 261–275.

[22] Cadée J, Van Luyn M, Brouwer L, Plantinga J, Van Wachem P, De Groot C, et al.: In vivo biocompatibility of dextran-based hydrogels. Journal of Biomedical Materials Research. 2000; 50: 397–404.

[23] De Backer L, Braeckmans K, Stuart M C, Demeester J, De Smedt S C, Raemdonck K: Bio-inspired pulmonary surfactant-modified nanogels: a promising siRNA delivery system. Journal of Controlled Release. 2015; 206: 177–186.

[24] Barba A A, Gaetano L, Sardo C, Dapas B, Abrami M, Grassi M, et al.: Novel lipid and polymeric materials as delivery systems for nucleic acid based drugs. Current Drug Metabolism. 2015; 16: 427–452. http://dx.doi.org/10.2174/1389200216666150812142557

[25] Caccavo D, Cascone S, Lamberti G, Barba A A: Controlled drug release from hydrogel-based matrices: experiments and modeling. International Journal of Pharmaceutics. 2015; 486: 144–152.

[26] Barba A A, Dalmoro A, d'Amore M, Vascello C, Lamberti G: Biocompatible nano-microparticles by solvent evaporation from multiple emulsions technique. Journal of Materials Science. 2014; 49: 5160–5170.

[27] Hamidi M, Azadi A, Rafiei P: Hydrogel nanoparticles in drug delivery. Advanced Drug Delivery Reviews. 2008; 60: 1638–1649.

[28] Gonçalves C, Pereira P, Gama M: Self-assembled hydrogel nanoparticles for drug delivery applications. Materials. 2010; 3: 1420–1460.

[29] Zhang H, Zhai Y, Wang J, Zhai G: New progress and prospects: the application of nanogel in drug delivery. Materials Science and Engineering: C. 2016; 60: 560–568.

[30] Bochicchio S, Dalmoro A, Barba A A, d'Amore M, Lamberti G: New preparative approaches for micro and nano drug delivery carriers. Current Drug Delivery. 2016; in press.

[31] Rao J P, Geckeler K E: Polymer nanoparticles: preparation techniques and size-control parameters. Progress in Polymer Science. 2011; 36: 887–913.

[32] Deepa G, Thulasidasan A K T, Anto R J, Pillai J J, Kumar G V: Cross-linked acrylic hydrogel for the controlled delivery of hydrophobic drugs in cancer therapy. International Journal of Nanomedicine. 2012; 7: 4077.

[33] Kai M P, Keeler A W, Perry J L, Reuter K G, Luft J C, O'Neal S K, et al.: Evaluation of drug loading, pharmacokinetic behavior, and toxicity of a cisplatin-containing hydrogel nanoparticle. Journal of Controlled Release. 2015; 204: 70–77.

[34] Ma D, Tian S, Baryza J, Luft J C, DeSimone J M: Reductively responsive hydrogel nanoparticles with uniform size, shape, and tunable composition for systemic siRNA delivery in vivo. Molecular Pharmaceutics. 2015; 12: 3518–3526.

[35] Kim K, Bae B, Kang Y J, Nam J-M, Kang S, Ryu J-H: Natural polypeptide-based supramolecular nanogels for stable noncovalent encapsulation. Biomacromolecules. 2013; 14: 3515–3522.

[36] Cavallaro G, Craparo E F, Sardo C, Lamberti G, Barba A A, Dalmoro A: PHEA–PLA biocompatible nanoparticles by technique of solvent evaporation from multiple emulsions. International Journal of Pharmaceutics. 2015; 495: 719–727. http://dx.doi.org/10.1016/j.ijpharm.2015.09.050

[37] An Z, Qiu Q, Liu G: Synthesis of architecturally well-defined nanogels via RAFT polymerization for potential bioapplications. Chemical Communications. 2011; 47: 12424–12440.

[38] Nagahama K, Mori Y, Ohya Y, Ouchi T: Biodegradable nanogel formation of polylactide-grafted dextran copolymer in dilute aqueous solution and enhancement of its stability by stereocomplexation. Biomacromolecules. 2007; 8: 2135–2141.

[39] Barba A, Bochicchio S, Lamberti G, Dalmoro A: Ultrasonic energy in liposome production: process modelling and size calculation. Soft Matter. 2014; 10: 2574–2581.

[40] Wang Y, Tu S, Pinchuk A N, Xiong M P: Active drug encapsulation and release kinetics from hydrogel-in-liposome nanoparticles. Journal of Colloid and Interface Science. 2013; 406: 247–255. http://dx.doi.org/10.1016/j.jcis.2013.05.081

[41] Ahmad M, Rai S M, Mahmood A: Hydrogel microparticles as an emerging tool in pharmaceutical field: a review. Advances in Polymer Technology. 2015; 35(2): 121–128.

[42] Kawaguchi H: Thermoresponsive microhydrogels: preparation, properties and applications. Polymer International. 2014; 63: 925–932.

[43] Bucatariu S, Fundueanu G, Prisacaru I, Balan M, Stoica I, Harabagiu V, et al.: Synthesis and characterization of thermosensitive poly (N-isopropylacrylamide-co-hydroxyethylacrylamide) microgels as potential carriers for drug delivery. Journal of Polymer Research. 2014; 21: 1–12.

[44] Kawaguchi H: Micro hydrogels: preparation, properties, and applications. Journal of Oleo Science. 2013; 62: 865–871.

[45] Lee E, Kim B: Preparation and characterization of pH-sensitive hydrogel microparticles as a biological on–off switch. Polymer Bulletin. 2011; 67: 67–76.

[46] Mundargi R C, Rangaswamy V, Aminabhavi T M: Poly (N-vinylcaprolactam-co-methacrylic acid) hydrogel microparticles for oral insulin delivery. Journal of Microencapsulation. 2011; 28: 384–394.

[47] Millon L, Mohammadi H, Wan W: Anisotropic polyvinyl alcohol hydrogel for cardiovascular applications. Journal of Biomedical Materials Research Part B: Applied Biomaterials. 2006; 79: 305–311.

[48] Lee K J, Yoon J, Lahann J: Recent advances with anisotropic particles. Current Opinion in Colloid & Interface Science. 2011; 16: 195–202.

[49] Shewan H M, Stokes J R: Review of techniques to manufacture micro-hydrogel particles for the food industry and their applications. Journal of Food Engineering. 2013; 119: 781–792.

[50] Han J-H, Kim J, Acter S, Kim Y, Lee H-N, Chang H-K, et al.: Uniform hollow-structured poly (vinyl amine) hydrogel microparticles with controlled mesh property and enhanced cell adhesion. Polymer. 2014; 55: 1143–1149.

[51] Barba A, Dalmoro A, d'Amore M, Lamberti G: Controlled release of drugs frommicro-particles produced by ultrasonic assisted atomization based on biocompatible polymers. Chemical and Biochemical Engineering Quarterly. 2012; 26: 345–353.

[52] Cascone S, Lamberti G, Titomanlio G, Barba A A, d'Amore M: Microencapsulation effectiveness of small active molecules in biopolymer by ultrasonic atomization technique. Drug Development and Industrial Pharmacy. 2012; 38: 1486–1493.

[53] Dalmoro A, Barba A A, Lamberti G, Grassi M, d'Amore M: Pharmaceutical applications of biocompatible polymer blends containing sodium alginate. Advances in Polymer Technology. 2012; 31: 219–230. 10.1002/adv.21276

[54] Dalmoro A, Barba A A, d'Amore M, Lamberti G: Single-pot semicontinuous bench scale apparatus to produce microparticles. Industrial & Engineering Chemistry Research. 2014; 53: 2771–2780. 10.1021/ie403308q

[55] Dalmoro A, Barba A A, Lamberti G, d'Amore M: Intensifying the microencapsulation process: ultrasonic atomization as an innovative approach. European Journal of Pharmaceutics and Biopharmaceutics. 2012; 80: 471–477.

[56] Barba A A, d'Amore M, Cascone S, Lamberti G, Titomanlio G: Intensification of biopolymeric microparticles production by ultrasonic assisted atomization. Chemical Engineering and Processing: Process Intensification. 2009; 48: 1477–1483.

[57] Dalmoro A, Barba A A, d'Amore M: Analysis of size correlations for microdroplets produced by ultrasonic atomization. The Scientific World Journal. 2013; 2013: 482910.

[58] Barba A A, Dalmoro A, d'Amore M, Lamberti G: Liposoluble vitamin encapsulation in shell–core microparticles produced by ultrasonic atomization and microwave stabilization. LWT-Food Science and Technology. 2015; 64: 149–156.

[59] Dalmoro A, Sitenkov A Y, Lamberti G, Barba A A, Moustafine R I: Ultrasonic atomization and polyelectrolyte complexation to produce gastroresistant shell–core microparticles. Journal of Applied Polymer Science. 2016; 133: 42976.

[60] Jamzad S, Tutunji L, Fassihi R: Analysis of macromolecular changes and drug release from hydrophilic matrix systems. International Journal of Pharmaceutics. 2005; 292: 75–85.

[61] Colombo P, Catellani P, Peppas N, Maggi L, Conte U: Swelling characteristics of hydrophilic matrices for controlled release new dimensionless number to describe the swelling and release behavior. International Journal of Pharmaceutics. 1992; 88: 99–109.

[62] Hewlett K O, L'Hote-Gaston J, Radler M, Shull K R: Direct measurement of the time-dependent mechanical response of HPMC and PEO compacts during swelling. International Journal of Pharmaceutics. 2012; 434: 494–501.

[63] Colombo P, Bettini R, Peppas N A: Observation of swelling process and diffusion front position during swelling in hydroxypropyl methyl cellulose (HPMC) matrices containing a soluble drug. Journal of Controlled Release. 1999; 61: 83–91.

[64] Huanbutta K, Terada K, Sriamornsak P, Nunthanid J: Advanced technologies for assessment of polymer swelling and erosion behaviors in pharmaceutical aspect. European Journal of Pharmaceutics and Biopharmaceutics. 2013; 83: 315–321.

[65] Zuleger S, Fassihi R, Lippold B C: Polymer particle erosion controlling drug release. II. Swelling investigations to clarify the release mechanism. International Journal of Pharmaceutics. 2002; 247: 23–37.

[66] Barba A A, d'Amore M, Cascone S, Chirico S, Lamberti G, Titomanlio G: On the behavior of HPMC/theophylline matrices for controlled drug delivery. Journal of Pharmaceutical Sciences. 2009; 98: 4100–4110.

[67] Chirico S, Dalmoro A, Lamberti G, Russo G, Titomanlio G: Analysis and modeling of swelling and erosion behavior for pure HPMC tablet. Journal of Controlled Release. 2007; 122: 181–188.

[68] Bettini R, Catellani P, Santi P, Massimo G, Peppas N, Colombo P: Translocation of drug particles in HPMC matrix gel layer: effect of drug solubility and influence on release rate. Journal of Controlled Release. 2001; 70: 383–391.

[69] Rajabi-Siahboomi A, Bowtell R, Mansfield P, Henderson A, Davies M, Melia C: Structure and behaviour in hydrophilic matrix sustained release dosage forms: 2. NMR-imaging studies of dimensional changes in the gel layer and core of HPMC tablets undergoing hydration. Journal of Controlled Release. 1994; 31: 121–128.

[70] Baille W E, Malveau C, Zhu X X, Marchessault R H: NMR imaging of high-amylose starch tablets. 1. Swelling and water uptake. Biomacromolecules. 2002; 3: 214–218.

[71] Abrahmsén-Alami S, Körner A, Nilsson I, Larsson A: New release cell for NMR microimaging of tablets: swelling and erosion of poly (ethylene oxide). International Journal of Pharmaceutics. 2007; 342: 105–114.

[72] Tajarobi F, Abrahmsén-Alami S, Carlsson A S, Larsson A: Simultaneous probing of swelling, erosion and dissolution by NMR-microimaging—effect of solubility of additives on HPMC matrix tablets. European Journal of Pharmaceutical Sciences. 2009; 37: 89–97.

[73] Caccavo D, Lamberti G, Barba A A, Abrahmsén-Alami S, Viridén A, Larsson A: Modeling of extended release matrix tablet performance: water up-take, polymer and drug release, and shape deformation. submitted.

[74] Barba A A, d'Amore M, Chirico S, Lamberti G, Titomanlio G: Swelling of cellulose derivative (HPMC) matrix systems for drug delivery. Carbohydrate Polymers. 2009; 78: 469–474. http://dx.doi.org/10.1016/j.carbpol.2009.05.001

[75] Lamberti G, Cascone S, Cafaro M M, Titomanlio G, d'Amore M, Barba A A: Measurements of water content in hydroxypropyl-methyl-cellulose based hydrogels via texture analysis. Carbohydrate Polymers. 2013; 92: 765–768.

[76] Cascone S, Lamberti G, Titomanlio G, d'Amore M, Barba A A: Measurements of non-uniform water content in hydroxypropyl-methyl-cellulose based matrices via texture analysis. Carbohydrate Polymers. 2014; 103: 348–354.

[77] Higuchi T: Rate of release of medicaments from ointment bases containing drugs in suspension. Journal of Pharmaceutical Sciences. 1961; 50: 874–875. 10.1002/jps.2600501018

[78] Peppas N A: Analysis of Fickian and non-Fickian drug release from polymers. Pharmaceutica Acta Helvetiae. 1985; 60: 110–111.

[79] Okay O. General properties of hydrogels. In: Gerlach G, Arndt K-F, editors. Hydrogel Sensors and Actuators: Engineering and Technology. Berlin, Heidelberg: Springer; 2010. p. 1–14. DOI: 10.1007/978-3-540-75645-3_1

[80] Yin L, Fei L, Cui F, Tang C, Yin C: Superporous hydrogels containing poly (acrylic acid-co-acrylamide)/O-carboxymethyl chitosan interpenetrating polymer networks. Biomaterials. 2007; 28: 1258–1266.

[81] Xu Y, Jia Y, Wang Z, Wang Z: Mathematical modeling and finite element simulation of slow release of drugs using hydrogels as carriers with various drug concentration distributions. Journal of Pharmaceutical Sciences. 2013; 102: 1532–1543. 10.1002/jps.23497

[82] Siepmann J, Kranz H, Bodmeier R, Peppas N A: HPMC-matrices for controlled drug delivery: a new model combining diffusion, swelling, and dissolution mechanisms and predicting the release kinetics. Pharmaceutical Research. 1999; 16: 1748–1756. 10.1023/A:1018914301328

[83] Fujita H. Diffusion in polymer-diluent systems. Fortschritte Der Hochpolymeren-Forschung. Advances in Polymer Science. 3/1: Berlin, Heidelberg: Springer; 1961. p. 1–47. DOI: 10.1007/BFb0050514

[84] Barba A A, d'Amore M, Chirico S, Lamberti G, Titomanlio G: A general code to predict the drug release kinetics from different shaped matrices. European Journal of Pharmaceutical Sciences. 2009; 36: 359–368. http://dx.doi.org/10.1016/j.ejps.2008.10.006

[85] Lamberti G, Galdi I, Barba A A: Controlled release from hydrogel-based solid matrices. A model accounting for water up-take, swelling and erosion. International Journal of Pharmaceutics. 2011; 407: 78–86. DOI 10.1016/j.ijpharm.2011.01.023

[86] Caccavo D, Cascone S, Lamberti G, Barba A A: Modeling the drug release from hydrogel-based matrices. Molecular Pharmaceutics. 2015; 12: 474–483. 10.1021/mp500563n

[87] Caccavo D, Cascone S, Lamberti G, Barba A A: Controlled drug release from hydrogel-based matrices: experiments and modeling. International Journal of Pharmaceutics. 2015; 486: 144–152. 10.1016/j.ijpharm.2015.03.054

[88] Drury J L, Mooney D J: Hydrogels for tissue engineering: scaffold design variables and applications. Biomaterials. 2003; 24: 4337–4351.

[89] Nicodemus G D, Bryant S J: Cell encapsulation in biodegradable hydrogels for tissue engineering applications. Tissue Engineering Part B: Reviews. 2008; 14: 149–165.

[90] Kuo C K, Ma P X: Ionically crosslinked alginate hydrogels as scaffolds for tissue engineering: part 1. Structure, gelation rate and mechanical properties. Biomaterials. 2001; 22: 511–521.

[91] Lee K Y, Mooney D J: Hydrogels for tissue engineering. Chemical Reviews. 2001; 101: 1869–1880.

[92] Treml H, Woelki S, Kohler H-H: Theory of capillary formation in alginate gels. Chemical Physics. 2003; 293: 341–353. 10.1016/s0301-0104(03)00336-7

[93] Caccavo D, Ström A, Larsson A, Lamberti G: Modeling capillary formation in calcium and copper alginate gels. Materials Science and Engineering: C. 2016; 58: 442–449.

[94] Bodde H, Aalten E, Junginger H: Hydrogel patches for transdermal drug delivery; in-vivo water exchange and skin compatibility. Journal of Pharmacy and Pharmacology. 1989; 41: 152–155.

[95] Mazzitelli S, Pagano C, Giusepponi D, Nastruzzi C, Perioli L: Hydrogel blends with adjustable properties as patches for transdermal delivery. International Journal of Pharmaceutics. 2013; 454: 47–57.

[96] Kaur A, Kaur G: Mucoadhesive buccal patches based on interpolymer complexes of chitosan–pectin for delivery of carvedilol. Saudi Pharmaceutical Journal. 2012; 20: 21–27.

[97] Peppas N A, Buri P A: Surface, interfacial and molecular aspects of polymer bioadhesion on soft tissues. Journal of Controlled Release. 1985; 2: 257–275. http://dx.doi.org/10.1016/0168-3659(85)90050-1

[98] Smart J D: The basics and underlying mechanisms of mucoadhesion. Advanced Drug Delivery Reviews. 2005; 57: 15561568.

[99] Caccavo D, Lamberti G, Cascone S, Barba A A, Larsson A: Understanding the adhesion phenomena in carbohydrate-hydrogel-based systems: water up-take, swelling and elastic detachment. Carbohydrate Polymers. 2015; 131: 41–49.

Chapter 9

Hydrogels with Self-Healing Attribute

Li Zhang, Moufeng Tian and Juntao Wu

Additional information is available at the end of the chapter

http://dx.doi.org/10.5772/64138

Abstract

Given increasing environmental issues and energy crisis, mimicking nature to confer materials with self-healing attribute to prolong their lifespan is highly imperative. As representative of soft matter with extensive applications, hydrogels have gained significant attention. In this chapter, a survey of the current strategies for synthesizing self-healing hydrogels based on inorganic-based, polymer and nanocomposite hydrogels is covered and highlighted. Several examples for non-autonomic and autonomic self-healing hydrogels, according to the trigger exerted, are presented. General mechanisms accounting for self-healing hydrogels are listed. Some typical instances to outline the emerging applications of self-healing hydrogels are also provided. Finally, a perspective on the current trends and challenges is briefly summarized.

Keywords: self-healing, hydrogel, hydrogen bond, interactions, application

1. Introduction

After billions of years of evolution, living organisms in nature possess special functions, such as super-hydrophobicity, self-healing, self-cleaning and anti-reflection [1]. Among these, self-healing, the ability of a system or material to repair itself and regenerate function upon the infliction of damage, gained numerous academic curiosity, for this fascinating property would extend the working lifespan and broaden their applications of creatures. Inspired by this, scientists have been trying their best to design and impart intriguing self-healing properties to desired materials with low production costs and improved safety. As representative of soft matter with widespread applications, hydrogels with a self-healing attribute have been developed rapidly as "smart" materials [2].

2. Self-healing process

Self-healing hydrogels can be divided into two different categories according to the required additional external energy or trigger exerted on the hydrogels during the self-healing process: non-autonomic and autonomic.

2.1. Non-autonomic self-healing process

Non-autonomic self-healing hydrogels require a modest external trigger, such as heat or light. Trigger-responsive hydrogels adapting to the surrounding environment and responding to external stimuli are emerging as 'smart hydrogels'. Ultrasound is used in sonophoresis to disrupt skin transiently for needle-free transdermal drug delivery, and damage caused by sonophoresis can be self-healed by skin after the removal of ultrasound stimulus. Similarly, Ca^{2+} cross-linked alginate smart hydrogels were fabricated and ultrasonicated to accelerate drug release, but the Ca^{2+} in physiological fluids would allow crosslinks to self-heal upon removal of the stimulus [3].

Graphene-poly(N,N-dimethylacrylamide) (PDMAA) hydrogel exhibits a thermally triggered (heated at 37°C in air for 12 h), analogous to the healing of wounds in human tissues (37°C), and near-infrared laser-triggered (near-infrared laser irradiation treatment for 2 h), similar to a minimal invasive surgery procedure, self-healing behaviour [4].

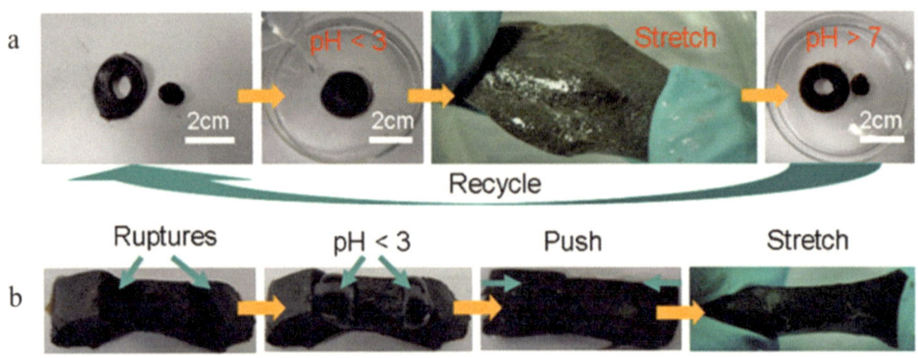

Figure 1. (a) Self-healing and recycling properties of GO/PAACA hydrogel; (b) self-healing process of the GO/PAACA hydrogel with two manual ruptures. The hydrogel was healed just by dripping two drops of acid solution into the ruptures and pushing the fresh surface into contact [5].

Besides ultrasound, heat and light, hydrogels can also self-heal through another important stimulation pH. After immersing in acid solution (pH<3) completely for several seconds, the separated parts of punched GO/poly(acryloyl-6-aminocaproic acid) (PAACA) hydrogel were recombined together, the healed hydrogel separated again when being reintroduced into the high-pH solution (**Figure 1a**) [5]. It was a reversible process, demonstrating fast self-healing capability and reversibility of healing-separation-rehealing process to the pH stimulus. Dropping two drops of acid solution instead of immersing into acid solution can heal the ruptures too (**Figure 1b**).

2.2. Autonomic self-healing process

Compared with non-autonomic self-healing hydrogels, hydrogels with autonomic self-healing capacity do not require any additional external trigger, the damage itself is the stimulus for the healing. Below we highlight some recent examples of autonomic self-healing hydrogels.

The multiple ionic interactions between Fe^{3+} ions and carboxyl groups of poly(acrylic acid) (PAA) produced a non-covalent network of polymer chains [6]. Inter-chain bonding between PAA and ferric ions across the interface autonomously self-heals the damage and rejoins the two halves of the hydrogel. Rheology measurement is used to quantitatively assess the extent of damage or self-healing of the PAA hydrogel. The gel-like character (G'>G") was recovered instantaneously, and G′ and G″ were recovered to the initial values immediately. This disruption and recovery of the PAA hydrogel properties under different oscillatory shears can be repeated several times.

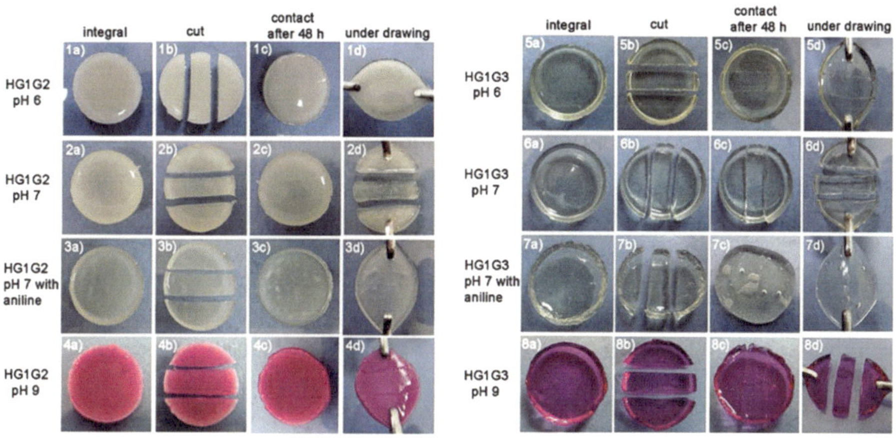

Figure 2. Self-healing properties of HG1G2 (10 wt%) and HG1G3(10 wt%) [7].

HG1G2 (prepared from aldehyde-terminated 3-armed PEO (G1) and dithiodipropionic acid dihydrazide (G2)) and HG1G3 (prepared from G1 and hexanedioic acid dihydrazide (G3)) hydrogels can automatically self-heal under both acidic (pH 3 and 6) and basic (pH 9) conditions through dynamic covalent bonds of acylhydrazone or disulphide (**Figure 2**). However, the hydrogels are not self-healable at pH 7 because both bonds are kinetically locked, whereas catalytic aniline added can accelerate the acylhydrazone exchange reaction and facilitate the hydrogels self-healing at pH 7. All of the self-healing processes are autonomic at room temperature in air [7].

3. Classification of self-healing hydrogels

The past decade has witnessed the rapid development of self-healing hydrogels. According to the materials utilized, self-healing hydrogels can be clarified into inorganic-based, polymer and nanocomposite hydrogels. The important aspects of and trends in self-healing hydrogels are discussed in the following section.

3.1. Self-healing inorganic-based hydrogels

Generally speaking, hydrogels consist of cross-linked networks of hydrophilic polymers swollen with an aqueous solution. Herein, inorganic-based hydrogels, with or without organic macromolecules or metal ion as the crosslinkers, are those whose main building blocks are inorganic [8]. It is different from classical nanocomposite hydrogels with polymer as main building blocks, and nanofillers usually provide physical cross-linking sites [9]. With one- or two-dimensional (2D) basic building blocks, inorganic-based hydrogels have attracted great attention because the fascinating properties of inorganics may bring additional functions to the macroscopic hydrogels [10–12].

Numerous intriguing properties of novel inorganic graphene have prompted scientists to translate its 2D sheets into complex three-dimensional (3D) macrostructures. Shi's group had synthesized the first graphene hydrogel via facile one-step hydrothermal reduction of a highly concentrated graphene oxide (GO) aqueous dispersion in 2010 [13]. Strong hydrophilicity and electrostatic repulsion effect result in random and uniform dispersing of separated GO sheets in GO aqueous dispersion before reduction. After hydrothermal reduction, the oxygenated functionalities, located on the basal planes and at the edges of GO, decreased significantly, whereas π-conjugation restored largely. Combined with hydrophobic effect, π-π stacking interaction promoted flexible rGO sheets to partially overlap and interlock with each other for self-assembling into hydrogel. Hydrothermal reduction strategy is one of the most direct and non-chemical methods to produce graphene hydrogels without introducing any other chemicals or further purification treatment. In addition to hydrothermal reduction, ultrasonication is another additive-free approach for preparing graphene hydrogels [14].

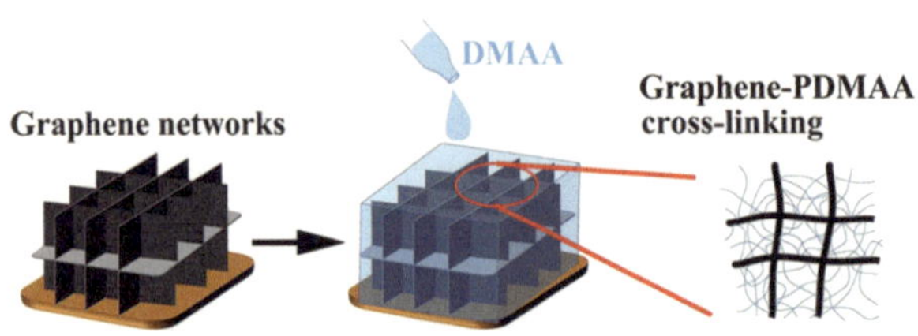

Figure 3. Preparation schematic of graphene-based hydrogel with cross-linking structure [4].

Besides, GO sheets can readily self-assemble into hydrogels in water with the assistance of various promoters, including multivalent metal ions (e.g. Ca^{2+}, Mg^{2+}, Fe^{3+}) and polymer [15].

After the monomer solution free-radical polymerized in the graphene hydrogel network, graphene-based hydrogel can be obtained (**Figure 3**). Graphene-PDMAA hydrogels prepared through this method demonstrate thermally triggered and near-infrared laser-triggered self-healing capacity. The electrical conductivity and compressive strength of the graphene–PDMAA hydrogels recover 60% and 82% approximately of the initial value through the thermally triggered self-healing process [4].

In another typical example, β-cyclodextrin (β-CD) functional graphene-PDMAA hydrogel with self-healing property can control drug linear release at the beginning of 7 h [16], making the double network graphene-based hydrogel potential candidate for anticancer drug carrier. However, single network β-CD functional graphene hydrogels can't self-heal. Therefore, the hydrogen bonding interaction between PDMAA and graphene attributed to self-healing capability of double network graphene-based hydrogels.

Clay, another typical inorganic widely spread on earth, distributes inhomogeneous charge on its disk-like structure. With this special structure and distributed ion, 'house-of-cards' structured hydrogel generates [17].

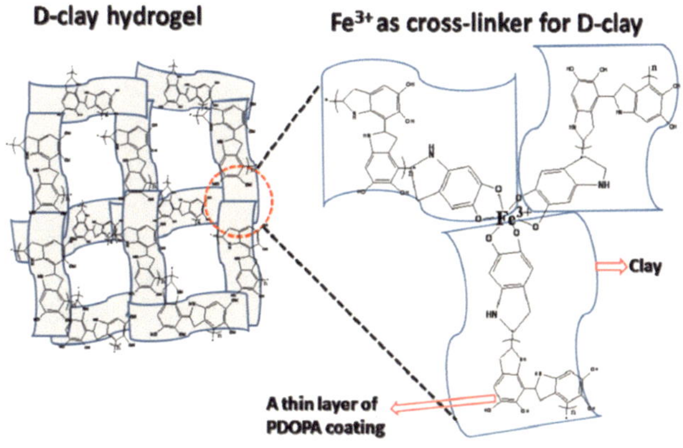

Figure 4. Schematic description of the formation mechanism for D-clay hydrogels [21].

Hydrogels with swollen clay as the basic building block and a small proportion of telechelic dendritic macromolecules as the crosslinker are fabricated readily by mixing the components in water at room temperature. Multiple adhesive termini in the telechelic dendritic macromolecules provide strong noncovalent interactions with clay in these hydrogels. Together with high mechanical strength, such supramolecular hydrogels are also featured by their rapid and complete self-healing behaviour [18]. Though the hydrogel can be fabricated facilely, key component telechelic dendritic macromolecules, physical crosslinker of the hydrogel, cannot be synthesized and obtained easily.

Catechol groups in 3,4-dihydroxy-L-phenylalanine (DOPA), a natural amino acid abundantly present in mussel adhesive proteins, are capable of forming metal-ligand complexes with various surfaces [19]. The reformation of catechol-Fe^{3+} complex after breaking can self-heal the damage [20]. Similar to catechol-ferric ion complexes in mussel-adhesive proteins fibres, clay-based hydrogels with polydopamine (PDOPA)-modified clay (D-clay) as the main building block and ferric ion (Fe^{3+}) as the physical crosslinker are constructed [21]. The coordination bonds between the PDOPA coating on clay nanosheets and Fe^{3+} drive the formation and self-heal the damage of the clay-based hydrogels (**Figure 4**).

3.2. Self-healing polymer hydrogels

Compared with inorganic-based hydrogels, polymer hydrogels are more biocompatible for they resemble biological systems in many ways. On the basis of this, self-healing polymer hydrogels gain much more attention. Li et al. [22] developed a kind of degradable and biocompatible poly(L-glutamic acid) (PLGA)-based self-healing hydrogels via host-guest interaction (**Figure 5a–h**). The self-healing ability of the hydrogels was confirmed by the macroscopic self-healing tests qualitatively (**Figure 5i**) and rheological measurements quantitatively. A 15% w/v PLGA30250 hydrogel in phosphate-buffered saline (PBS) at 37°C can totally degrade in 72 days. Cytotoxicity investigation had shown that the hydrogels could provide a suitable environment for cell growth. Taken together, the self-healing hydrogels could potentially be applied in biomedical fields.

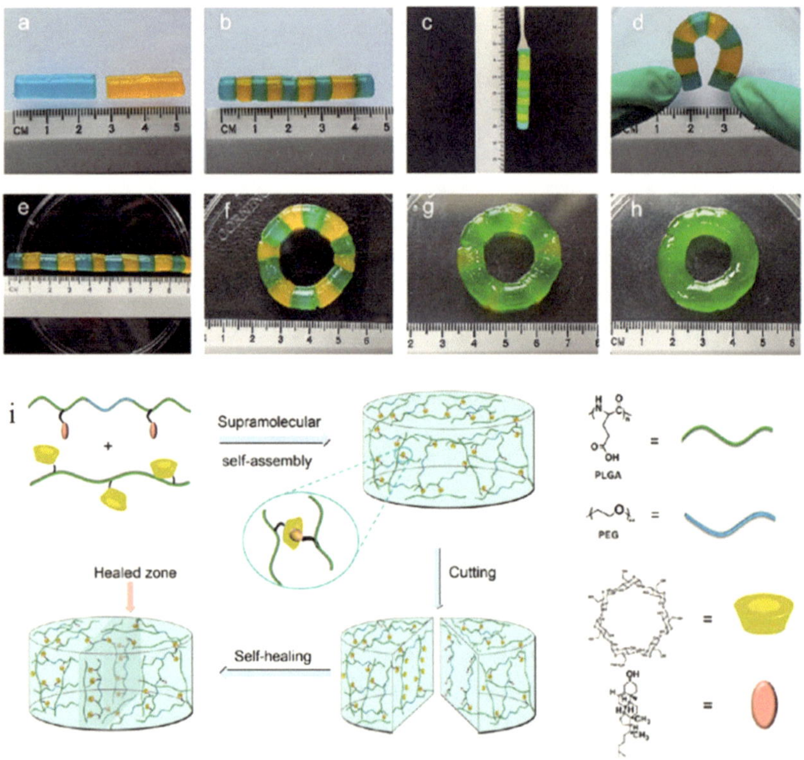

Figure 5. Flexibility of self-healing hydrogels: (a) two rod-like hydrogels stained with erioglaucine disodium salt (EDS) and tartrazine (TAR), respectively; (b and e) colour alternating hydrogel columns; (c) the self-healed hydrogel column held vertically by forceps; (d and f) the healed hydrogel rod bent to a semicircle or a circle. The evolution process of the colorant diffusion: Photo images of colour alternating hydrogel column stored for 1 (f), 12 (g) and 36 h (h). Preparation and self-healing process of supramolecular hydrogels (i) [22].

As a typical biocompatibility and nontoxicity macromolecule with crystalline nature, poly(vinyl alcohol) (PVA) has long been used to fabricate hydrogels via freezing/thawing methods [23, 24]. After self-healing for 12 h at room temperature without any external stimulus,

PVA hydrogel can withstand bending without failure at the interface (**Figure 6**). Tensile tests confirmed that the fracture stress after 48 h healing is ~200 kPa, and the healing efficiency is ~72% [24].

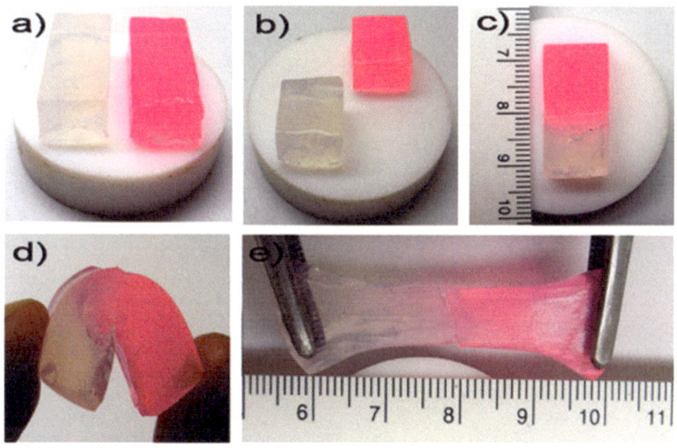

Figure 6. Self-healing behaviour of PVA hydrogel: (a) two pieces of original hydrogel with and without rhodamine B for coloration; (b) two halves of the original hydrogels cut from the middle; (c) self-healed hydrogel upon bringing the two separate halves in contact for 12 h in air at room temperature without any external stimulus; (d) bending of the self-healed hydrogel; (e) stretching of the self-healed hydrogel to about 100% extension [24].

Self-healing attributes to the hydrogen bonding between PAA chains. However, hydrogen bonding is commonly a weak noncovalent bond. In order to solve the issue, diaminotriazine-diaminotriazine (DAT-DAT) hydrogen bonding is formed, and both the tensile and compressive strength can be significantly enhanced; after heating at 90°C for 3 h, healing efficiency (HE) is able to reach 84% and the strength of the mended hydrogel was up to 1 MPa, quite amazing in self-healing hydrogels [25]. Though mechanically robust, lack of effective energy dissipating made these hydrogels brittle. What's more, they don't have the antibacterial property, thus limiting their implantation application [26].

Polyimide (PI) is an important high-performance plastic, and there are many hydrogen bonding, π-π stacking interactions between the chains of its traditional precursor poly(amic acid) (PAA). These non-covalent interactions may confer PAA special attributes, such as self-healing, as macroscopic material. To verify our hypothesis, we aim to synthesize a PAA hydrogel. However, as is known to all, PAA is hydrophobic. In order to overcome this limitation, triethylamine (TEA) was utilized to react with the carboxyl of PAA, and a hydrophilic polyelectrolyte poly(amic acid) ammonium salt (PAS) was obtained (**Figure 7a**) [27]. The as-prepared PAS hydrogel is temperature responsive and robust. In addition, multiple intermolecular interactions, including hydrogen bonds, π-π stacking and polyanion-polycation interactions, served as a combined driving force for the self-healing of hydrogel networks (**Figure 7b,c**). On the basis of this pioneering work, PI aerogels with multifunction were fabricated [28]. Our strategy could open a pathway and possibility for the fabrication of other self-healing functional hydrogels through transformation of hydrophobic polymers to hydrophilic.

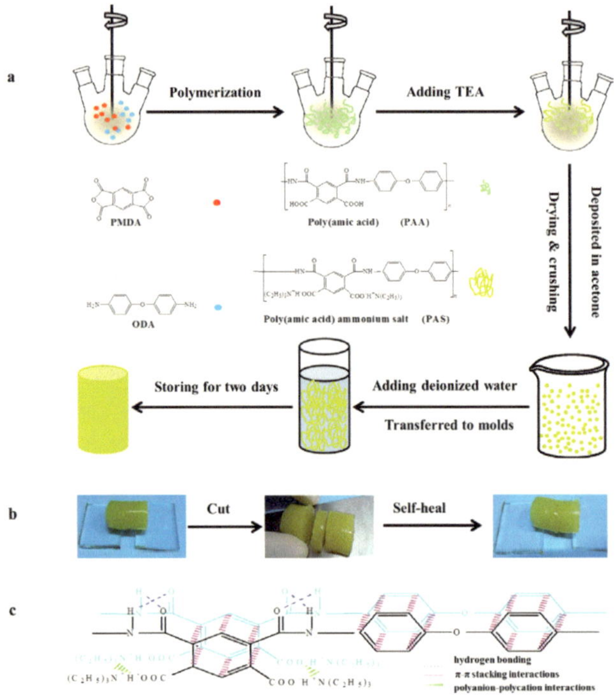

Figure 7. (a) Schematic diagram for the synthesis of PAS hydrogel; (b) self-healing process of PAS hydrogel; (c) possible interactions between PAS chains of PAS hydrogels [27].

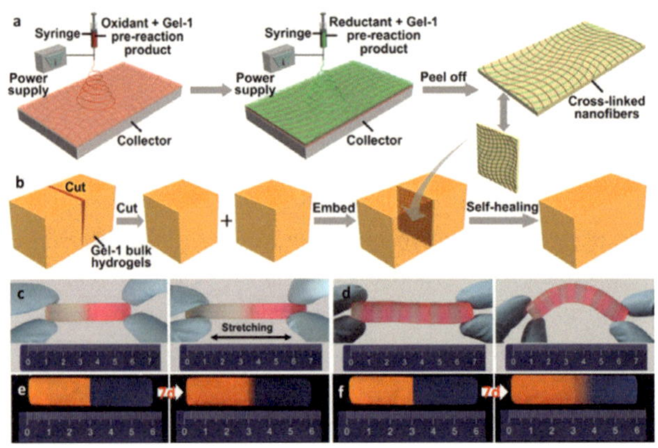

Figure 8. (a) Fabrication of gel-1 healing layer formed of nanofibre networks; (b) enhanced self-healing procedure upon embedding a healing layer into the damage area of gel-1 bulk hydrogels; a sample constructed by juxtaposing a healing layer between two hydrogel blocks exhibits extraordinary mechanical toughness: (c) before and after stretching; (d) a hydrogel column constructed from the connection of 12 hydrogel blocks alternating with six healing layers can be held horizontally and bended by hands; fluorescent images of hydrogels constructed by connecting together two hydrogel blocks (e) without and (f) with 0.05 wt% rhodamine B, respectively, stored for 7 days [29].

http://dx.doi.org/10.5772/64138

Substances	Healing conditions	Healing efficiency	Self-healing mechanisms	Ref.
PVA	RT, 48 h	~72%	Hydrogen bonds	[24]
PNAGA	90 °C, 3 h	84%	Hydrogen bonds	[25]
A6ACA	pH ≤3, <2 s	Hydrogen bonds	[30]	
PAMPSA	water, 72 h	~100%	Hydrogen bonds	[31]
PU	RT,10 min	>60%	Quadruple hydrogen bonds	[32]
DNODN, BNOBN	RT, instantly	57%	Multiple intermolecular interactions	[33]
PAS	RT, instantly	Multiple intermolecular interactions	[27]	
P(AM-co-DMAPS)		~100%	Multiple intermolecular interactions	[34]
Poly(L-glutamic acid)	60 s		Host-guest interaction	[22]
pAA-6βCD/pAA-Fc	GSH aq, 24 h	84%	Host-guest interaction	[35]
PAAm	Press, few seconds	~100%	Hydrophobic interactions	[36]
Agar/HPAAm	RT, 24 h	40%	Hydrophobic interactions	[37]
PNIPAAm	24°C, 30 min	~100%	Hydrophobic interactions	[38]
PAAm	35°C, 3 min	98%	Hydrophobic interactions	[39]
Ferrocene-modified chitosan	RT, 4 h	~100%	Hydrophobic interaction	[40]
Dex-l-PEG	37°C, 12 h	98.7%	Diels-Alder reaction	[41]
GOX/CAT	RT, 5 h	100 %	Imine bonds	[42]
CSMA/SC	Moisture, RT, 2 h	Imine bonds	[43]	
Chitosan-PEG	RT, 2 h		Imine bonds	[44]
cPEG/BDBA	Rapidly		Boronate-catechol	[45]
Poly(acrylic acid)	RT, ~6 h	~88%	Fe^{3+}-carboxylic	[6]
HG1G2	RT, 48 h		Acylhydrazone(acid) disulphide (base)	[7]

Table 1. Examples of various self-healing polymer hydrogels and their self-healing properties.

Noticeably, Fang et al. reported a new strategy to prepare robust, repeatable self-healing hydrogels assisted by a healing layer composed of electrospun nanofibre containing redox agents (**Figure 8a,b**) [29]. Self-healed hydrogels embedded with the electrospun nanofibre exhibit extraordinary mechanical toughness (**Figure 8c, d**). With the same composition of fibre layer as the bulk hydrogels, electrospun nanofibre with redox agents as healing layer can accelerate the molecular diffusion (**Figure 8e,f**), resulting in half of the healing time compared with the one without healing layer, and the healing efficiency is about 80%. This work provides promising avenues to endow electrospun cross-linked nanofibre networks with more potential applications.

Limited by the length of the chapter, other fascinating works cannot be introduced. To make up for this, herein, examples of various self-healing polymer hydrogels and their self-healing properties are summarized in **Table 1** [6, 7, 22, 24, 25, 27, 30–45].

3.3. Self-healing nanocomposite hydrogels

In general, nanocomposite hydrogels refer to cross-linked polymer networks swollen with water in the presence of nanoparticles or nanostructures [9]. The combination of self-healing hydrogels with nanostructures can lead to nanocomposite hydrogels with unique properties. Nanocomposite hydrogels with Fe_3O_4 nanoparticles make them magnetic hydrogels [46], and graphene nanosheets will afford the nanocomposite hydrogels high conductive, mechanical, absorption properties [47, 48] and shortened self-recovery time [49].

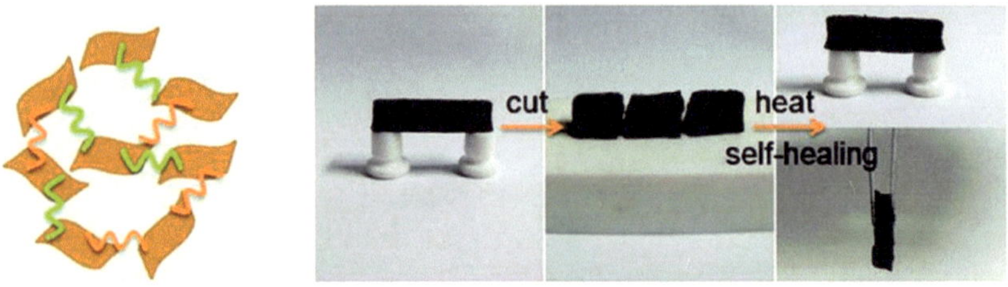

Figure 9. Self-healing nature of graphene-based hydrogel: left, schematic diagram of graphene-DNA hydrogel; and right, demonstration of self-healing properties [55].

One of the most frequently used nanoplatelets (NPs) in nanocomposite hydrogels over the past few years is clay nanoplatelets. Kazutoshi Haraguchi et al. [50], Haraguchi and Li [51] and Haraguchi [52] had conducted many pioneering works on clay-containing nanocomposite hydrogels through in situ free-radical polymerization. In 2011, his group reported two kinds of nanocomposite hydrogels, PDMAA-nanocomposite (D-NC) and PNIPA-nanocomposite (N-NC) [53]. Mechanical damage in NC hydrogels healed autonomously after 48 h at 37°C. While in contrast to the D-NC hydrogels, the N-NC hydrogels hardly self-heal at high temperature (50 or 80°C) after mechanical damage, which can be attributed to the coil-globule transition of PNIPAAm chains above lower-critical-solution temperature (LCST) (32°C). Using the same method, recently, Gao et al. [54] reported polyacrylamide/montmorillonite nanocomposite hydrogels (MnA hydrogels) with unprecedented stretchability and toughness, i.e.

high fracture elongation up to 11800% and fracture toughness up to 10.1 $MJ{\cdot}m^{-3}$. The MnA hydrogels demonstrated interesting self-healing properties. After self-healing directly at room temperature, the self-healed hydrogels became weak and easily disrupted at the interface. While drying the preliminarily jointed hydrogels to water content of about 10 wt% and reswollen at room temperature to its original water content, the fracture strains of self-healed hydrogels were very close to the as-prepared hydrogels. During the drying-reswelling procedure, the preliminary contact of the cut segments facilitated the diffusion of polymers across the interface. Subsequent drying is the driving force for diffused polymers to closely contact the clay platelets and vital to establish strong PAAm/MMT adhesion.

In addition to nanoclay, other nanosheets, such as GO, have also been used to develop self-healing nanocomposite hydrogels. Until now, there are mainly two approaches to introduce GO into the nanocomposite hydrogels: self-assembly and in situ polymerization.

Shi's group provided a new insight for the self-assembly graphene oxide and DNA into a composite hydrogel with high mechanical strength, stability and absorption capacity. This hydrogel was also self-healable after heating at 90°C for 3 min (**Figure 9**) [55]. After heating treatment, the blocks adhered to each other and the resulting self-healed hydrogel was strong enough to bridge two posts horizontally or allow vertical handle.

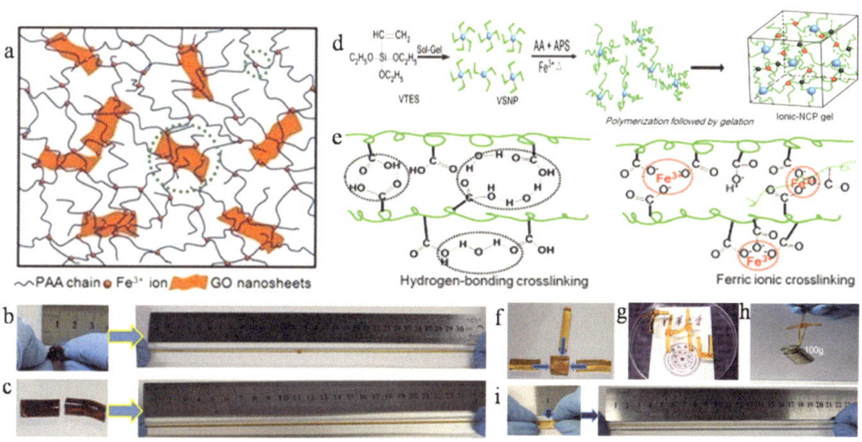

Figure 10. a) A scheme of the three-dimensional network structure of GO-PAA nanocomposite hydrogels facilitated by Fe^{3+} ions with dual cross-linking effects (the smaller green dotted circle represents the first cross-linking points that are Fe^{3+} ions creating ionic cross-linking among PAA chains, and the larger green dotted circle represents the second cross-linking points that are GO nanosheets linking PAA chains through coordination); the GO-PAA nanocomposite hydrogels can be (b) stretched over 30 times from a knotted state and (c) stretched over 20 times after being healed (45°C, 48 h) from a cut-off state [57]; (d) preparation of ionic-NCP gels; (e) illustration of the physical crosslinks formed by hydrogen bonding and Fe^{3+} ionic interactions; self-healing properties of ionic-NCP gels; (f) images of severed cylinder samples of different sizes that were connected and healed as indicated; (g) free-standing healed samples after incubation at 50°C for 48 h; and (h) the healed hydrogel suspending a 100 g weight; (i) stretching of the healed ionic-NCP gel to more than 18 times of its initial length [58].

GO/PAACA nanocomposite hydrogels with double network had been prepared through in situ polymerization. Ca^{2+} acted as crosslinkers through coordination interactions with both oxygen-containing groups of GO nanosheets and polar groups of PAACA side chains, thus inducing the formation of the 3D cross-linked network. Meanwhile, Ca^{2+} content was an important factor to influence the mechanical properties of the fracture stress of the hydrogels. The as-prepared hydrogels are pH triggered and self-healable [5].

Zhong et al. [56–58] proposed a 'single network, dual (or hierarchical) cross-linking' strategy to prepare various tough and highly stretchable nanocomposite hydrogels. For example, nanocomposite hydrogels based on GO prepared by in situ free radical polymerization achieved remarkable synergistic effect (**Figure 10a**): (i) the reinforced-GO nanosheets provide excellent mechanical strength and the cross-linking site between GO nanosheet and PAA chain which triggered by Fe^{3+} ions; (ii) the matrix-PAA chains provide excellent flexibility; (iii) the linker-Fe^{3+} ions provide dynamic and recoverable cross-linking [57]. GO/PAA nanocomposite hydrogels demonstrated superior stretchability and exceptional self-healing properties (**Figure 10b,c**). When stretched, the ionic interaction among PAA chains can dynamically break and recombine to dissipate energy and homogenize the gel network simultaneously. Coordinated with PAA chains, GO nanosheets can maintain the configuration of the hydrogels and act as stress transfer centres transferring the stress to the polymer matrix. The healed GO-PAA nanocomposite hydrogels treated at 45°C for 48 h have the best mechanical properties, with a tensile strength of 777 kPa, elongation at break of 2980%, and the healed nanocomposite hydrogels have a tensile strength of ~495 kPa and elongation at break of ~2470%. The self-healing stemmed from dynamic ionic interactions.

Besides two-dimensional nanosheets, 0-dimensional nanoparticles are also frequently introduced into hydrogels to endow hydrogel multifunction. Silica nanoparticles were hybridized with vinyl to graft an acrylic acid monomer on the surface for the growth of poly(acrylic) acid (PAA) [58]. Network structure of the obtained ionic cross-linked VSNP/PAA nanocomposite physical hydrogels (Ionic-NCP gels) consisted of ferric ion-mediated reversible physical cross-linking, intra- and inter-polymer chain hydrogen bonding, physical entanglement of the polymer chains and multivalent covalent cross-linking through the vinyl hybrid silica nanoparticles (VSNPs) (**Figure 10 d,e**). The prepared Ionic-NCP gels have excellent mechanical properties, with the tensile strength being approximately 860 kPa. In contrast, the hydrogel without ferric ions is much weaker, with a tensile strength less than 50 kPa at fracture, illustrating the importance of ionic cross-linking. In addition, these Ionic-NCP gels exhibit excellent self-healing properties, and the healed hydrogel could be stretched to approximately 15 times its initial length and possessed prominent tensile strength (**Figure 10f–i**).

Different from VSNPs who served as multivalent covalent cross-linking points, poly(2-dimethylaminoethyl methacrylate) (PDMAEMA) brush-modified silica nanoparticles (SiO_2@PDMAEMA) acted as multifunctional noncovalent crosslinkers in a poly(acrylic acid) (PAA) network structure, dissipating energy, whereas the electrostatic interactions between cationic PDMAEMA and anionic PAA render the hydrogel self-healing properties [59]. Grafting of high-density PDMAEMA brushes onto the surface of SiO_2 nanoparticles contributed to the efficient self-healing property significantly. Self-healing efficiency of the healed SiO_2@PDMAEMA/PAA hydrogel is nearly 100%, whereas that of the SiO_2/PAA is only about

50% after 12 h, indicating that the physical adsorption between bare SiO_2 nanoparticles and PAA chains is relatively weak.

From the abovementioned examples, we can see that modification is necessary and vital to nanoparticles. Fe_3O_4 nanoparticles with carboxyl modification could be dispersed well in chitosan solution to form a stable ferrofluid due to the interaction between NH_2 on chitosan and carboxyl groups. External magnetic field exerted could help the as-prepared magnetic hydrogel to stay at the target position and bring hydrogel pieces together to ensure the self-healing process [46].

Moreover, there were other remarkable reports on the self-healing nanocomposite hydrogels: Liu et al. [60] reported self-healed graphene oxide composite hydrogels with very high tensile strengths (up to 0.35 MPa) and extremely high elongations (up to 4900%). Peng et al. [47] presented nanocomposite hydrogels with excellent electrical self-healing properties (completely healed after 20 s) and ultra-high water-absorption ability (up to 350 times). Du et al. [61] reported that multifunctional PPA/carbon nanotube (CNT) hydrogels could respond to a variety of environmental stimuli and exhibit autonomous self-healing properties in both wet and dried states, pointed out that hierarchical hydrogen bond manipulation is a general but powerful strategy to design and synthesize a host of multifunctional materials for versatile applications.

4. Self-healing mechanisms of hydrogels

As previously mentioned, research on self-healing hydrogels is fruitful. In general, self-healing mechanisms of these hydrogels can be illustrated physically and chemically. From the physical viewpoint, during self-repairing, flow of molecular segments across the interface allows the hydrogel networks to be broken at the interface to be rebuilt. As is known, this mechanism is the so-called molecule diffusion. To visualize the molecule diffusion mechanism, digital monitoring [6, 7, 42, 46, 62], scanning electron microscopy (SEM) and ultraviolet (UV) excitation are ideal methods and characterization techniques [29, 63]. Chemically, re-bonding of cleaved non-covalent interactions and dynamic covalent bonds after mechanical damage is particularly critical [2]. As depicted in **Figure 11**, non-covalent interactions between molecules or polymer chains include hydrogen bonds, hydrophobic interactions, crystallization, host-guest interactions, polymer-nanocomposite interactions and multiple intermolecular interactions. Dynamic covalent bonds include cyclohexenes (reversible Diels-Alder cycloaddition), sulphur-sulphur bonds (disulphide), boron-oxygen bonds (phenylboronate ester), carbon-nitrogen bonds (imine, acylhydrazone) and carbon-carbon/carbon-sulphur bonds (reversible radical reaction). Compared to dynamic covalent bonds, non-covalent interactions show a higher dynamic behaviour. Molecule diffusion physically and re-bonding of cleaved bonds chemically combine with each other to facilitate the self-healing of hydrogels.

Factors which accelerate polymer diffusion lead to the acceleration of self-healing [53]. In order to look into the underlying mechanism, the impact of many parameters on the self-healing efficiency, including separation time [24], self-healing time [31, 53], temperature [30, 53], chain length [30, 64] and nanomaterial content [49, 60], should be deeply investigated. With long separation time, polymer chains at the interface could rearrange to minimize the surface energy. Thus the number of free groups on each surface decreases over time, which reduces the number of cleaved bonds that can be reformed across the interface when the two surfaces are brought together [24]. The longer self-healing time, the deeper polymer chains diffuse across the interface, and hence stronger interactions are formed. Temperature strongly affects the diffusion rate of the polymer chains. Short chains facilitate fast molecular diffusion but cannot diffuse deeper, while long chains confer high strength recovery at the interface to hydrogels [60]. Introducing nanomaterials, such as clay or GO, may prevent the polymer chains from diffusing deeper and shorten self-recovery time of the hydrogel, and chain length between nanosheets becomes smaller, with the increase of nanomaterial content [49, 53, 60]. What's more, self-healing properties of hydrogels can be tuned by parameters such as ratio of strong and weak hydrogen bonds [61] and number of freezing/thawing cycles [24].

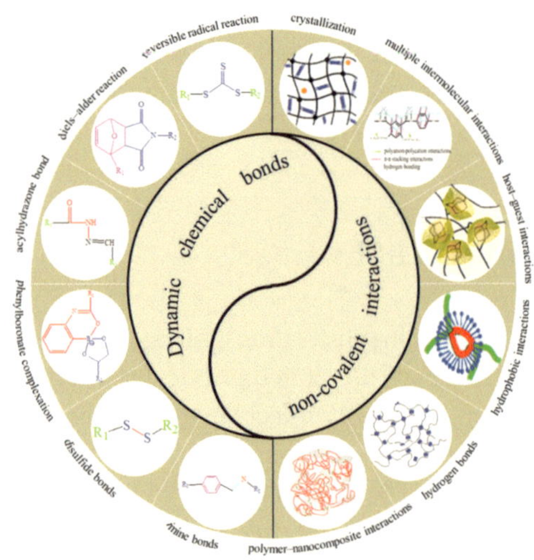

Figure 11. Various dynamic covalent bonds (left) and non-covalent interactions (right) used for self-healing of hydrogels.

Self-healing mechanisms of hydrogels are complicated, and serious effort should be made. Only if many impact parameters, including but not being limited to the abovementioned ones, on self-healing hydrogels are investigated deeply and systematically, we can design and utilize hydrogels well to a wide range of areas.

5. Potential applications of self-healing hydrogels

Hydrogels with self-healing attribute have aroused increasing interest for this fascinating property and would extend the working lifespan of materials. Consequently, they have been widely explored for applications in the fields of biomedicine (scaffolds for tissue engineering, drug/cell carriers and tissue adhesives) and industry (coatings, sealants, sensors, superoleophobic and antibiofouling interfacial materials, enhanced oil recovery(EOR) and remote actuation) (**Figure 12**). Actually, all the superior contributions cannot be included in this chapter because of space constraint. Herein, some typical instances to outline the potential applications of self-healing hydrogels are summarized.

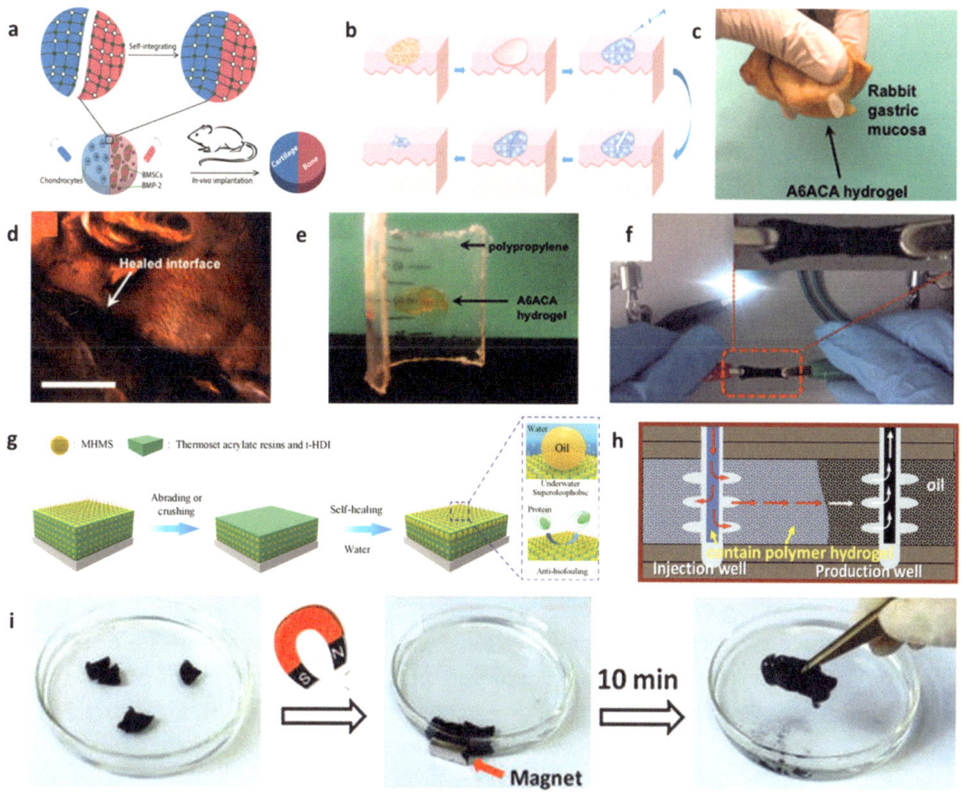

Figure 12. Applications in the fields of biomedicine (a, scaffolds for tissue engineering [67]; b, drug/cell carriers [43] and c, tissue adhesives [30]) and industry (d, coatings [30]; e, sealants [30]; f, sensors [47]; g, superoleophobic and antibiofouling interfacial materials [74]; h, enhanced oil recovery (EOR) [34] and i, remote actuation [46]).

The above mentioned studies illustrate that hydrogels with self-healing attribute confer a possible solution to a wide range of biomedicine and industrial applications in the near future.

5.1. Biomedicine applications

5.1.1. Scaffolds for tissue engineering

Injectable hydrogels are particularly appealing because they provide homogeneous cell distribution within the desired tissue and are emerging as promising materials for tissue engineering scaffolds [65, 66]. For instance, self-healing hydrogels encapsulating chondrocytes and bone marrow stem cells (BMSCs)/bone morphogenetic protein 2 (BMP-2) were merged together and then implanted subcutaneously in a nude mouse to form the cartilage-bone tissue complex (**Figure 12a**) [67]. The complex was successfully regenerated with cartilage occupying a larger volume (~70%) than bone (~30%), displaying capability of the self-integrating hydrogel as scaffold for growth of both bone and cartilage tissues.

5.1.2. Drug/cell carriers

After injection, the broken hydrogel fragments, encapsulated with pharmaceutical drugs/cells, could merge together at the target site, avoiding the risk of catheter clogging by premature polymerization [30, 68, 69]. Cells were encapsulated inside the chondroitin sulphate multiple aldehyde/N-succinyl-chitosan (CSMA/SC) injectable and self-healing hydrogel and retained their proliferative capacity in the hydrogel microenvironment (**Figure 12b**), demonstrating that the hydrogel can be used as a vehicle for the delivery of therapeutic cells [43].

5.1.3. Tissue adhesives

Acryloyl-6-aminocaproic acid (A6ACA) hydrogels adhere well to the gastric mucosa, and the adhesion is strong enough to support the weight of the hydrogel, and the hydrogels could potentially be used as tissue adhesives for stomach perforations (**Figure 12c**) [30].

5.2. Industrial applications

5.2.1. Coatings/sealants

When casting or coating onto substrates, self-healing hydrogels can be used as coatings and ultimately improve the performance and lifetime of implantable biomaterials [6, 70]. The rapid self-healing A6ACA hydrogels had been explored as coatings and could self-heal the imparted crack within seconds when exposed to (or sprayed with) low-pH buffers (**Figure 12d**) [30]. With another important application, A6ACA hydrogels are capable of sticking to various plastics, such as polypropylene and polystyrene, as sealants (**Figure 12e**) [30]. As a proof of concept, the hole of a polypropylene container could be sealed instantly after being coated with A6ACA hydrogel, and leakage of HCl acid in the container was prevented.

5.2.2. Sensors

As mentioned above, self-healing hydrogels with nanostructures could afford new properties to the original hydrogels. For example, with high electrical healing efficiency, self-healed rGO/superabsorbent polymer (SAP) hydrogels can light up light-emitting diode (LED) again (**Figure 12f**) and will find more practical applications in sensor systems [47].

5.2.3. Superoleophobic and antibiofouling interfacial materials

Marine biofouling and oil leakage accidents limit the application of marine vessels and plague people for thousands of years. To significantly improve the safety and prolong the lifetime of marine vessels, self-healing underwater superoleophobic and antibiofouling interfacial materials through the self-assembly of hydrophilic polymeric chain-modified hierarchical microgel, miniature hydrogels with a size ranging from tens of nanometres to several micron [71–73], spheres had been prepared [74]. Once the surface is mechanically damaged, the surface materials can recover oil- and biofouling-resistant properties in water (**Figure 12g**), similar to the skins of some marine organisms such as sharks or whales and will serve in the underwater environment for a long time.

5.2.4. Enhanced oil recovery (EOR)

As is known, the development of enhanced oil recovery (EOR) techniques is vital to deal with the increasing energy crisis. An injection of polymer hydrogel would increase the viscosity of the injection fluid, which could efficiently prevent from bypassing the oil and enhance oil recovery (**Figure 12h**) [34]. Hydrogels with self-healing ability could repair the damage induced by injecting and transporting and maintain their functionalities for a longer period of time.

5.2.5. Remote actuation

The use of magnetic fields can allow remote actuation of self-healing nanocomposite hydrogels. Under an external magnetic field, ruptured sections of magnetic self-healing hydrogels can be facilely directed and automatically combined together, illustrating self-healing and remote actuation (**Figure 12i**) [46].

6. Conclusions

To summarize, mimicking nature to confer hydrogels with self-healing attribute has been a topic of growing interest over the past decade, and remarkable progress has been made. As depicted, brand new methods, such as dual-network, 'single network, dual (or hierarchical) cross-linking' strategy and DAT-DAT hydrogen bonding, are reported to endow self-healing of hydrogels with high strength. Hierarchical hydrogen bonds are constructed and flexibly modulated to tailor the properties of resultant hydrogels. Mussel serves as an important source of inspiration to design and fabricate catechol-containing hydrogels with self-healing properties. Nanostructures such as clay, graphene and Fe_3O_4 were introduced to confer appealing properties and functionality to hydrogel systems.

However, there is still significant room for improvement of self-healing hydrogels to meet industrial and application requirements: (i) first and foremost, when designing self-healing hydrogels, particular attention should be paid to cost, processability, toxicity, biocompatibility, mechanical properties and multi-functionalities; (ii) novel fabrication strategy should be developed; (iii) theoretical research on self-healing mechanism at the nanoscale and assessment systems for quantitatively evaluating self-healing performance should be improved; (iv) more potential applications of self-healing hydrogels should be explored.

To confront the aforementioned challenges and boom the self-healing of hydrogel field, researchers from material, chemistry, physics, biology and engineering should be cooperative multidisciplinarily. Given the prolonging lifespan and efficient utilizations of resources and energy, novel multifunctional and commercial applications of self-healing hydrogels are awaited in the near future.

Acknowledgements

The work is financially supported by the National Natural Science Foundation of China (Grants 51373007 and 51003004), the Beijing Natural Science Foundation (Grant 2142019), the National Basic Research Program of China (Grants 2010CB934700), the Fundamental Research Funds for the Central Universities and the SRF for ROCS, SEM.

Author details

Li Zhang[1,2], Moufeng Tian[2] and Juntao Wu[1*]

*Address all correspondence to: wjt@buaa.edu.cn

1 Key Laboratory of Bio-Inspired Smart Interfacial Science and Technology of Ministry of Education, School of Chemistry and Environment, Beihang University, BeiJing, China

2 Beijing Composite Materials Co., Ltd. Badaling Economic Development Zone, BeiJing, China

References

[1] Kesong Liu and Lei Jiang. Multifunct ional integration: from biological to bio-inspired materials. ACS NANO. 2011;5(9):6786–6790. DOI: 10.102 1/nn203250y

[2] Zhao Wei, Jian Hai Yang, Jinxiong Zhou, Feng Xu, Miklo's Zr'nyi, Patrick H. Dussault, Yoshihito Osada, and Yong Mei Chen. Self-healing gels based on constitutional dynamic chemistry and their potential applications. Chem. Soc. Rev. 2014;43(23):8114–8131. DOI: 10.1039/c4cs00219a

[3] Nathaniel Huebsch, Cathal J. Kearney, Xuanhe Zhao, Jaeyun Kim, Christine A. Cezar, Zhigang Suo, and David J. Mooney. Ultrasound-triggered disruption and self-healing of reversibly cross-linked hydrogels for drug delivery and enhanced chemotherapy. PNAS. 2014;111(27):9762–9767. DOI: 10.1073/pnas.1405469111

[4] Chengyi Hou, Yourong Duan, Qinghong Zhang, Hongzhi Wang, and Yaogang Li. Bio-applicable and electroactive near-infrared laser-triggered self-healing hydrogels based on graphene networks. J. Mater. Chem. 2012; 22(30):14991–14996. DOI: 10.1039/c2jm32255b

[5] Huai-Ping Cong, Ping Wang, and Shu-Hong Yu. Stretchable and self-healing graphene oxide – polymer composite hydrogels: a dual-network design. Chem. Mater. 2013;25(16):3357–3362. DOI: 10.1021/cm401919c

[6] Zengjiang Wei, Jie He, Tony Liang, Hyuntaek Oh, Jasmin Athas, Zhen Tong, Chaoyang Wang, and Zhihong Nie. Autonomous self-healing of poly(acrylic acid) hydrogels induced by the migration of ferric ions. Polym. Chem. 2013;4(17):4601–4605. DOI: 10.1039/c3py00692a

[7] Guohua Deng, Fuya Li, Hongxia Yu, Fuyong Liu, Chenyang Liu, Weixiang Sun, Huanfeng Jiang, and Yongming Chen. Dynamic hydrogels with an environmental adaptive self-healing ability and dual responsive sol – gel transitions. ACS Macro Lett. 2012;1(2):275–279. DOI: 10.1021/mz200195n

[8] Hua Bai, Chun Li, Xiaolin Wang, and Gaoquan Shi. A pH-sensitive graphene oxide composite hydrogel. Chem. Commun. 2010;46(14):2376–2378. DOI: 10.1039/c000051e

[9] Patrick Schexnailder and Gudrun Schmidt. Nanocomposite polymer hydrogels. Colloid. Polym. Sci. 2009;287(1):1–11. DOI: 10.1007/s00396-008-1949-0

[10] Tomoki Ogoshi, Yoshinori Takashima, Hiroyasu Yamaguchi, and Akira Harada. Chemically-responsive sol-gel transition of Supramolecular Single-Walled Carbon Nanotubes (SWNTs) hydrogel made by hybrids of SWNTs and cyclodextrins. J. Am. Chem. Soc. 2007;129(16):4878–4879. DOI: 10.1021/ja070457

[11] Nina I. Kovtyukhova, Thomas E. Mallouk, Ling Pan, and Elizabeth C. Dickey. Individual single-walled nanotubes and hydrogels made by oxidative exfoliation of carbon nanotube ropes. J. Am. Chem. Soc. 2003;125(32):9761–9769. DOI: 10.1021/ja0344516

[12] Hai-Wei Liang, Q ing-Fang Guan, Li-Feng Chen, Zhu Zhu, Wen-Jun Zhang, and Shu-Hong Yu. Macroscopic-scale template synthesis of robust carbonaceous nanofiber hydrogels and aerogels and their applications. Angew. Chem. Int. Ed. 2012;51:5101–5105. DOI: 10.1002/anie.201200710

[13] Yuxi Xu, Kaixuan Sheng, Chun Li, and Gaoquan Shi. Self-assembled graphene hydrogel via a one-step hydrothermal process. ACS NANO. 2010;4(7):4324–4330. DOI: 10.1021/nn101187z

[14] Owen C. Compton, Zhi An, Karl W. Putz, Bong Jin Hong, Brad G. Hauser, L. Catherine Brinson, and SonBinh T. Nguyen. Additive-free hydrogelation of graphene oxide by ultrasonication. Carbon. 2012;50:3399–3406. DOI: 10.1016/j.carbon.2012.01.061

[15] Chun Li and Gaoquan Shi. Functional gels based on chemically modified graphenes. Adv. Mater. 2014;26:3992–4012. DOI: 10.1002/adma.201306104

[16] P. Chen, X. Wang, G. Y. Wang, Y. R. Duo, X. Y. Zhang, X. H. Hu, and X. J. Zhang. Double network self-healing graphene hydrogel by two step method for anticancer drug delivery. Mater. Technol. Adv. Perform. Mater. 2014;29(4):210–213. DOI: 10.1179/1753555714Y.0000000128

[17] Dayong Yang, Songming Peng, Mark R. Hartman, Tiffany Gupton-Campolongo, Edward J. Rice, Anna Kathryn Chang, Zi Gu, G. Q. (Max) Lu and Dan Luo. Enhanced transcription and translation in clay hydrogel and implications for early life evolution. Sci. Rep. 2013;3(3165):1–6. DOI: 10.1038/srep03165

[18] Qigang Wang, Justin L. Mynar, Masaru Yoshida, Eunji Lee, Myongsoo Lee, Kou Okuro, Kazushi Kinbara, and Takuzo Aida. High-water-content mouldable hydrogels by mixing clay and a dendritic molecular binder. Nature. 2010;463(7279):339–343. DOI: 10.1038/nature08693

[19] Haeshin Lee, Bruce P. Lee, Phillip B. Messersmith. A reversible wet/dry adhesive inspired by mussels and geckos. Nature. 2007;448(7151):338–341. DOI: 10.1038/ature0 968

[20] Devin G. Barrett, Dominic E. Fullenkamp, Lihong He, Niels Holten-Andersen, Ka Yee C. Lee, and Phillip B. Messersmith. pH-based regulation of hydrogel mechanical properties through mussel-inspired chemistry and processing. Adv. Funct. Mater. 2013;23(9):1111–1119. DOI: 10.1002/adfm.201201922

[21] Shu Huang, Liping Yang, Ming Liu, Si Lei Phua, Wu Aik Yee, Wanshuang Liu, Rui Zhou, and Xuehong Lu. Complexes of polydopamine-modifi ed clay and ferric ions as the framework for pollutant-absorbing supramolecular hydrogels. Langmuir. 2013;29(4):1238–1244. DOI: 10.1021/la303855t

[22] Guifei Li, Jie Wu, Bo Wang, Shifeng Yan, Kunxi Zhang, Jianxun Ding, and Jingbo Yin. Self-healing supramolecular self-assembled hydrogels based on poly(L-glutamic acid). Biomacromolecules. 2015;16(11):3508–3518. DOI: 10.1021/acs.biomac.5b01287

[23] Christie M. Hassan and Nikolaos A. Peppas. Structure and applications of poly (vinyl alcohol) hydrogels produced by conventional crosslinking or by freezing/ thawing methods. In: J.Y. Chang, D.Y. Godovsky, M.J. Han, C.M. Hassan, J. Kim, B. Lee, Y. Lee, N.A.Peppas, R.P. Quirk, T. Yoo (Eds.), Biopolymers PVA hydrogels, anionic polymerisation nanocomposites. Berlin: Springer; 2000. p. 37–65. DOI: 02/3020hu - 54320

[24] Hongji Zhang, Hesheng Xia, and Yue Zhao. Poly(vinyl alcohol) hydrogel can autonomously self-heal. ACS Macro Lett. 2012;1(11):1233–1236. DOI: 10.1021/mz300451r

[25] Xiyang Dai, Yinyu Zhang, Lina Gao, Tao Bai, Wei Wang, Yuanlu Cui, and Wenguang Liu. A mechanically strong, highly stable, thermoplastic, and self-healable supramo-

lecular polymer hydrogel. Adv. Mater. 2015;27(23):3566–3571. DOI: 10.1002/adma.201500534

[26] Bing Xu, Yongmao Li, Fei Gao, Xinyun Zhai, Mengge Sun, William Lu, Zhiqiang Cao, and Wenguang Liu. High strength multifunctional multiwalled hydrogel tubes: ion-triggered shape memory, antibacterial, and anti-inflammatory efficacies. ACS Appl. Mater. Interfaces. 2015;7(30):16865–16872. DOI: 10.1021/acsami.5b05074

[27] Li Zhang, Juntao Wu, Na Sun, Xiaomin Zhang, and Lei Jiang. A novel self-healing poly(amic acid) ammonium salt hydrogel with temperature-responsivity and robust mechanical properties. J. Mater. Chem. A. 2014;2(21):7666–7668. DOI: 10.1039/C3TA1549G

[28] Li Zhang, Juntao Wu, Xiaomin Zhang, Guangming Gong, Jingang Liu, and Lin Guo. Multifunctional, marvelous polyimide aerogels as highly efficient and recyclable sorbents. RSC Adv. 2015;5(17):12592–12596. DOI: 10.1039/c4ra15115a

[29] Yuan Fang, Cai-Feng Wang, Zhi-Hong Zhang, Huan Shao, and Su Chen. Robust self-healing hydrogels assisted by cross-linked nanofiber networks. Sci. Rep. 2013;3:1–7. DOI: 10.1038/srep02811

[30] Ameya Phadke, Chao Zhang, Bedri Arman, Cheng-Chih Hsu, Raghunath A. Mashelkar, Ashish K. Lele, Michael J. Tauber, Gaurav Arya, and Shyni Varghese. Rapid self-healing hydrogels. PNAS. 2012;109(12):4383–4388. DOI: 10.1073/pnas.1201122109

[31] Xinmi ng Xing, Lianwei Li, Tao Wang, Yanwei Ding, Guangming Liu, and Guangzhao Zhang. A self-healing polymeric material: from gel to plastic. J. Mater. Chem. A. 2014;2(29):11049–11053. DOI: 10.1039/c4ta02079k

[32] Yinlei Lin and Guangji Li. An intermolecular quadruple hydrogen-bonding strategy to fabricate self-healing and highly deformable polyurethane hydrogels. J. Mater. Chem. B. 2014;2(39):6878–6885. DOI: 10.1039/c4tb00862f

[33] Lin Li, Bin Yan, Jingqi Yang, Lingyun Chen, and Hongbo Zeng. Novel mussel-inspired injectable self-healing hydrogel with anti-biofouling property. Adv. Mater. 2015;27(7):1294–1299. DOI: 10.1002/adma.201405166

[34] He Liu, Chunming Xiong, Zhen Tao, Yujiao Fan, Xiaofen Tang, and Haiyang Yang. Zwitterionic copolymer-based and hydrogen bonding-strengthened self-healing hydrogel. RSC Adv. 2015;5(42):33083–33088. DOI: 10.1039/c4ra15003a

[35] Masaki Nakahata, Yoshinori Takashima, Hiroyasu Yamaguchi, and Akira Harada. Redox-responsive self-healing materials formed from host–guest polymers. Nat. Commun. 2011;2(511):1–6. DOI: 10.1038/ncomms1521

[36] Deniz C. Tuncaboylu, Murat Sari, Wilhelm Oppermann, and Oguz Okay. Tough and self-healing hydrogels formed via hydrophobic interactions. Macromolecules. 2011;44(12):4997–5005. DOI: 10.1021/ma200579v

[37] Qiang Chen, Lin Zhu, Hong Chen, Hongli Yan, Lina Huang, Jia Yang, and Jie Zheng. A novel design strategy for fully physically linked double network hydrogels with tough, fatigue resistant, and selfhealing properties. Adv. Funct. Mater. 2015; 25(10): 1598–1607. DOI: 10.1002/adfm.201404357

[38] Umit Gulyuz and Oguz Okay. Self-healing poly(N-isopropylacrylamide) hydrogels. Eur. Polym. J. 2015 ;72:12–22. DOI: 10.1016/j.eurpolymj.2015.09.002

[39] Gizem Akay, Azadeh Hassan-Raeisi, Deniz C. Tuncaboylu, Nermin Orakdogen, Suzan Abdurrahmanoglu, Wilhelm Oppermann, and Oguz Okay. Self-healing hydrogels formed in catanionic surfactant solutions. Soft Matter. 2013;9(12):2254–2261. DOI: 10.1039/c2sm27515e

[40] Ya-Kun Li, Cheng-Gong Guo, Liang Wang, Youqian Xu, Chen-yang Liu, and Cai-Qi Wang. A self-healing and multi-responsive hydrogel based on biodegradable ferrocene-modified chitosan. RSC Adv. 2014;4(98):55133–55138. DOI: 10.1039/c4ra10694f

[41] Zhao Wei, Jian Hai Yang, Xiao Jing Du, Feng Xu, Miklos Zrinyi, Yoshihito Osada, Fei Li, and Yong Mei Chen. Dextran-based self-healing hydrogels formed by reversible Diels–Alder reaction under physiological conditions. Macromol. Rapid Commun. 2013;34(18):1464–1470. DOI: 10.1002/marc.201300494

[42] Yuzhou Gao, Quan Luo, Shanpeng Qiao, Liang Wang, Zeyuan Dong, Jiayun Xu, and Junqiu Liu. Enzymetically regulating the self-healing of protein hydrogels with high healing efficiency. Angew. Chem. 2014;126(35):9497–9500. DOI: 10.1002/anie.201404 531

[43] Shaoyu Lü, Chunmei Gao, Xiubin Xu, Xiao Bai, Haogang Duan, Nannan Gao, Chen Feng, Yun Xiong, and Mingzhu Liu. Injectable and self healing carbohydrate based hydrogel for cell encapsulation. ACS Appl. Mater. Interfaces. 2015;7(23):1302913037. DOI: 10.1021/acsami.5b03143

[44] Yaling Zhang, Lei Tao, Shuxi Li, and Yen Wei. Synthesis of multiresponsive and dynamic chitosan-based hydrogels for controlled release of bioactive molecules. Biomacromolecules. 2011;12(8):2894–2901. DOI: 10.1021/bm200423f

[45] Lihong He, Dominic E. Fullenkamp, Jose' G. Rivera, and Phillip B. Messersmith. pH responsive self-healing hydrogels formed by boronate–catechol complexation. Chem. Commun. 2011;47(26):7497–7499. DOI: 10.1039/c1cc11928a

[46] Yaling Zhang, Bin Yang, Xiaoyong Zhang, Liangxin Xu, Lei Tao, Shuxi Li, and Yen Wei. A magnetic self-healing hydrogel. Chem. Commun. 2012;48(74):9305–9307. DOI: 10.1039/c2cc34745h

[47] Rengui Peng, Yang Yu, Sheng Chen, Yingkui Yang, and Youhong Tang. Conductive nanocomposite hydrogels with self-healing property. RSC Adv. 2014;4(66):35149–35155. DOI: 10.1039/c4ra05381h

[48] Wei Cu i, Jin Ji, Yi-Feng Cai, Hang Li, and Rong Ran. Robust, anti-fatigue, and self-healing graphene oxide/hydrophobically associated composite hydrogels and their use

as recyclable adsorbents for dye wastewater treatment. J. Mater. Chem. A. 2015;3(33): 17445–17458. DOI: 10.1039/c5ta04470g

[49] Subhasish Roy, Abhishek Baral, and Arindam Banerjee. An amino-acid-based self-healing hydrogel: modulation of the self-healing properties by incorporating carbon-based nanomaterials. Chem. Eur. J. 2013;19(44):14950–14957. DOI: 10.1002 /chem. 2013 01655

[50] Kazutoshi Haraguchi, Mariko Ebato, and Toru Takehisa. Polymer–clay nanocomposites exhibiting abnormal necking phenomena accompanied by extremely large reversible elongations and excellent transparency. Adv. Mater. 2006;18(17):2250–2254. DOI: 10.1002/adma.200600143

[51] Kazutoshi Haraguchi, Huan-Jun Li. Hydrophobic surface characteristics of nanocomposite hydrogels. Macromol. Symp., 291–292, 2010;29(1):159–167. DOI: 10.1002/masy. 201050519

[52] Kazutoshi Haraguchi. Nanocomposite hydrogels. Curr. Opin. Solid State Mater. Sci. 2007;11(3):47–54. DOI: 10.1016/j.cossms.2008.05.001

[53] Kazutoshi Haraguchi, Kazuhisa Uyama, Hisashi Tanimoto. Self-healing in nanocomposite hydrogels. Macromol. Rapid Commun. 2011;32(16):1253–1258. DOI: 10.1002/ marc.201100248

[54] Guorong Gao, Gaolai Du, Yuanna Sun, and Jun Fu. Self-healable, tough, and ultra-stretchable nanocomposite hydrogels based on reversible polyacrylamide/montmorillonite adsorption. ACS Appl. Mater. Interfaces. 2015;7(8):5029–5037. DOI: 10.1021/ acsami.5b00704

[55] Yuxi Xu, Qiong Wu, Yiqing Sun, Hua Bai, and Gaoquan Shi. Three-dimensional self-assembly of graphene oxide and DNA into multifunctional hydrogels. ACS NANO. 2010;4(12):7358–7362. DOI: 10.1021/nn1027104

[56] Ming Zhong, Fu-Kuan Shi, Yi-Tao Liu, Xiao-Ying Liu, and Xu-Ming Xie. Tough superabsorbent poly(acrylic acid) nanocomposite physical hydrogels fabricated by a dually cross-linked single network strategy. Chin. Chem. Lett. 2016;27(3):312–316. DOI: 10.1016/j.cclet.2015.12.020

[57] Ming Zhong, Yi-Tao Liu, and Xu-Ming Xie. Self-healable, super tough graphene oxide–poly(acrylic acid) nanocomposite hydrogels facilitated by dual cross-linking effects through dynamic ionic interactions. J. Mater. Chem. B. 2015;3(19):4001–4008. DOI: 10.1039/c5tb00075k

[58] Ming Zhong, Xiao-Ying Liu, Fu-Kuan Shi, Li-Qin Zhang, Xi-Ping Wang, Andrew G. Cheetham, Honggang Cui, and Xu-Ming Xie. Self-healable, tough and highly stretchable ionic nanocomposite physical hydrogels. Soft Matter. 2015;11(21):4235–4241. DOI: 10.1039/c5sm00493d

[59] Jing Zheng, Peng Xiao, Wei Liu, Jiawei Zhang, Youju Huang, Tao Chen. Mechanical robust and self-healable supramolecular hydrogel. Macromol. Rapid Commun. 2016;37(3): 265–270. DOI: 10.1002/marc.201500571

[60] Jiaqi Liu, Guoshan Song, Changcheng He, Huiliang Wang. Self-healing in tough graphene oxide composite hydrogels. Macromol. Rapid Commun. . 2013;34(12):1002–1007. DOI: 10.1002/marc.201300242

[61] Ran Du, Juanxia Wu, Liang Chen, Huan Huang, Xuetong Zhang, and Jin Zhang. Hierarchical hydrogen bonds directed multi-functional carbon nanotube-based supramolecular hydrogels. Small. 2014;10(7):1387–1393. DOI: 10.1002/smll.201302649

[62] Ting-Chen Tseng, Lei Tao, Fu-Yu Hsieh, Yen Wei, Ing-MingChiu, and Shan-hui Hsu. An injectable, self-healing hydrogel to repair the central nervous system. Adv. Mater. 2015;27(23):3518–3524. DOI: 10.1002/adma.201500762

[63] Zhiyan Xu, Junxia Peng, Ni Yan, Hang Yu, Shasha Zhang, Kaiqiang Liu and Yu Fang. Simple design but marvelous performances: molecular gels of superior strength and self-healing properties. Soft Matter. 2013;9(4):1091–1099. DOI: 10.1039/c2sm27208c

[64] Shibaji Basak, Jayanta Nanda and Arindam Banerjee. Multi-stimuli responsive self-healing metallohydrogels: tuning of the gel recovery property. Chem. Commun. 2014;50(18):2356–2359. DOI: 10.1039/c3cc48896a

[65] Deepti Dyondi, Thomas J. Webster, and Rinti Banerjee. A nanoparticulate injectable hydrogel as a tissue engineering scaffold for multiple growth factor delivery for bone regeneration. Int. J. Nanomed. 2013;8:47–59. DOI: 10.2147/IJN.S37953

[66] Hoi Ki Cheung, Tim Tian Y. Han, Dale M. Marecak, John F. Watkins, Brian G. Amsden, and Lauren E. Flynn. Composite hydrogel scaffolds incorporating decellularized adipose tissue for soft tissue engineering with adipose-derived stem cells. Biomaterials. 2014;35(6):1914–1923. DOI: 10.1016/j.biomaterials.2013.11.067

[67] Sen Hou, Xuefei Wang, Sean Park, Xiaobing Jin, and Peter X. Ma. Rapid self-integrating, injectable hydrogel for tissue complex regeneration. Adv. Healthc. Mater. 2015;4(10):1491–1495. DOI: 10.1002/adhm.201500093

[68] Bin Yang, Yaling Zhang, Xiaoyong Zhang, Lei Tao, Shuxi Li, and Yen Wei. Facilely prepared inexpensive and biocompatible self-healing hydrogel: a new injectable cell therapy carrier. Polym. Chem. 2012;3(12):3235–3238. DOI: 10.1039/c2py20627g

[69] Murat Guvendiren, Hoang D. Lu, and Jason A. Burdick. Shear-thinning hydrogels for biomedical applications. Soft Matter. 2012;8(2):260–272. DOI: 10.1039/C1SM06513K

[70] Antoinette B. South and L. Andrew Lyon. Autonomic self-healing of hydrogel thin films. Angew. Chem. Int. Ed. 2010;122(4):779–783. DOI: 10.1002/ange.200906040

[71] Ying Guan and Yongjun Zhang. Boronic acid-containing hydrogels: synthesis and their applications. Chem. Soc. Rev. 2013;42(20):8106–8121. DOI: 10.1039/C3CS60152H

[72] Ying Guan and Yongjun Zhang. PNIPAM microgels for biomedical applications: from dispersed particles to 3D assemblies. Soft Matter. 2011;7(14):6375–6384. DOI: 10.1039/C0SM01541E

[73] Pablo Froimowicz, Daniel Klinger, and Katharina Landfester. Photoreactive nanoparticles as nanometric building blocks for the generation of self-healing hydrogel thin films. Chem.-A Eur. J. 2011;17(44):12465–12475. DOI: 10.1002/chem.201100685

[74] Kunlin Chen, Shuxue Zhou, and Limin Wu. Self-healing underwater superoleophobic and antibiofouling coatings based on the assembly of hierarchical microgel spheres. ACS NANO. 2016;10(1):1386–1394. DOI: 10.1021/acsnano.5b06816